The Man Who Dammed
Hetch Hetchy

Public Lands History

GENERAL EDITORS

Ruth M. Alexander
Adrian Howkins
Jared Orsi
Sarah Payne

The Man Who Dammed Hetch Hetchy

San Francisco's Fight for a Yosemite Water Supply

Donald C. Jackson

University of Oklahoma Press : Norman

Library of Congress Control Number: 2024059291
ISBN: 978-0-8061-9557-5 (hardcover)

The Man Who Dammed Hetch Hetchy: San Francisco's Fight for a Yosemite Water Supply is Volume 8 in the Public Lands History series.

The paper in this book meets the guidelines for permanence and durability of the Committee on Production Guidelines for Book Longevity of the Council on Library Resources, Inc. ∞

The manufacturer's authorized representative in the EU for product safety is Mare Nostrum Group B.V., Mauritskade 21D, 1091 GC Amsterdam, The Netherlands, email: gpsr@mare-nostrum.co.uk.

To the memory of

David P. Billington and Norris Hundley Jr.

and to Donald J. Pisani

Friends, Colleagues, Scholars

Postcard of Hetch Hetchy in Yosemite National Park, circa 1925, showing the valley before and after initial construction of O'Shaughnessy Dam. Author's collection.

Senate passed Hetchy Bill at 11:57 PM.
A great victory for the city and for my plans of development.
—John R. Freeman, diary entry for December 6, 1913

John R. Freeman

Contents

FREEMAN PLAN 1912

SAN FRANCISCO

(Receiving Reservoir) (San Miguel)

Future booster to San Andreas Reservoir
5.5 billion gallons storage

Future bypass 5 miles
CRYSTAL SPRINGS RESERVOIR
22 Billion gallons storage

Maximum Head - 375 ft.

SAN FRANCISCO BAY
About 7000 ft submerged
6.4 miles on embankment
or concrete piers

IRVINGTON GATE HO.

CALAVERAS RES.
SAN ANTONIO RES.
VALLE RESERVOIR, Possible future additional supply

TO SUPPLEMENT PRESENT SPRING VALLEY
SUPPLY UNTIL AQUEDUCT IS COMPLETED

Branch line from here to Oakland and other Bay Cities

30.8 Miles of tunnel 0.1 miles of siphon (about)

Maximum Head - 490 ft.

SAN JOAQUIN RIVER - About 600 ft.
Submerged

POWER HOUSE No.1
AT MOCCASIN CR.
Not needed to be built at the start

SAN JOAQUIN VALLEY CROSSING
45 Miles of steel pipe, 7 ft.
6 in. diam, cement lined.
Only one pipe at first,
240 million gallons per day.
Second pipe added later,
400 million gallons per day.
By larger second pipe
500 million gallons per day.

No power is needed for the delivery
of water to SAN FRANCISCO. This large
power is simply held in reserve for
the city's future benefit

TUOLUMNE RIVER SIPHON, 10 ft. Steel Pipe
1250 ft drop, 70,000 h.p.

15.8 MILES of TUNNEL,
ABT 0.3 MILES
OF SIPHON

18.8 Miles of
tunnel, abt.0.6
miles of siphon

Future tunnel and power No.2
135 ft drop

South Fork
TUOLUMNE
EARLY INTAKE
DIVERSION DAM

Flows down TUOLUMNE RIVER unt'l
tunnel is built

7500 H.P.

300 ft HIGH

HETCH HETCHY DAM

Future Cherry-Eleanor
Development: Eleanor
Reservoir Dam 60 ft. high,
11 Billion gallons storage.
Cherry Reservoir 8 Billion
gallons storage. Dam 50 ft.
high. These Dams can be
increased to 150 ft high as
needed in the future

HETCH HETCHY RESERVOIR
EL. 3800 - 110 Billion
gallons storage

ELEVATION

4000
3000
2000
1000
MAIN SEA LEVEL

180
175
170
165
160
155
150
145
140
135
130
125
120
115
105
100
95
90
85
80
75
70
65
60
55
50
45
40
35
30
25
15
10

MILES FROM SAN FRANCISCO

CITY HALL

Hydraulic Gradient

19.1 Miles of steel pipe, 6.5 ft. Diam.
Lined with Cement.
Only one pipe at First,
second pipe added later

Steel pipe alternative,

About 18.5 miles of tunnel

The FREEMAN LINE is shortened still
more by tunneling straight through
instead of following contour in
canal on canyon side

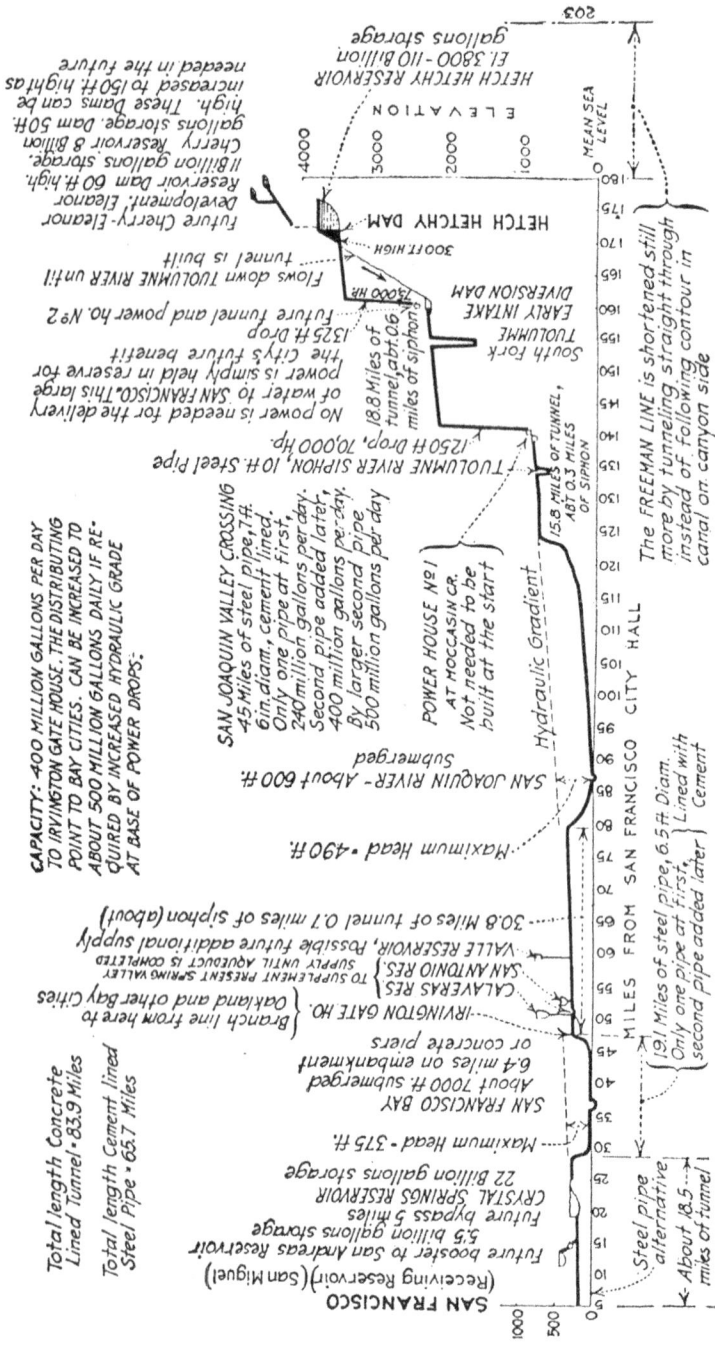

CAPACITY: 400 MILLION GALLONS PER DAY
TO IRVINGTON GATE HOUSE. THE DISTRIBUTING
POINT TO BAY CITIES. CAN BE INCREASED TO
ABOUT 500 MILLION GALLONS DAILY IF RE-
QUIRED BY INCREASED HYDRAULIC GRADE
AT BASE OF POWER DROPS:-

Total length Concrete
Lined Tunnel - 83.9 Miles

Total length Cement lined
Steel Pipe - 65.7 Miles

203-

Illustrations

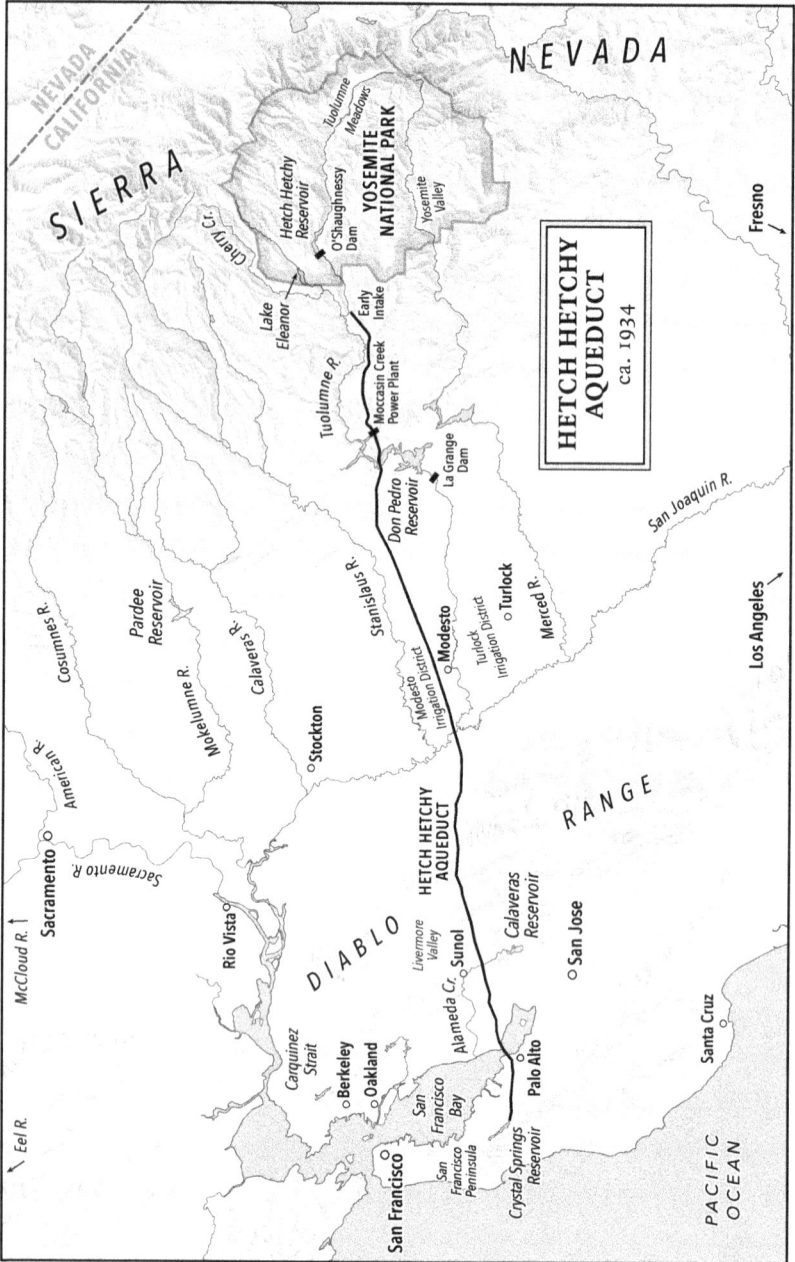

Map of Hetch Hetchy Aqueduct, circa 1934. Cartography by Erin Greb.

The Man Who Dammed
Hetch Hetchy

Introduction

John R. Freeman and "Something Better for the City"

On November 6, 2012, San Francisco's citizens acted upon one of the most ambitious environmental initiatives ever proposed to a municipal electorate. Endorsed by more than 16,000 petition signers, Ballot Proposition F polled voters on their willingness to "require the City to prepare a two-phase plan to evaluate how to drain the Hetch Hetchy Reservoir and identify replacement water and power sources."[1] A seemingly straightforward proposal, the notion that the city might empty a major water supply reservoir high in the Sierra Nevada (capable of storing 117 billion gallons at an elevation more than 3,500 feet above sea level) had deep and contentious roots.

A century earlier, naturalist John Muir and a cadre of followers vigorously fought San Francisco's plans to build a dam in the northern reaches of Yosemite National Park, a dam that would inundate Hetch Hetchy Valley under a massive reservoir. This remarkable campaign led by Muir sparked a national outcry against "park invaders" who threatened the sanctity of the Yosemite preserve. In this environmental crusade, Muir famously drew upon religious imagery to rally the nation against San Francisco's presumed perfidy, extolling the valley as "A grand landscape garden, one of Nature's rarest and most precious mountain mansions. . . . It is one of God's best gifts, and ought to be carefully guarded."[2] He also raged against the forces of mammon with the legendary jeremiad "Dam Hetch Hetchy! As well dam for water-tanks the people's cathedrals and churches, for no holier temple has ever been consecrated by the heart of man."[3]

1

Hetch Hetchy Valley, circa 1905. Tueeulala Falls (*left*) and Wapama Falls (*right*)
plunge down to the Tuolumne River as it winds through a verdant mountain meadow.
The valley floor lies about 3,500 feet above sea level. (*Sierra Club Bulletin* 4 [January
1908], 210)

John Muir's impassioned pleas brought the damming of Hetch Hetchy Valley
into the core of an emerging environmental consciousness, a perspective that—
for at least some Americans—questioned the social value of economic growth
founded upon the unchecked exploitation of natural resources. But despite a
welter of political protest, this battle to protect Hetch Hetchy Valley was lost in
December 1913, when Congress and President Woodrow Wilson green-lighted
San Francisco's plans and, through legislation known as the Raker Act, authorized construction of what (in 1923) became O'Shaughnessy Dam.[4] However,
bitter memories of this defeat had long festered and now, almost one hundred
years later, hopes were high among a coterie of environmentalists that an
ancient evil might at last be rectified. Times had changed, and many Americans
were willing to question the costs exacted by large-scale engineering projects.
So why couldn't the citizens of twenty-first-century San Francisco take action
to free Hetch Hetchy Valley from the concrete tyranny of O'Shaughnessy Dam
and restore its splendor as a mountain meadow?

The idea of tearing down O'Shaughnessy Dam did not abruptly arise in
the fall of 2012. In the 1960s and 1970s, environmental activists picked up

Muir's banner and began to rail against the damage dams can bring to rural landscapes, to fragile riparian ecosystems, to indigenous people, and to populations of spawning fish. In the late twentieth century, dam removal became a touchstone for many conservationists, and when, in 1999, the 160-year-old Edwards Dam across the Kennebec River in southern Maine was demolished under the auspices of the Federal Energy Regulatory Commission, the possibility that large-scale dams might be breached in service to environmental progress gained widespread attention.[5] If the Edwards Dam could be removed, then why couldn't the notorious dam that flooded a once pristine valley in Yosemite National Park be erased from the Sierra landscape? By 2012 the time seemed ripe for change, and many Sierra Club members and their allies pushed hard to get Proposition F before city voters.[6]

The residents of San Francisco, encompassing a locus of leftist Democrats and libertarians dedicated to social causes of all sorts, might have seemed a welcoming audience for the message of dam removal, especially if it meant protecting one of America's iconic national parks. A closer look at the political dynamics of the city's relationship to the O'Shaughnessy Dam, however, reveals an entrenched antagonism toward any plans to drain Hetch Hetchy and reconfigure the city's municipal water supply system. Many people in modern-day San Francisco and parts of the southern Bay Area (including much of "Silicon Valley") rely upon the Hetch Hetchy system as a source of safe drinking water. They are skeptical of any need to drastically modify the status quo.

Alarmed by the possibility that Proposition F might pass, the city's political leadership took a strong stand, resolutely opposing any proposals that would help explore the feasibility of removing O'Shaughnessy Dam. President of the board of supervisors David Chui and Mayor Ed Lee branded the ballot initiative as "a Trojan horse that threatens irreparable harm to our economy and our environment."[7] Former San Francisco mayor and US Senator Dianne Feinstein echoed this view, declaring that "Hetch Hetchy provides critical water supplies to 2.5 million people and thousands of businesses, and any effort to jeopardize that water supply is simply unacceptable."[8] Beyond the political class, city newspapers such as the *San Francisco Bay Guardian* ("Vote this [proposition] down and let's focus our attention on dealing with real environmental and social problems") and *SFGate* ("It doesn't make sense environmentally . . . to eliminate a system that produces clean power and delivers high-quality water to 2.4 million people through the force of gravity") also chimed in, castigating any plans that would expend city funds and civic energy on efforts to remove the dam.[9]

Completed in 1938 to a height 430 feet above its deepest foundations, the massive O'Shaughnessy Dam creates a seven-mile-long reservoir inundating Hetch Hetchy Valley in northern Yosemite National Park. (LC-USZ62-134607, Library of Congress, Washington, DC)

Many people in the Bay Area liked the fact that they consumed water from a protected mountain reservoir, and they were loath to rely entirely on storage at lower elevations in the San Joaquin Valley and coastal environs. In addition, draining the Hetch Hetchy reservoir would reduce the system's ability to generate greenhouse gas–friendly hydroelectric power. Environmentalists certainly had support in their crusade to reconceive the city's long-standing hydraulic infrastructure, but there were two sides to the question of what role the O'Shaughnessy Dam/Hetch Hetchy reservoir should play in San Francisco's future. So what would the people decide?

When the votes were counted for Ballot Proposition F, the verdict was clear. The San Francisco electorate registered a resounding "no" to funding plans for the drainage of Hetch Hetchy. By a margin of more than three to one (No: 249,304—Yes: 74,885), the initiative's proponents failed to attract even 25 percent of the vote. Perhaps someday O'Shaughnessy Dam will be removed and the valley restored, but as of 2025, that time is not now.[10]

Something Better for the City

Some people reading this book may have never heard of Hetch Hetchy Valley and the battles over its use as a reservoir. But most people interested in the history of San Francisco, Yosemite National Park, dams and water in the American West, or environmental history writ large will likely have some awareness of the Hetch Hetchy controversy. Principally because of Muir's poetically strident opposition, the controversy has become embedded in the canon of the American environmental movement, a lesson highlighting the movement's struggle to defend what appears to be wilderness terrain against urban encroachment. Less known (or at least less appreciated in the environmental community) are the historical claims, justifications, and strategies used to validate the city's cause and win authorization of the dam. The origin, evolution, and advocacy of these latter arguments constitute the core subject of this book.

In the contemporary debate over Hetch Hetchy, arguments on both sides are generally structured in terms of what constitutes the greatest public good. Dam opponents point to the environmental glories of a restored valley that will benefit generations to come; conversely, advocates for keeping the city's existing infrastructure praise the high quality of water stored in the mountain reservoir and point to the environmentally friendly hydroelectric power it generates. Claims of what constitutes the best use of Hetch Hetchy are hardly new and can be traced back to the start of the twentieth century when battles over the city's plans to dam Hetch Hetchy first erupted on the political landscape.

Within the historiography of American environmentalism, the Hetch Hetchy controversy is most commonly framed as a clash between the preservationist ideals of John Muir and the utilitarian beliefs of Gifford Pinchot. Muir helped bring the wonders of Yosemite into the national consciousness and, at least rhetorically, led the fight against San Francisco's plans to inundate Hetch Hetchy. In sharp contrast, the famed conservationist Pinchot, as head of the US Forest Service and confidant of President Theodore Roosevelt, helped promote the city's desire to flood Hetch Hetchy for a reservoir, espousing that a municipally owned dam would best serve the Progressive ideal of providing the greatest good for the greatest number for the longest time.

Following a path forged by Holway Jones in his 1965 history of Yosemite National Park (*John Muir and the Sierra Club: The Battle for Yosemite*) and followed by Roderick Nash in his widely read *Wilderness and the American Mind* published in 1967, the "Muir vs. Pinchot" confrontation has come to dominate contemporary discussion of the political conflict over Hetch Hetchy.[11] However, a closer look at Pinchot's pro-dam activism reveals that he held his support for San Francisco's plans at some distance—notably, during his lifetime Pinchot never visited Hetch Hetchy Valley, and he never offered any independent defense for why San Francisco needed to dam the valley rather than seek out other supply sources. He simply accepted the city's claims of the need to dam Hetch Hetchy and relied upon other parties to justify these claims. Unquestionably Pinchot played a significant role in championing the city's Hetch Hetchy cause in the years from 1906 through 1909 (as detailed in chapter 1), but from January 1910, after he was dismissed as head of the Forest Service by President William Howard Taft, through the summer of 1913, Pinchot essentially disappeared from the controversy. Although he did return in June 1913 to testify for about half an hour in support of the city at a House congressional hearing, during the prior three and a half years he had played no role in developing, defending, or promoting the city's Hetch Hetchy initiative. And it was within these critical years that the city's plans for the Hetch Hetchy Dam and Aqueduct experienced a radical transformation.[12]

If the nationally known Pinchot is not the central figure responsible for San Francisco's campaign to win the battle for an expansive Hetch Hetchy project, then how to account for the city's success? Here, the focus shifts to the efforts of John R. Freeman, an MIT-trained, New England–based consulting engineer first hired by the city in the spring of 1910, who eventually conceived and promoted the essential form of what became the modern-day Hetch Hetchy Dam and Aqueduct. Although Freeman is scarcely known to the general public in any way comparable to Muir or Pinchot, during his lifetime he was one of the most important American engineers of the Progressive Era. Born in rural Maine to a farming family of modest means, from the 1890s and into the new century, he achieved prominence as a consulting engineer for water supply systems in Boston, New York City, and Los Angeles. He also consulted on major hydroelectric power projects from New York to California and became a major figure in the world of fire insurance and fire suppression. In many ways, no one better exemplified the precepts of progressive conservation than Freeman, as he drew upon his scientific training and hard-earned engineering expertise to help create what he believed to be a better world. That is not to say that he was selfless in his work habits or disinterested in financial compensation. Far from it. But he was energized by something more than generous consulting fees when undertaking an engineering project and bringing it to completion. He wanted to conceive and create water and power systems that, despite costs borne by some, would overall enhance the life of the nation and, from his perspective, improve society.[13]

John R. Freeman, circa 1905. A graduate of the Massachusetts Institute of Technology (class of 1876), Freeman worked as a consultant for municipal water supply systems in Boston, New York, and Los Angeles before agreeing to support San Francisco's Hetch Hetchy project in April 1910. Author's collection.

When Freeman signed on to assist then–City Engineer Marsden Manson, eight years had passed since San Francisco had first proposed a municipally owned dam and aqueduct connecting the Bay Area to the Tuolumne River. But the city's initial plan consisted of a relatively modest-sized system, offering only 60 million gallons a day (mgd) with limited hydroelectric power capacity. To Freeman's experienced eye, this fell far short of what was possible—and what would provide the greatest good for the greatest number in terms of using the river's resources. Looking to the future, he envisioned a system serving millions of people; it would feature a supply capacity of 400 mgd and include hydro-electric power plants with a combined capacity exceeding 150,000 horsepower.

For almost a year in his capacity as a consultant for San Francisco, Freeman, unimpressed with Manson's capabilities and engineering acumen, held back from making any great commitment to the city's cause. Then turnover at the federal Department of the Interior offered him an opening. Meeting with the newly appointed secretary of the interior, Walter Fisher, in April 1911, Freeman boldly proclaimed that he had resisted being a "subservient counselor" to the city and had instead "been recommending the policy of opening this case wide-open," believing there was "something better for the city" to be had. This book is the story of that "better" plan and how Freeman made it possible.[14]

John R. Freeman: Center Stage

Some historians have appreciated the overriding importance of Freeman's massive 421-page *Hetch Hetchy Report* detailing his plans for the Hetch Hetchy Dam and Aqueduct (published in the fall of 1912), which he wielded to win the valley for San Francisco.[15] For example, environmental historian Robert Righter called the report "a turning point for the city" in its battle for the Hetch Hetchy Dam, and renowned California historian Kevin Starr proclaimed that it "envision[ed] and help[ed] materialize the future."[16] But overall, major works in the hydraulic history of the American West have largely ignored Freeman—for example, he goes unmentioned in Donald Worster's *Rivers of Empire,* Marc Reisner's *Cadillac Desert,* and Norris Hundley's *The Great Thirst*—a historiographical lacuna that may well speak to our still imperfect understanding of the water history of California, the arid West, and America as a whole.[17] At the very least, this absence reveals that there is still much to investigate about how engineers in the Progressive Era operated in creating the great water supply systems that continue to define key elements of America's hydraulic infrastructure. In

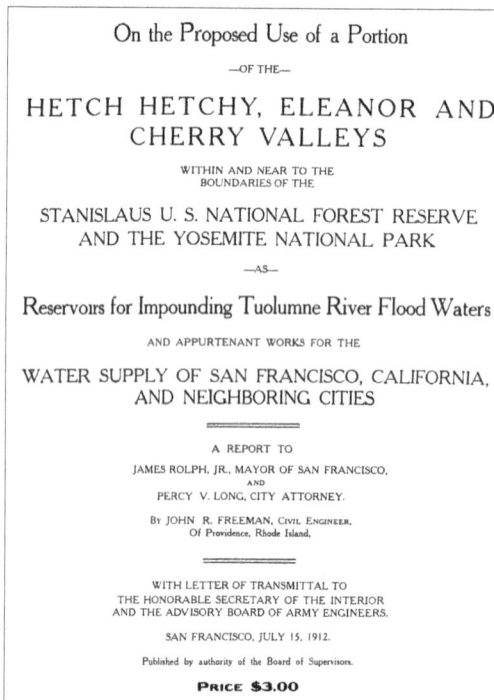

On the Proposed Use of a Portion

—OF THE—

HETCH HETCHY, ELEANOR AND CHERRY VALLEYS

WITHIN AND NEAR TO THE
BOUNDARIES OF THE

STANISLAUS U. S. NATIONAL FOREST RESERVE
AND THE YOSEMITE NATIONAL PARK

—AS—

Reservoirs for Impounding Tuolumne River Flood Waters

AND APPURTENANT WORKS FOR THE

WATER SUPPLY OF SAN FRANCISCO, CALIFORNIA,
AND NEIGHBORING CITIES

A REPORT TO
JAMES ROLPH, JR., MAYOR OF SAN FRANCISCO,
AND
PERCY V. LONG, CITY ATTORNEY.

BY JOHN R. FREEMAN, CIVIL ENGINEER,
Of Providence, Rhode Island,

WITH LETTER OF TRANSMITTAL TO
THE HONORABLE SECRETARY OF THE INTERIOR
AND THE ADVISORY BOARD OF ARMY ENGINEERS.

SAN FRANCISCO, JULY 15, 1912.

Published by authority of the Board of Supervisors.

PRICE $3.00

Title page of John Freeman's 421-page *Hetch Hetchy Report,* published in 1912.

The Man Who Dammed Hetch Hetchy, I hope to enhance and broaden understanding of this important subject.

Environmental preservationists and historians of the twenty-first century may consider the crafting of Freeman's *Hetch Hetchy Report,* and his unabashed advocacy and lobbying for the Hetch Hetchy Dam, to constitute little more than blasphemous subterfuge, a cruel trick foisted upon an unwitting American public. They can deplore his claims that the beauty and accessibility of the national park would be enhanced by the city's dam and reservoir. And they can rail against the heartbreak he brought to Muir and his fellow travelers when his report was used to justify passage of the Raker Act and served as a blueprint for what became (and still remains) San Francisco's Yosemite water supply system. But beyond historical hand-wringing, readers interested in the full and complicated story of Hetch Hetchy will benefit from taking time to consider exactly how Freeman's report came to be, the experiences that prepared him to create such an influential document, and how he operated as an engineer and Capitol Hill lobbyist in championing the city's cause. In essence, this book presents the story of how San Francisco, under Freeman's guidance and leadership, won the battle to dam Hetch Hetchy. It is a story that has only been hinted at—and never told—in the existing historiography of the Hetch Hetchy controversy.

Holding that view in mind, I wrote this book with the goal of documenting the origins of the Hetch Hetchy Dam and Aqueduct by placing John R. Freeman at center stage. No biography of Freeman has ever been written, so in chapter 2 I have taken care to sketch in some detail his life before Hetch Hetchy; this is important because his professional experiences were key to both the technical skills he brought to his work for San Francisco and his ability to operate effectively in the political realm. The story of Hetch Hetchy, of course, does not begin with Freeman's arrival on the scene in the spring of 1910. Thus, chapter 1 discusses the history of the valley, its inclusion in the reserve that became Yosemite National Park, the creation of the Modesto and Turlock Irrigation Districts within the Tuolumne River watershed, the expansion of San Francisco's water supply in the nineteenth and early twentieth centuries, and the origins of the city's plans for a Yosemite water supply. While other books on Hetch Hetchy cover much of this ground, *The Man Who Dammed Hetch Hetchy* brings its own perspective to these relatively familiar events in the story of the controversy. Beginning in chapter 3, the book details Freeman's three-and-a-half-year Hetch Hetchy odyssey starting in the spring of 1910, extending through the extraordinary summer of 1912 and the creation of the *Hetch Hetchy Report,* and

A provocative illustration published in Freeman's 1912 *Hetch Hetchy Report* (p. 16) showing how an auto road circling the reservoir might facilitate tourist access to Hetch Hetchy Valley and Wapama Falls.

into 1913, when the "great victory" for his vision of the Hetch Hetchy Dam and Aqueduct was won in the nation's capital.

As readers will become aware, the path traveled by Freeman in advancing the city's cause was not a simple one; it followed a complex trajectory rife with contingency and unexpected turns—this is especially true in regard to his inter-actions with the often vilified Spring Valley Water Company and with northern California's investor-owned electric power companies. Underlying everything presented in this book, however, is a basic argument that, absent Freeman and the work he undertook for the city—especially following Marsden Manson's mental collapse in the spring of 1912 and his subsequent resignation as city engineer—no Hetch Hetchy Dam and related large-scale aqueduct featuring a huge hydroelectric power capacity would have been authorized for construction by Congress (and approved by President Wilson) in December 1913. Perhaps some municipal dam and aqueduct tapping into the Tuolumne River would have won approval at a later date, but if so, that would have been years in the future. And it may be that no dam flooding Hetch Hetchy would have ever been built. In the phrasing of Kevin Starr, Freeman did indeed "materialize the future" of Hetch Hetchy in a way that no one else—whether engineer (such as Marsden Manson), politician (such as San Francisco's various mayors or members of the board of supervisors), or conservationist Gifford Pinchot—could have accomplished in pre–World War One America.[18]

Progressivism and Conservation

Through the course of this book, I often refer to Freeman as a "progressive engineer" who operated within a professional milieu where he sought to create a better world based upon the fruits of scientific and technological advancement. Of course, it is a staple of American history to denote the period from the late nineteenth century through the 1920s as the Progressive Age or the Progressive Era, a time in which great efforts were made to "reform" society and, depending upon one's interests and perspective, make it a better place for the American people. Such reforms were incredibly wide ranging in their foci and include (but are hardly limited to) such diverse subjects as child labor, factory work and the "eight-hour day," women's suffrage, prohibition, urban political and social reform, direct election of US senators, tariff reform and the income tax, antitrust and banking regulation, fire safety, food safety, forest conservation, and public health and water supply.[19]

For many progressive leaders, a key concern involved the profligate waste of natural resources—which could include untapped rivers emptying into salty estuaries or land-locked lakes—and a concomitant desire to use these resources for human purposes as efficiently and effectively as possible. Their goal was to conserve nature's bountiful gifts, protect them from wasteful exploitation by short-sighted profiteers, and utilize them in a way that provided "the greatest good for the greatest number for the longest time." In his classic book *Conservation and the Gospel of Efficiency: The Progressive Conservation Movement, 1890–1920,* Samuel P. Hays analyzes this movement and observes how "[c]onservation above all, was a scientific movement, and its role in history arises from the implications of science and technology in modern society. . . . The new realms of science and technology, appearing to open up unlimited opportunities for human achievement, filled conservation leaders with intense optimism. . . . They displayed that deep sense of hope which pervaded all those at the turn of the century for whom science and technology were revealing visions of an abundant future." In his work, Hays makes no reference to Freeman, as he focuses on federal engineers, scientists, and bureaucrats—Gifford Pinchot looms large in his analysis—who guided conservation policy related to water resources, forestry, and public land use. But Hays's emphasis on the role of scientific technology in stimulating the work of progressive conservationists is a theme that fully resonates with Freeman's efforts to advance the field of hydraulic engineering.[20]

As detailed in chapter 2, the MIT-trained Freeman built upon his early work with New England textile mills and learned how to construct hydropower systems that, based upon carefully recorded streamflow records and an understanding of terrain and topography, could enhance the growth of urban industrial centers and spur the creation of far-flung hydroelectric power networks. His remarkable interest in fire safety and factory insurance was centered on the deployment of systematic protocols and operational regimes (including automatic sprinklers) that were calculated to reduce the possibility of devastating conflagrations and thus increase economic efficiency (and lower insurance premiums). His efforts in developing municipal water supply systems for Boston and New York City were intended to provide ever-growing populations of urban consumers with huge quantities of safe and reliable water that would improve peoples' lives, reduce the risk of widespread fires, and spur economic growth on a regional scale. Similarly, he took on the task of planning a dam across the lower reaches of the Charles River between Boston and Cambridge that would inundate malodorous mud flats and create a spectacular artificial lake within the heart of one of America's iconic urban centers. All of these accomplishments did not come by happenstance but were founded upon hard work and an understanding of the precepts of science and modern engineering. In this, Freeman's career pre–Hetch Hetchy aligns easily with Hays's view of how elite progressive engineers were motivated by "visions of an abundant future" as they sought to maximize use of natural resources and improve society.

In his own mind, Freeman possessed an awareness of how his plans for Hetch Hetchy correlated with a progressive vision of environmental improvement. In the fall of 1913, he traveled to Washington, DC, to lobby for the Raker Act, meeting with numerous senators, testifying at a key Senate committee hearing, and working to win authorization for his vision of the Hetch Hetchy Dam and Aqueduct. Describing one of his visits to the nation's capital that September, he framed his work as promoting a progressive cause in service to the people of San Francisco: "I was spending most of the past weekend in Washington at the home of William Kent, Progressive member of Congress, and in a general atmosphere of 'progressivism.' It is a fine thing for the country that we have Progressives of one aim or another to suit the times."[21]

To be clear, as an engineer Freeman did not always, or necessarily, have the "right" answer to questions about what was the most appropriate technological solution for a given problem. As I describe in my book *Building the Ultimate Dam,* he had no interest in embracing the innovative multiple-arch dam designs proposed by the California engineer John S. Eastwood, designs that, by reducing concrete quantities, were intended to reduce dam construction costs and thus facilitate increased storage of floodwaters across the arid West.[22] In his communications with the leadership of the Great Western Power Company, Freeman objected to Eastwood's designs ostensibly on the grounds that taking chances with a new form of dam technology was too risky given that the consequences of failure were so great. But there is also no question that he saw Eastwood as a threat to his own authority over the building of massive gravity dams. For Freeman, this threat needed to be countered not because Eastwood's progressive designs—progressive because they were economically efficient and minimized material requirements—were necessarily deficient in terms of "scientific technology" but because in large part they posed a threat to the East Coast engineer's professional stature and livelihood. While advancing a progressive vision of water resources development, Freeman was also fully capable of acting in his own self-interest.

Freeman's work as a consultant was no doubt vital to San Francisco's success in winning approval of the Hetch Hetchy project, and this included the engineering expertise he brought to his design of a 400 mgd gravity-flow aqueduct capable of generating more than 150,000 horsepower of hydroelectricity. But one of the key truths about the Hetch Hetchy saga revealed in this book is that Freeman did not act simply as a dispassionate engineer in his work for San Francisco. He also served as a politically savvy advocate and Capitol Hill lobbyist in advancing his, and the city's, plan for Hetch Hetchy. Of course, he was not alone in projecting a political agenda as, to some degree or another, all progressive conservationists of the early twentieth century acted within the political realm. Freeman may not have been unique in his standing as a politically motivated engineer, but in light of what he accomplished at Hetch Hetchy, it is difficult to call forth any other individual who was more successful in taking on and implementing an engineering project of such far-reaching consequence and political complexity.

Lest Anyone Wonder . . .

This book is not intended to provide a guide to what will or should happen to Hetch Hetchy Valley in the years to come. Instead, the intent is to provide a context for understanding how San Francisco, in the face of wide-ranging opposition by John Muir and his allies—and also in the face of politically potent antagonism from irrigation farmers in the San Joaquin Valley—obtained the right to build a huge concrete dam within the confines of Yosemite National Park. The arguments made by Freeman in promoting the city's plans were not grounded in some idea that the "nature" of Hetch Hetchy Valley held no value. Instead, he believed that the greater public good could be advanced by using the natural attributes of the mountain valley to create an expansive and valuable reservoir, improve the quality and reliability of the city's water supply, and help meet the region's hydraulic engineering needs for a century (or more) to come. Personal interests and motives aside, Freeman believed, to the great consternation of Muir and fellow preservationists, that his plans for San Francisco offered "something better for the city" in its search to ensure a future water supply for millions of people. As it turned out, more than a few of these people showed up at the polls in November 2012, voting to keep the dam that Freeman conceived a century earlier in place and operational, providing high-quality water from the Sierra Nevada for residents of the City by the Bay.

CHAPTER 1

Hetch Hetchy—Park or Reservoir?

By the time Hetch Hetchy Valley attracted the professional interest of John R. Freeman, the Sierra Nevada landform possessed a long history. In geologic time, the valley's origins stretched back 200 million years, when magma along the western edge of the North American plate began to solidify. Some 65 million years ago the granite core had been exposed, but another 40 million years passed before tectonic uplift created a towering escarpment that, at its peaks, reached elevations more than 10,000 feet above sea level. As the range rose, the slope of rivers draining the highlands increased, with "faster flowing streams cut[ting] deeper and deeper canyons into the mountain block." Volcanic mudflow then covered the highland terrain, a time in which "the present river courses and drainage patterns throughout the Sierra became well established." The final stage in the physical creation of what became Yosemite National Park started about 3 million years ago, with "a mountain icefield form[ing] along the mountain crests." As colossal glaciers scoured the Sierra landscape, the future park assumed its modern-day geologic form. At Hetch Hetchy, glacial action carved out the final U-shape of the valley, where steep granite cliffs—adorned by two spectacular waterfalls—soar above the course of the now-inundated Tuolumne River.[1]

Not long after glaciers largely receded from the Tuolumne watershed, humans began migrating across the Bering Strait land-bridge into what became North America. About three thousand years ago, ancestors of Miwok and Paiute Indians started inhabiting and cultivating the Hetch Hetchy Valley in seasonal mountain journeys; notably, the name Hetch Hetchy is believed to be derived from the Miwok language.[2] During the early twentieth century, the valley was often denoted by park proponents as "wild" or an expanse of "wilderness."

However, this appellation ignored how California's indigenous people had already transformed the valley's flora. As historian Char Miller notes, at a minimum "they used fire to maintain meadows and inhibit the growth of some plants and [to] select for those they prized, including deergrass (for baskets), California black oak (for acorns) and Pinyons (for seeds)."[3]

This ecological transformation of the Hetch Hetchy environs might appear relatively modest—at least compared to the inundation brought by a massive concrete dam. Nonetheless, thousands of years have elapsed since the valley landscape stood in a condition unchanged or uninfluenced by the hand of human culture. When naturalist John Muir rhapsodized over Hetch Hetchy as a "grand landscape garden" and "one of Nature's rarest and most precious mountain mansions," he was speaking of a valley long nurtured by California's first peoples.[4]

Spanish settlers moved north from Mexico into Alta California beginning in the late eighteenth century, but they made little effort to explore the interior mountain ranges. Major settlements ringed the Pacific coastline, and the small community of Yerba Buena (renamed San Francisco in 1847) lay close by the Golden Gate that welcomed seafaring vessels to the shelter of San Francisco Bay.[5] The vast expanse of the Sierra Nevada, including the main Yosemite Valley (along the Merced River) and Hetch Hetchy Valley (along the Tuolumne), remained largely unknown to Euro-American settlers prior to the discovery of gold east of Sacramento in January 1848. All changed during the ensuing "Gold Rush," when tens of thousands of migrants fervently scoured the rivers and watersheds of the Sierra Nevada in search of undiscovered placer goldfields.[6]

The expanding presence of Anglo-American culture in the early 1850s brought more than the ecological changes wrought by large-scale hydraulic mining. It also forced indigenous people from land that had been a source of sustenance for centuries. The so-called discovery of Yosemite Valley and Hetch Hetchy was sparked by both the search for gold and a concerted effort to drive Miwok and Paiute tribes from their mountain sanctuaries. Precisely when Anglo-Americans first visited Yosemite Valley and Hetch Hetchy is uncertain, but by 1850 the gold-seeking Screech brothers had traversed the Tuolumne into Hetch Hetchy, and in 1851 the so-called Mariposa Battalion had raided Miwok settlements in Yosemite Valley, seeking to enslave Native peoples. A new era in Sierra Nevada history was under way.[7]

After the Mexican-American War ended in 1848, California became a territory of the United States. With a rapidly rising population driven by the Gold Rush, California achieved statehood in 1850; in area it was the nation's second-largest state (after Texas). Although far distant from major East Coast cities, a

Postcard view of Yosemite Valley, circa 1920, with the Merced River in the foreground and Bridalveil Fall on the right. By the early 1860s, lithographs and photographic images were bringing the scenic splendor of Yosemite into America's national consciousness. Author's collection.

remarkably sophisticated communication system centered around the US Post Office quickly brought the new Pacific Coast state into the national consciousness. While Hetch Hetchy Valley received little or no notice in the 1850s and '60s, the visual splendor of Yosemite Valley attracted a growing cadre of admirers. The most important of these was James Hutchings, an Anglo-American who, in 1855, began publishing *Hutchings' California Magazine,* a periodical that celebrated the scenic virtues of Yosemite and other parts of the state's remarkable landscape. In the 1850s, cabins appeared in Yosemite to accommodate a nascent tourist trade, and in 1864, Hutchings took ownership of the most prominent of these rural hostels. Yosemite Valley rapidly became a destination for hundreds, and eventually thousands, of visitors every year.[8]

By the mid-1850s, photography had become an essential feature of American culture, and as word of the wondrous spectacle of Yosemite began to spread, photographers such as Carleton Watkins soon visited the valley. Through widely distributed lithographs, stereo views, and commercial photographs, evocative images of the Yosemite landscape were brought into parlor rooms and public venues across America. In 1862, Watkins presented an exhibit of large-format Yosemite photos in New York City, drawing acclaim and helping evince the splendor of Yosemite as a national treasure.[9] Even the horrors of the Civil War could not stop recognition of Yosemite Valley as a special part of the American geography, a recognition that sparked a movement to protect it from commercial degradation. Thus, in June 1864, while Union and Confederate soldiers were locked in combat outside of Richmond and General Sherman was beginning his assault on Atlanta, the US Congress, with President Lincoln's signature and assent, approved the creation of a "Yosemite Valley Reserve." This unprecedented federal law transferred ownership of the scenic valley to the State of

By the end of the nineteenth century, Yosemite Valley had become a favored destination for tourists who could afford a vacation in the Sierra Nevada. Here, a fashionably dressed visitor takes to the trail. Author's collection.

California so that it could be publicly administered and—while promoting tourism—protected from private exploitation.[10]

Encompassing about sixty square miles, the Yosemite reserve did not include the watershed that fed the Merced River above the main valley, and it included no land within the Tuolumne watershed. In fact, it was not a *national* park at all because control over the valley had been transferred from the federal government to California state officials (operating through a board of politically appointed commissioners). Nonetheless, Yosemite Valley had achieved recognition as an extraordinary scenic treasure worthy of protection. The next year, landscape architect Frederick Law Olmsted, famed co-designer of New York City's Central Park, drafted a report attesting to the valley's value as both a tourist destination (akin to the Swiss Alps) and a source of moral regeneration for the nation's urban citizenry.[11] Tourism became central to how the park would be operated in the public interest, but the new park commissioners opposed Hutchings's presence in the valley, demanding that he forfeit his claims in deference to state authority. Private concessionaires would be called on to serve visitor needs, but on terms codified and regulated by government officials.[12]

John Muir and Robert Underwood Johnson

A few years passed before a US Supreme Court ruling forced Hutchings out of Yosemite in 1873, but in the meantime a new epoch in the park's history came with the arrival of John Muir in California in the spring of 1868. When he first visited Yosemite Valley later that summer, the thirty-year-old Scotch immigrant by way of Wisconsin attracted scant attention. After working for several months as a farmhand in the San Joaquin Valley, Muir spent the following summer in the High Sierra tending a flock of sheep; soon after, he signed on as a hired hand at Hutchings's hotel in Yosemite Valley. During this time, he became enthralled with the Sierra landscape and began writing eloquent essays that, in time, earned him the sobriquet "John of the Mountains."

As part of his Sierra explorations, Muir visited Hetch Hetchy and, in March 1873, published an account of a fall 1871 ramble to the Tuolumne; this offered the American public one of the first printed descriptions of the remote valley.[13] As Muir detailed in the *Boston Weekly Transcript,* Hetch Hetchy "is a Yosemite Valley in depth and in width, and is over twenty miles in length, abounding in falls and cascades . . . [offering] one of the very grandest [views] I ever beheld. . . . On the north side of the valley there is a vast perpendicular rock front 1800 feet high. . . . In spring a large stream pours over its brow with a clear fall of at least one thousand feet [modern-day Tueeulala Falls]. East of this, on the same side, is the Hetch Hetchy Fall [modern-day Wapama Falls]." Enchanted by Hetch Hetchy's lofty cliffs and tumbling waterfalls, Muir also reveled in the valley's flora: "[D]iversified with groves and meadows . . . the crystal river glides between sheltering groves of alder and poplar and flowering dogwood[;] . . . river nooks are gloriously bordered with ferns and sedges and drooping willow."[14]

Acknowledging that the valley was attracting few tourists, Muir observed how it both retained artifacts of its Native American history and was an outpost of Anglo-American sheep herding: "The whole valley is at present claimed by the 'Smith brothers' as a summer sheep range. Sheep are driven into Hetch Hetchy every spring, about the same time that a nearly equal number of tourists are driven into Yosemite. . . . At present there are a couple of shepherds' cabins and a group of Indian huts in the valley, which I believe is all that will come under the head of improvements."[15]

In calling out the natural features of Hetch Hetchy as well as human artifice, Muir was aware of the dynamic nature of the valley environs. And in ways that foretold how the natural flow of the river could be altered by blockage of the outlet chasm, he further witnessed how "[a]t the end of the valley the river enters a narrow cañon which cannot devour spring floods sufficiently fast to prevent the lower half of the valley from becoming a lake."[16] Perhaps it would not be so far-fetched as to imagine the possibility of expanding upon this seasonal impoundment and build a more permanent structure to store the Tuolumne's floodwaters. Muir would never hold any interest in such a possibility, but as the decades passed, others would.

No great groundswell of interest in Hetch Hetchy Valley followed publication of Muir's March 1873 essay, but the valley did not remain totally unknown.[17] For example, in the 1870s Albert Bierstadt painted a florid rendering of the valley as part of a series of California mountain views; a report from an 1878–79 geographical survey described Hetch Hetchy as "worthy of special remark as perhaps the most special feature of the great Tuolumne Canon"; and an 1885 article in *Outing: An Illustrated Monthly Magazine of Recreation* described it as "one perfectly-cut little gem" that could be of interest to "genuine out-door people." Compared to the Merced Yosemite, however, Hetch Hetchy languished in obscurity, with sheepherding dominating the valley. By the 1880s the valley meadow was being "patented" through General Land Office sales; eventually

720 acres of Hetch Hetchy bottomland (more than a square mile) was removed from the public domain and sold into private hands.[18]

Through the 1870s Muir continued writing and proselyting about Yosemite and the Sierra Nevada, attaining a national prominence that laid a foundation for his fame as a naturalist. His first essay for *Harper's New Monthly* appeared in 1875 ("Living Glaciers of California"), and between 1878 and 1882 he published twelve articles in *Scribner's Monthly* and its successor, *Century Magazine*.[19] But in April 1880 his life assumed a new direction when he married Louisa Strentzel and took charge of his father-in-law's farm in Martinez, a few miles inland from San Francisco Bay. For most of the 1880s Muir devoted his energies to raising a family and maintaining the orchard estate. After a hiatus of several years, in 1887 Muir started writing about nature again in the regional magazine *Picturesque California,* where he published a series of essays on the state's natural features and attractions.[20] His return to the public arena caught the attention of Robert Underwood Johnson, an associate editor of *Century Magazine* in New York City.

Born in Ohio in 1853, Johnson had graduated from Earlham College before going east to work as an editor/writer for Charles Scribner's publishing empire.[21] Ever in search of good copy, in June 1889 he journeyed to San Francisco and met with Muir to see if the naturalist might be induced to write some articles on the Sierra Nevada. Presumably Muir's essays could appeal to a national readership that was becoming ever more interested in conservation and fearful of the impact that industrial/economic growth was inflicting upon the American landscape. The meeting proved fortuitous, and their blossoming friendship helped elevate both to leadership in the nascent movement to preserve the nation's natural resources.[22]

What proved most memorable from the Muir/Johnson partnership was not simply the written words that Muir produced to advance the progressive cause of conservation, but the way that Muir's association with *Century Magazine* helped spawn Yosemite National Park, a vast tract of 1,500 square miles that would encompass the entire upper Merced and Tuolumne watersheds. The creation of this new national reserve in 1890 depended on more than the shared advocacy of Muir and Johnson—the lobbying muscle of the Southern Pacific Railroad also played an essential role.[23] But in no small part the legislation underlying the park's formation derived from a trip the two men took to Yosemite soon after they met, a visit enshrined as a centerpiece of the national park's origin story.[24] It was a tale oft told, and here is one version that Johnson shared in November 1912 during a hearing on San Francisco's plans to dam Hetch Hetchy:

As no doubt many of you know it fell to my lot to be the originator of this park. In 1889 I was with Mr. Muir, in what is now the national park, and we went from Yosemite over along the Tuolumne Canyon to Soda Springs, camped out one night after having seen a great deal of the most wonderful scenery there. I said to him "Muir, why is it that this has never been reserved?" "Well," he said, "I do not know. It ought to be reserved." I said . . . "I propose to you, Muir, that we start a movement here and now to make a national park. . . ." And we talked it over and he agreed.[25]

Johnson was certainly fond of placing himself at the center of the park's genesis, but it is true that, at his behest, Muir wrote two important articles espousing the park in *Century Magazine*.[26] And it is also true that in the lead-up to passage of the law creating the new park Johnson appeared before the House Committee on Public Lands and testified as to the importance of including the upper Tuolumne watershed in the protected reserve.[27] The bill passed Congress

With support from John Muir and Robert Johnson, in 1890 Hetch Hetchy Valley was included within the boundaries of what became Yosemite National Park. But the valley and the narrow gorge at its western end (highlighted in this photograph ca. 1905) also offered an ideal reservoir site lying 3,500 feet above sea level. (Ray W. Taylor, *Hetch Hetchy* [San Francisco, 1926], 35)

on September 30, 1890, with President Benjamin Harrison signing it the next day.[28] Although making no specific reference to a "Yosemite National Park" (the bill instead refers to a "forest reservation"), the legislation does make clear that the mountainous area covered by the law, which included Hetch Hetchy Valley (minus, of course, the 720 acres of meadow land already in private ownership), was to be given a special protective status "under the exclusive control of the Secretary of the Interior, whose duty it shall be . . . to make and publish such rules and regulations [covering its use] as he may deem necessary or proper." These regulations were far-ranging and provided "for the preservation from injury of all timber, mineral deposits, natural curiosities, or wonders within said reservation, and their retention in their natural condition."[29]

Regardless of whether or not the phrase "national park" appears in the act's text, the interior secretary is clearly charged with preserving from injury those "natural curiosities, or wonders" lying within the designated reservation and

The valley floor at Hetch Hetchy looking upstream with Kolana Rock at right, circa 1900. Prior to the creation of Yosemite National Park, 720 acres of land in the valley (including the meadow shown here) were in private ownership. (From panoramic photograph, digital ID pan 6a19577, Library of Congress, Washington, DC)

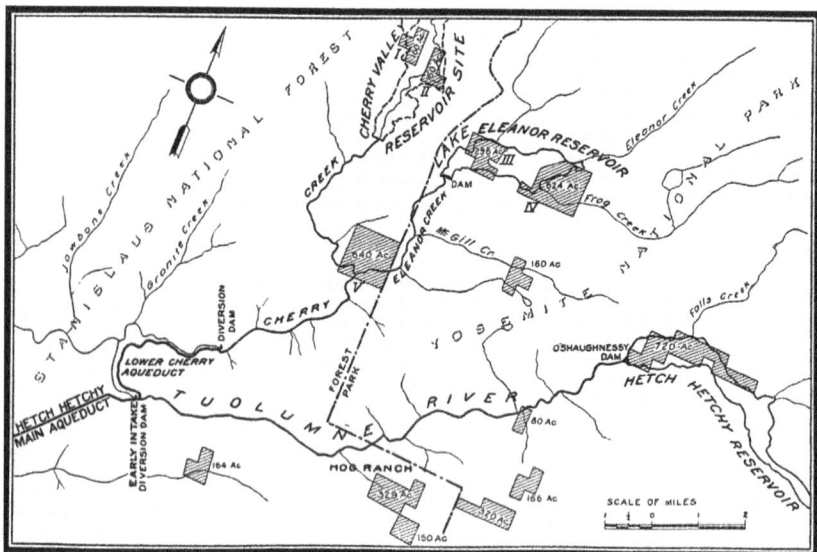

Map of the upper Tuolumne River watershed showing the relationship of Hetch Hetchy Valley to nearby Cherry Creek and Eleanor Creek (*on left*). This map also shows numerous land parcels that, by the start of the twentieth century, were privately owned and not a part of either Yosemite National Park or Stanislaus National Forest. (Ray W. Taylor, *Hetch Hetchy* [San Francisco, 1926], 120)

ensuring their "retention in their natural condition." Almost certainly the wonder that was Hetch Hetchy Valley would be considered as requiring protection in its "natural condition." Or at least that is how park supporters believed the bill should be interpreted.

Passage of the 1890 act protecting "wonders" present in the Sierra Nevada also helped energize many Californians to take a more active role in both defending the state's natural resources and making them more accessible for public enjoyment. In 1892 a group of influential men in the Bay Area organized the Sierra Club, asking John Muir to serve as the organization's first president. With this, preservation of the Sierra Nevada landscape and its flora and fauna (but not the preservation of Native tribes with claims to mountain land) assumed a new political dimension. Although not every member of the Sierra Club would later castigate San Francisco as a "park invader" because of its desire to flood Hetch Hetchy, many—including Muir, the Sierra Club's first president—would adopt that view. By the 1890s, the foundation for a later anti-dam coalition was in place and dedicated to defending what, in 1906, would officially become Yosemite National Park.[30]

Irrigation and Reservoirs

The Hetch Hetchy controversy is commonly posed as a battle between those wishing to preserve the sanctity of Yosemite National Park and San Francisco boosters seeking a municipally owned mountain water supply. Left out of this framing are farmers in the San Joaquin Valley who wished to take the water from the Tuolumne River and, as irrigation advocates commonly phrased it, "make the desert bloom." Interest in using the region's scarce water resources to support agriculture was by no means confined to the Tuolumne watershed. It extended across the American West and, in the late nineteenth century, sparked a politically potent "irrigation crusade." This movement looked to populate the vast arid tracts lying west of the 100th meridian with Anglo-American farming colonies dependent upon dams and canals and lateral ditches to water dry fields and sustain long-term crop production.[31]

Early irrigated agriculture in the Anglo-American West (such as that initiated by Mormon settlers in Utah in 1847) relied on relatively small diversion dams and canals.[32] These systems were simple to build and, if damaged by flooding, could be rebuilt without major hardship. However, westerners had little difficulty imagining that much larger irrigation projects might be possible if only they could capture seasonal spring floods behind large storage dams. By releasing copious quantities of stored water over the course of the growing season, it would be possible to irrigate larger tracts of land and produce more crops. In the Tuolumne basin, large-scale Anglo agriculture took root with wheat farming in the 1860s that, within a few years, produced annual yields reaching five million bushels. But the drought years of 1870/71 demonstrated the danger of relying upon natural precipitation for sustained agriculture in a semiarid land. As more of the San Joaquin Valley came into private ownership and the Southern Pacific Railroad extended its trackage in search of agricultural markets, the desire to tap into the flow of the Tuolumne River for crop irrigation took on greater urgency.[33]

Dams and canals cost money, and a stream the size of the Tuolumne requires a substantial political structure that can finance and operate a large-scale water control system. The organization of irrigation districts capable of developing the Tuolumne became possible in 1887, when California enacted the Wright Irrigation District Act. Under this law, groups of private California landowners were empowered to pool their land as collateral and, acting collectively as an irrigation district, sell bonds to investors to pay for dams, canals, and distribution laterals. As irrigated land in the district became more productive, farmers could pay for this increased productivity through tax assessments. In turn, these taxes would be used to repay the bond holders; over a span of twenty or thirty years (sometimes more), a district's debt could be extinguished. And then, in theory, further expansion could be financed under a new bond and a new structuring of assessment and repayment.[34]

Along the Tuolumne, two districts were formed under the Wright Act. To the north lay the Modesto Irrigation District (encompassing about 82,000 acres) and, to the south, the larger Turlock Irrigation District (about 176,000 acres).[35] Starting in 1890, the districts jointly financed construction of the 127-foot-high La Grange Dam lying across the Tuolumne at a crest elevation about 300 feet above sea level. A massive masonry overflow dam with a maximum thickness of 90 feet, the La Grange Dam took three years to build and cost almost $600,000. Providing little storage capacity, the dam was designed to raise and divert the natural flow of the Tuolumne into distribution canals and laterals extending down both sides of the valley. In tandem, the La Grange Dam and associated canals were costly to construct, and the two irrigation districts struggled to remain financially solvent during the economic hard times of the 1890s. However, the districts escaped bankruptcy and by the turn of the twentieth century had jointly established water rights to 2,350 cubic feet per second of the Tuolumne's natural flow.[36] Actual irrigation came slowly, with a lawyer for the Modesto district estimating in 1907 that only about 60,000 acres total in the two districts were presently under cultivation. But key infrastructure was in place, and after almost twenty years of struggle, by 1910 the combined districts seemed poised for dramatic growth.[37]

Construction of the La Grange Dam, and the subsequent diversion of natural streamflow, legally established the Turlock and Modesto Irrigation Districts as key players in the development of the Tuolumne. Nonetheless, the districts knew that their diversion dam provided little storage and that seasonal spring floods largely escaped over the dam and down to the San Joaquin River (and thence into San Francisco Bay). In the latter part of the nineteenth century, it was no secret that much of the seasonal flooding in California—and the arid West as a whole—was being "lost" or "wasted" in terms of nourishing crop land. This had been a major theme highlighted in John Wesley Powell's famous *Report on the Lands of the Arid Region*, first printed in 1878 and made widely available the following year.[38] Later, in his capacity as director of the US Geological Survey (USGS), Powell convinced the US Congress to authorize a federally sponsored irrigation survey designed to explore the American West and locate reservoir storage sites that could help maximize economic use of the region's scarce

Completed by the Modesto and Turlock Irrigation Districts in 1893, the La Grange Dam diverts water from the lower Tuolumne River and irrigates more than 200,000 acres of farm land in the San Joaquin Valley. The canal serving the Modesto District is visible on the left in this ca. 1910 postcard view. Author's collection.

water resources. This irrigation survey began in 1888, with USGS engineers and surveyors visiting several California streams, including the Tuolumne, in search of possible reservoir sites. Thus, at almost the same time that Muir and Johnson were seeking to protect the upper Tuolumne basin as a "forest reservation" (and while the Turlock and Modesto districts were developing plans for the La Grange diversion dam), Hetchy Hetch Valley was officially designated as California "Reservoir Site No. 33" in a widely distributed USGS report. Specifically, this report noted that "the height of [a prospective Hetch Hetchy] dam is 100 feet . . . [and] the approximate content of [the] reservoir is 25,000 acre feet. . . . The reservoir site is partly settled. The irrigable lands are in the San Joaquin Valley, 70 miles to the west."[39]

With this brief description, Hetch Hetchy Valley officially achieved recognition as a reservoir site capable of helping irrigate land some seventy miles downstream. In addition, the report acknowledged that hundreds of acres in Hetch Hetchy were already "partly settled" and had been "patented" into private ownership during the 1880s. In its concluding sentence, the report noted that all land in the valley (minus that already in private hands) "has been reserved as forest reservation, October 1, 1890." Absent was any discussion of how a dam at Hetch Hetchy might be financed or what its relationship to the Turlock and Modesto districts might be. Additionally, other than acknowledging that a "forest reservation" encompassing the valley had been established by federal legislation, there was no consideration of how such a reservation—one specifically intended to protect "natural curiosities or wonders"—might affect future dam construction plans.

Although Congress curtailed funding for Powell's irrigation survey in late 1890, USGS subsidies for streamflow measurement continued through the remainder of the decade. In 1899 a further USGS investigation of the Tuolumne River watershed was initiated, this time under the supervision of consulting engineer J. B. Lippincott and his assistant John H. Quinton.[40] Building upon streamflow measurements recorded at the La Grange Dam gauging station, most of this 1899 report focused specifically on Hetch Hetchy Valley ("a mountain meadow at an elevation of 3,700 feet above the sea") and how storage at the site could support irrigation of "about 250,000 acres in the Turlock and Modesto districts." Such storage was possible because of "rugged granite walls . . . that seem to rise almost perpendicularly upon all sides to a height of 2,500 feet above this beautiful emerald meadow. . . . [T]he granite walls on either side approach so as to confine the river at low water to a width of about 20 feet, and this has been selected as a dam site."[41]

In the 1899 report, Hetch Hetchy is exalted as a "beautiful emerald meadow," but the focus quickly shifts to the valley's "narrow outlet" that—true to the prevailing mission of the USGS—had been "selected as a dam site." No mention is made of the fact that the valley lies within the borders of a federally protected forest reservation. Instead, the bulk of the investigation considers how the storage capacity provided by a curved gravity dam 150 feet high (impounding more than 40 billion gallons) could support increased agricultural production.[42]

The 1899 USGS report seems to constitute a rather straightforward delineation of the viability of storing water at Hetch Hetchy to enhance the economic fortunes of Turlock and Modesto farmers. But buried in the text is a short paragraph positing a different use for the reservoir, whereby it could "furnish the city of San Francisco with an unfailing supply of pure water. Without entering into details, it will suffice to say that the dam and reservoir as proposed would ensure a supply in the driest years of 250 gallons per diem per capita for 1,000,000 people."[43] Beyond this, no further explanation is offered on how water stored in a Hetch Hetchy reservoir might be delivered to a million citizens in San Francisco. Also left unstated is exactly how a municipal use of

Tuolumne Cascade above Hetch Hetchy, circa 1920. The upper Tuolumne watershed was included within the boundaries of Yosemite National Park, but the waterscape also offered the possibility of providing an abundant and remarkably pure water supply for greater San Francisco. Author's Collection.

Tuolumne streamflow might transgress upon the rights of Turlock and Modesto farmers. Nonetheless, the idea that greater San Francisco could draw upon the Tuolumne for a future water supply entered the arena of public discourse. And this possible use had come with the seeming endorsement of USGS engineers.

So questions arise: What was the status of San Francisco's water supply system at the turn of the century? And would city officials welcome the possibility that—as briefly articulated in the 1899 USGS report—a new water supply system originating in the Sierra Nevada might serve the needs of a rapidly growing urban community?

San Francisco's Water Supply

In 1847, when US soldiers first occupied San Francisco (formerly known as Yerba Buena), the settlement had a population of about 1,000 people. This quickly changed after the Gold Rush brought tens of thousands of Euro-American settlers into California. By 1850 the port of San Francisco supported approximately 25,000 residents; a decade later, the number had risen to almost 57,000. At the start of the twentieth century, the US Census tallied a city population of more than 342,000.[44]

The San Francisco community always struggled to find reliable sources of fresh water. Located at the northern end of a rocky peninsula, the city was surrounded by salt water on three sides (Pacific Ocean to the west, the Golden Gate channel to the north, San Francisco Bay to the east), and as the population skyrocketed, local wells and cisterns quickly proved inadequate. In the early 1850s, a private company built a small conduit delivering water from Los Lobos Creek (near the Golden Gate) to the city center. But it was not until 1858 that a state law created the legal means for organizing private water companies such that investors might have confidence that water rates would be regulated in a manner protecting the long-term viability of their investments. In 1858 both the existing San Francisco Water Company and the nascent Spring Valley Water Works organized under the new law, with city residents receiving service from these privately-financed enterprises. In January 1865, the two companies merged under the name Spring Valley Water Works, marking the beginning of the city's modern-day hydraulic infrastructure.[45]

The key to the Spring Valley company's early success was its ability to tap water supply sources lying within the coastal range escarpment covering the San Francisco Peninsula. In the early 1860s, Spring Valley began building a lengthy supply line originating at a storage reservoir along Pilarcitos Creek (a coastal stream flowing into the Pacific Ocean) lying almost 700 feet above sea level. With a storage capacity of 950 million gallons, Lake Pilarcitos fed water into a thirteen-mile-long, 30-inch-diameter pipe connecting to the Lake Honda Distributing Reservoir inside the city limits. At 365 feet above sea level, Lake Honda could, by gravity flow, serve the city's main business district as well as many urban residents. Construction of the pathbreaking Pilarcitos system (completed in 1866) was overseen by the company's hydraulic engineer Hermann Schussler, who, over the next forty years, engineered an expansive water supply system that first extended down the San Francisco Peninsula and then moved into the Alameda Creek watershed lying east of the San Francisco Bay. Over several decades,

Located on the San Francisco Peninsula south of the city, the Spring Valley Dam (commonly called Crystal Springs Dam) was constructed in the 1880s by the corporate predecessor of the Spring Valley Water Company. Since 1930 the massive concrete gravity dam has been owned by the City of San Francisco and operated as part of the municipal water supply system. Author's collection.

the company invested millions of dollars into Schussler's system, including $2 million to construct the 145-foot-high Crystal Springs Dam (also called Lower Crystal Springs Dam or Spring Valley Dam). Upon completion in 1888, this massive concrete gravity structure stood as one of the largest dams in America. By 1900, Spring Valley delivered on average more than 25 million gallons of water per day to the people and businesses of San Francisco.[46]

The success of Spring Valley did not come without controversy, as city officials perceived it to be a monopoly more interested in generating profits than in sustaining a valuable public service. In the 1870s, city leaders set out to purchase the assets of Spring Valley, but while the company thought a fair price would be about $16 million, the city considered $11 million to be more reasonable. This failure to reach a "meeting of the minds" initiated a protracted struggle that in 1900 spurred changes to the San Francisco City Charter, mandating a municipally owned water supply system.[47]

Even after Spring Valley came to dominate San Francisco's water infrastructure in the mid-1860s, other companies sought to develop water supply systems to serve the city. The most famous of these was the Lake Tahoe and San Francisco Water Works Company organized by the engineer/entrepreneur Alexis Von Schmidt. With a bit of bombast, Von Schmidt proposed bringing water from Lake Tahoe (on the eastern slope of the Sierra Nevada) to Oakland and San Francisco via "the Grandest Aqueduct in the World." Exactly how practical such a project was when first proposed in 1870 is subject to debate. But Von Schmidt had previously worked for the Spring Valley Water Works and, in 1870, had enhanced his engineering reputation by skillfully removing an imposing navigational hazard in San Francisco Bay. Although the Lake Tahoe scheme never bore fruit, it did raise the possibility that water supplies in the Sierra Nevada could someday slake the thirst of greater San Francisco.[48]

Through the 1880s and into the 1890s, the Spring Valley company continued to expand its system on the San Francisco Peninsula and within the Alameda Creek watershed. But a broader awareness of the water resources of the Sierra Nevada remained ever present, fostered by the work of the USGS, by local irrigation advocates, and by a growing interest in hydroelectric power. In San Francisco, a desire to supplant Spring Valley with a municipal water supply system gained significant momentum following the election of James D. Phelan as mayor in 1897.

Born in San Francisco in 1861, Phelan was the son of an Irish immigrant who had journeyed west in the Gold Rush and attained financial success as founder of San Francisco's First National Gold Bank. Following in his father's footsteps as a prominent and savvy banker, young James also became a leading member of the Democratic Party, an opponent of monopolistic "trusts," and a proponent of municipally owned utilities.[49] Upon becoming mayor, he joined with other Democrats, including the newly elected city attorney Franklin K. Lane, to push for a new city charter that advanced such causes.[50] More than 300 pages long, this charter became legally effective in January 1900. Notably, the charter specifically addressed the issue of public utilities, including water supply: "It is hereby declared to be the purpose and intention of the city and county of San Francisco that its public utilities shall be gradually acquired and ultimately owned by the city and the county."[51]

To fulfill the charter's mandate, future water supplies could be obtained by "condemnation" (i.e., eminent domain action through state courts), by "negotiations for the permanent acquisition" of assets owned by existing water companies, or by "original construction" of completely new projects. Precisely how the city was to take control over a municipally owned water supply was not detailed in the charter, but the city was authorized both to build a new system and/or to take ownership of an existing system (such as the Spring Valley enterprise). The only significant constraint in carrying out this mandate was that any project or purchase that incurred "municipal bonded indebtedness" would have to be submitted to city voters and approved by "at least two-thirds of the electors voting thereon." A simple majority would not suffice when it came to taking on debt and issuing municipal bonds.[52]

With authorization of the new city charter and with the Quinton/Lippincott USGS report endorsing the viability of a large dam at Hetch Hetchy, Mayor Phelan was ready to move on a project that would bring water from the Sierra Nevada to San Francisco. But action on the part of Phelan and the city awaited passage of a new federal law, one intended to facilitate economic use of land lying within the public domain and, quite explicitly, within the "forest reservation" known as Yosemite National Park.

The 1901 Right-of-Way Act

The 1890 federal law creating the "forest reservation" later designated as Yosemite National Park seemingly blocked commercial development within the confines of the reserve. Specifically, the act obligated the secretary of the interior to preserve "natural curiosities and wonders" in their "natural condition." But a Congress that ostensibly authorized such protection could, at some later time, also pass laws allowing uses of the park's resources that people like John Muir or Robert Underwood Johnson would detest. In fact, not long after completion of the Quinton/Lippincott USGS report on the Tuolumne River, Congress enacted the Right-of-Way Act of February 15, 1901, a law creating the foundation for San Francisco's initial attempts to dam Hetch Hetchy.[53]

In the abstract, the Right-of-Way Act was seemingly quite reasonable, as it provided a mechanism for individuals, corporations, irrigation districts, and municipalities to apply to the secretary of the interior for authorization to use public land controlled by the federal government. Specifically, the law allowed for the construction of "water plants, dams, and reservoirs, to promote irrigation . . . or the supplying of water for domestic, public, or any other beneficial public use." Adhering to the precepts of progressive conservation, such legal prescriptions aligned with the idea that the government should work to develop natural resources in the public domain to provide the greatest good to the greatest number of people. Federally owned resources were to be used, but they were to be used wisely under the direction of capable government officials.

Complicating the authority accorded by the act, however, was the provision that any permit issued by a secretary could, at some future time, be revoked or amended by a subsequent interior secretary, someone who might have a different view of the need for or efficacy of a previously issued permit. In essence, no permit issued under the act was necessarily permanent or inviolate. As a further complication—and one that specifically affected Hetch Hetchy—the Right-of-Way Act expressly permitted "the use of rights of way through the public lands, forest and other reservations of the United States, *and the Yosemite, Sequoia, and General Grant National Parks* [emphasis added]."[54]

Therein lay the genesis of the Hetch Hetchy controversy. A federal law had created a national park (or reserve) with boundaries that included Hetch Hetchy Valley. Subsequently, a different federal law authorized the interior secretary to allow municipal and commercial development of resources within this same park. The question thus became: Which law should take precedence? The guidance given by the Right-of-Way Act to resolve such conflict was that any permits approved by the interior secretary were not to be "incompatible with the public interest." So, in the context of Hetch Hetchy Valley, what possible uses would serve the "public interest" and what criteria were appropriate for determining what uses were "incompatible" with the ideals of a national park? As it turned out, the battle over these questions relative to Hetch Hetchy stretched out for more than a decade.

Passage of the 1901 Right-of-Way Act initially prompted little excitement or attention, with Muir and his compatriots in the Sierra Club taking no notice of it.[55] But San Francisco's Mayor Phelan was aware that the new law might serve the interests of the city. Using the recent USGS report as a springboard, in July 1901 Phelan personally filed a water rights claim to 250 cubic feet per second of Tuolumne River flow at the Hetch Hetchy gorge.[56] Later in the fall, and under the authority of the Right-of-Way Act, he followed up with an application to Secretary of the Interior E. A. Hitchcock seeking permission to build storage dams at Hetch Hetchy and Lake Eleanor (located along Eleanor Creek, a tributary of the Tuolumne River entering downstream from Hetch Hetchy).[57] During the course of the next year, City Engineer Carl E. Grunsky prepared a "Report on the Tuolumne River Water Supply Project," which he submitted to the city's board of public works in July 1902. Here, Grunsky described a reservoir and aqueduct system with a capacity of 60 million gallons a day and costing about $39 million. Hydroelectric power would be developed as part of the Grunsky plan, but essentially all of this power would be used to pump aqueduct water some six hundred feet over the Mount Diablo Coastal Range separating the San Joaquin Valley from the lowlands surrounding San Francisco Bay.[58]

As a follow-up, in November Grunsky also prepared "Report on the Availability of Water Supply Sources for San Francisco," which offered a justification for the city's need to dam Hetch Hetchy. After assessing the capacity and availability of rivers draining into the Sacramento and San Joaquin watersheds, the possibility of filtering water pumped from the Sacramento River, and the resources of the Spring Valley Water Works, Grunsky concluded that "in light of information now at hand [the Tuolumne River] is the most available source of supply for an independent system of municipal water works."[59]

By 1903 Phelan had transferred his Hetch Hetchy/Lake Eleanor application to the City of San Francisco, with hopes that Secretary Hitchcock would, under authority of the 1901 Right-of-Way Act, authorize a permit. The city soon discovered, however, that the secretary held little interest in such plans. Taking the provisions of the 1890 law to heart and, recognizing that there was nothing in the 1901 act that *required* him to issue any permits, in December 1903 Hitchcock determined, "If natural scenic attractions of the grade and character of Lake Eleanor and Hetch Hetchy valley are not of the class which the law commands the Secretary to preserve and retain in their natural condition, it would seem difficult to find any in the Park that are." He further emphasized how the 1890 law had been specifically intended to protect natural wonders such as Lake Eleanor and Hetch Hetchy and suggested that "it is the aggregation of such natural scenic features that makes the Yosemite Park a wonderland which the Congress of the United States sought by law to preserve for all coming time." Lest there be any doubt, Hitchcock made his position clear: "It is inconceivable that it was intended by the Act of February 15, 1901, to confer any authority to be exercised [by the Secretary] for the subversion of those natural conditions which are essential to the very purposes for which the park was established. . . . [The law] clearly defines my duty . . . I am constrained to deny the application."[60]

Hoping that Hitchcock's mind might be changed, City Attorney Franklin Lane requested a rehearing, pleading that "the building of the dams would not detract from the natural beauties and wonders of the Yosemite National Park. . . . Hetch Hetchy Valley would assume the character of a mountain lake of unusual beauty."[61] But Hitchcock was unmoved. In February 1905 he again rejected the city's application, averring, "I do not see what good end would be subserved by especially setting aside reservoir sites on the Government lands in the Yosemite National Park. . . . In my judgement, this wonder of nature [at Hetch Hetchy] could not be used for reservoir purposes except through further legislation by Congress."[62] With Hitchcock's second denial, the Hetch Hetchy issue appeared to be dead, at least so long as he remained interior secretary or until Congress approved further legislation.

A Plan Dropped and Revived

In November 1901, James Phelan chose not to run for reelection as mayor, and the office was taken by Eugene Schmitz, the Union Labor candidate and protégé of urban political boss Abe Ruef. At first, the Schmitz/Ruef administration supported the Phelan-driven plan for a Hetch Hetchy water supply system. However, once Secretary Hitchcock rejected the city's application for a second time in 1905, Schmitz and the board of supervisors began to search for other water sources. In February 1906, the supervisors formally signaled this policy change with "Resolution No. 6949": "Whereas, the city has expended thousands upon thousands of dollars in tentative efforts to secure the Tuolumne or Hetch Hetchy system . . . [and] it is said that the Tuolumne supply cannot be acquired for years to come, if at all. . . . Resolved, that the City refrain from spending any further money, energy, or time in the futile attempt to acquire the so called Tuolumne system."[63]

As of early 1906, Hetch Hetchy appeared to be permanently off the table as a supply source for the city. But while the supervisors and the mayor were focused on pursuing other options, not everyone in San Francisco shared their perspective. The most important person who refused to give up the dream of a Hetch Hetchy reservoir was the engineer Marsden Manson. Born in Leewood, Virginia, in 1850, Manson graduated as an engineer from the Virginia Military Institute in 1873. His introduction to hydraulic engineering came the next year when he began working for the James River and Kanawha Canal. In 1877, he relocated to California, joining the State Engineering Department and, a few years later, receiving a PhD from the University of California at Berkeley. With degree in hand, in the 1890s he became chief engineer of the California State Board of Harbor Commissioners. In 1899 he tied his professional fortunes to the city of San Francisco, taking charge of the municipal sewer commission and, between 1900 and 1904, serving on the San Francisco Board of Public Works. During this time, Manson became a champion of the Hetch Hetchy project.[64]

It was also during this time that Gifford Pinchot, head of the US Forest Service, began to take an interest in promoting Hetch Hetchy as a San Francisco reservoir site. Exactly what prompted him to take up the city's cause is unclear, but apparently it drew in no small part from a desire to oppose the "monopoly" controlled by the privately owned Spring Valley Water Company (in 1904, this company became the legal successor to the Spring Valley Water Works).[65] Pinchot was a conservationist who believed in government regulation of natural resources while also advocating for policies that provided the greatest good to the greatest number for the longest time. As early as February 1905, he had recommended to President Theodore Roosevelt "the reservation of the Hetch Hetchy and Tuolumne Meadows reservoir sites for the eventual use of San Francisco"; he also told William Colby, secretary of the Sierra Club, "that I feel strongly that San Francisco must have an adequate water supply, and it has seemed to me that the action of [Hitchcock] was based purely on a technicality and entirely failed to meet the needs of the situation."[66]

In the face of opposition offered by the Schmitz administration, there seemed little likelihood that, despite what Manson or Pinchot might desire, the Hetch Hetchy initiative would be revived any time soon. Two events, however—one natural and the other political—opened the door for reconsideration. The first came on April 18, 1906, when San Francisco was jolted by a major earthquake measuring almost 8.0 on the Richter Scale. In the immediate aftermath of the shock, property damage seemed relatively modest. But then massive fires ignited, raging across the urban landscape and taking a deadly toll. A key reason the conflagration proved so destructive was that the hydrants and pipes operated by the Spring Valley system ran dry, with many supply lines broken by the shifting terrain. The problem was not that water was unavailable in Spring Valley's reservoirs, it was that pipes needed to deliver and distribute this water were breached and broken by the severe temblor.[67]

The horrific fires sparked by the earthquake brought attention to the failings of the Spring Valley company and, rather incongruously, resurrected the idea that perhaps it would be in the city's best interest to develop a municipally owned Sierra water supply. Of course, a reservoir at Hetch Hetchy, even if filled to the brim, could have done nothing more to ameliorate the effect of the earthquake-induced fires than water stored in Spring Valley's Crystal Springs reservoir located about twenty miles south of the city. Nonetheless, the flames that ravaged San Francisco upended the status quo relative to water supply. Despite the board of supervisors' resolution passed just a few months prior, the idea of a municipally owned Hetch Hetchy system came back in play.

Postcard view of ruins along Montgomery Street; the powerful earthquake that shook San Francisco on April 18, 1906, caused some buildings to collapse, but most damage resulted from fires that spread unchecked because of breaks in water supply mains. Author's collection.

About three weeks after the earthquake, Marsden Manson penned a brief handwritten note to Pinchot raising the possibility that changes might be afoot: "My Dear Mr. Pinchot, it is possible that the application for reservoir rights of way will be renewed by S.F. . . . Be ready to help us if this matter is taken up again."[68] Upon hearing the news, Pinchot happily offered a helping hand: "I was very glad to learn . . . that the earthquake had damaged neither your activity nor your courage. I hope sincerely that in the regeneration of San Francisco its people may be able to make provision for a water supply from the Yosemite National Park, which will probably be equal to any in the world. I will stand ready to render any assistance which lies in my power."[69]

If the earthquake-induced fires had been the only factor at play working to revive the Hetch Hetchy scheme, then Manson's hopes might well have languished unfulfilled. But the Schmitz/Ruef administration did not fare well in the aftermath of the disaster, and charges of graft and corruption soon emerged. The grievances were far-ranging; one line of inquiry concerned efforts by the Bay Cities Water Company to collude with Ruef and his cronies in the purchase of

One of the Spring Valley Water Company supply pipes that ruptured during the April 1906 earthquake. Such breaks prevented water in reservoirs south of the city from reaching urban fire hydrants. (Hermann Schussler, *The Water Supply of San Francisco* [New York, 1906], photo #6, 80)

a purported water supply tied to the Cosumnes and American Rivers. The Bay Cities scheme, with a reported one-million-dollar kickback marked for Ruef, never received any authorization or approval. But with Ruef's and Schmitz's indictment on other charges at the end of 1906, the board of supervisors' anti–Hetch Hetchy/Tuolumne River resolution soon faded into irrelevance. What had seemed impossible after Hitchcock had rejected the city's Hetch Hetchy application for a second time in 1905 had become, if not likely, at least conceivable.[70]

In March 1907, Secretary Hitchcock had been in office eight years. Then, at the age of 71, he resigned his post. Exactly why he stepped down is not known, although his health was such that he died barely two years later. Regardless of the reason, by resigning he gave President Roosevelt an opportunity to appoint a new secretary of the interior, one who might have a different view of how best to conserve, and use, federal land in the Sierra Nevada. With the concurrence of Pinchot, Roosevelt chose as his new interior secretary the 42-year-old James Garfield, the namesake son of the president who had been assassinated in 1881. The younger Garfield was a graduate of Williams College, had studied law at Columbia University, and was a prominent member of the Cleveland, Ohio, bar. Since 1904 he had served as commissioner of the US Bureau of Corporations (a predecessor of the Federal Trade Commission).[71] Upon taking office as interior secretary on March 5, 1907, Garfield began to consider permitting San Francisco to use Hetch Hetchy. In late July, he held a hearing in San Francisco during which he heard arguments in favor of such a plan offered by Manson, the newly appointed mayor Edward Taylor (a replacement for the disgraced Schmitz), and former mayor James D. Phelan, among others; he also heard lengthy objections raised by irrigation lawyers from Turlock and Modesto who considered the Tuolumne to be the exclusive domain of the two irrigation districts. In the end, nothing official was decided in the two-and-a-half-hour-long meeting, but it did give public notice that Hetch Hetchy was back under consideration as a reservoir site and that Garfield considered it plausible that the resources of the Tuolumne were sufficient to meet the needs of both the irrigation districts and the city.[72]

In helping resuscitate San Francisco's application for a Hetch Hetchy dam, Garfield no doubt acted in a manner amenable to the beliefs of his boss, President Roosevelt. In 1903, Roosevelt had toured Yosemite National Park in the company of Muir, and by all accounts, the two men had gotten along famously. Seemingly, they were almost brothers-in-arms, sharing in the ideals of conservation and the joys of camping and outdoor life.[73] After that joyous meeting, Roosevelt might have been expected to oppose San Francisco's plans for flooding Hetch Hetchy. But Roosevelt also had strong ties to Pinchot, and he looked to the bureaucratic forester for counsel in formulating government policy on how to best utilize the nation's natural resources. When it came to whether Hetch Hetchy should be used as either park or reservoir, Roosevelt needed to choose either Muir or Pinchot as a guiding north star. He picked Pinchot.[74]

In November 1907, Democratic Party stalwart Edward Taylor was formally elected San Francisco mayor and, as Taylor's city engineer, Manson formally revived the city's Hetch Hetchy application under the 1901 Right-of-Way Act.[75] Garfield and the Roosevelt administration were ready to respond, and in an April 1908 letter to the editor of the nationally renowned *Collier's Weekly*, Pinchot laid out the basic justification to be used when the government approved the city's plans. In this letter, Pinchot opened with a complaint calling out the private Spring Valley company: "The present water supply of the city of San Francisco is both inadequate and unsatisfactory. . . . The water companies, interested in supplying the city with water for business reasons, have taken advantage of the long delay since it was first proposed to bring water from the Sierra Nevada to San Francisco."[76]

Pinchot believed that a safe and voluminous municipal water supply represented the best use of the Tuolumne River: "Municipal supply is the highest use to which water can be put. . . . Practical questions of the comfort, health, safety, and convenience of great bodies of people must have the consideration to which they are fairly entitled. . . . The demands of a great population such as that of San Francisco Bay must be met, and . . . if they cannot be met without taking Hetch Hetchy then Hetch Hetchy should be devoted to this use." Although he had made no study of alternative sources, Pinchot's support for damming Hetch Hetchy came with the notable caveat that the valley should only be used as a reservoir if all other possibilities proved insufficient. In addition, if San Francisco was granted a permit to dam Hetch Hetchy, Pinchot believed that the city should be obligated to fully develop the Eleanor Creek watershed as a supply source before expanding the system to include Hetch Hetchy: "First, the city of San Francisco should be given the Lake Eleanor reservoir site and should be required to develop and use it to its fullest capacity. Second, the Hetch Hetchy reservoir site should be reserved for the city, but it should not be used until Lake Eleanor has been used to its full capacity."[77]

While acknowledging that he had "never seen Hetch Hetchy" in person, Pinchot felt confident that the reservoir would inflict no grave harm on the national park. In his words to *Collier's*: "If it becomes necessary to transform the floor of Hetch Hetchy into a lake instead of its present condition as a ranch, I do not believe that serious injury to th[is] scenic valley of the Sierras will be done."[78]

The Garfield Permit

Less than two weeks after Pinchot sent his letter to *Collier's*, Secretary Garfield issued a permit to the city under the authority of the 1901 Right-of-Way Act. Taking a very different tack than his predecessor Hitchcock, Garfield saw no reason that the 1890 law authorizing Yosemite National Park required him to protect the natural wonders of Hetch Hetchy (and Lake Eleanor) from use by San Francisco. In issuing what has come to be known as the Garfield Permit, the new secretary followed a path closely aligned with Pinchot's missive sent but a few days before. In fact, the entire second paragraph of Pinchot's letter, starting with the sentence "The present water supply of the city of San Francisco . . . ," and most of the third paragraph appears verbatim in the Garfield Permit.[79]

In his edict, Garfield averred that "the 'public interest' will be much better conserved by granting the permit," and, like Pinchot, he declined to independently investigate whether San Francisco could meet its future water supply needs from other sources: "In considering the reinstated application of the City of San Francisco, I do not need to pass upon the claim that this is the only practicable and reasonable source of water supply for the city. It is sufficient that after careful and competent study the officials of the City insist such is the case." But in allowing the city to proceed with its plans for damming Hetch Hetchy, Garfield included the all-important caveat requiring complete development of Eleanor Creek before work at Hetch Hetchy could proceed: "San Francisco will develop the Lake Eleanor site to its full capacity before beginning the development of the Hetch Hetchy site." In addition, San Francisco was not to "interfere in the slightest particular with the right of the Modesto Irrigation District and the Turlock Irrigation District to use the natural flow of the Tuolumne River . . . to the full extent of their claims." And lest there be any question as to the districts' prevailing water rights, the permit specifically acknowledged that the districts possessed rights to a flow of 2,350 cubic feet per second under California law.[80]

In another key stipulation, Garfield took notice of how "San Francisco practically owns all the patented land in the floor of the Hetch Hetchy reservoir site and sufficient areas adjacent areas in the Yosemite National Park and the Sierra National Forest to equal the remaining of that reservoir area. The city will surrender to the United States equivalent areas outside of the reservoir sites and within the National Park and adjacent reserves in exchange for the remaining land in the reservoir sites, for which the authority from Congress shall be obtained if necessary."[81] There did not appear to be anything of particular consequence in this latter proviso, as it seemed to provide a relatively straightforward means for the city to take ownership of all land to be inundated by the proposed reservoirs. To the city's dismay, however, the proposed land swap subsequently became the focus of virulent political opposition.

Lest there be any doubt that President Roosevelt approved of the Garfield Permit, one need only consider the timing of the secretary's action. The permit was publicly announced in Washington, DC, on May 11. Just two days later, President Roosevelt convened a gala three-day conference of governors at the White House, organized under the theme "Conservation of Natural Resources." More than two hundred politicians, industrial magnates (including Andrew Carnegie), and conservation leaders from around the nation participated in the gathering. At the conference, the Garfield Permit stood as a symbol of how Roosevelt viewed the proper goals of conservation. Sustaining the economic and social growth of San Francisco took precedence over protection of a remote valley in the northern reaches of Yosemite National Park. Furthermore, to help preclude pernicious grumbling, Roosevelt and Pinchot left Muir off the conference invitation list. In

1903, Roosevelt and Muir may have shared the pleasures of Yosemite Valley, but by 1908 the president was walking a different path.[82]

With the requirement that Eleanor Creek be fully developed as a water supply before damming Hetch Hetchy, City Engineer Manson set out to reconfigure the aqueduct design so that the use of an enlarged Lake Eleanor came first. Adjustments to Grunsky's 1902 plan were needed, but by locating the aqueduct intake near where the flow of Eleanor Creek entered the Tuolumne River, Manson's reconfigured design could, when the time came years hence, readily accommodate water stored in a future Hetch Hetchy reservoir. The aqueduct design below the intake did not vary much from the 1902 precedent, and the Manson plan, like that of Grunsky, projected an ultimate capacity of only 60 million gallons per day.[83]

Although the actual construction of a dam at Hetch Hetchy lay many years in the future, Manson and city leaders wanted to formalize control over the reservoir site by taking ownership of all land to be inundated in the valley. As part of this effort, the city sought citizen approval for a $600,000 bond issuance to facilitate the proposed land purchases in the Sierra. In November 1908 a referendum to this effect was brought to city voters; any question as to the electorate's interest in the Lake Eleanor/Hetch Hetchy project was quickly answered. By a margin of more than six-to-one (34,950 yeas vs. 5,647 nays) the city's Tuolumne River initiative won resounding approval. There were opponents of the city's plan among the local citizenry, but they were a decided minority. While not always monolithic or unquestioning, citizen support for the damming of Hetch Hetchy would remain steadfast in San Francisco in the months and years ahead.[84]

With funding assured, the city sought to initiate the land swap that would exchange privately owned parcels in and near the park for government-controlled land in the Hetch Hetchy and Lake Eleanor reservoir zones. And, as the Garfield Permit postulated, such a trade could be authorized by legislation passed by Congress. In the abstract, it all seemed quite straightforward. But the need to bring the issue of damming Hetch Hetchy to Capitol Hill would offer Muir and fellow defenders of Yosemite National Park an opportunity to rally a nationwide phalanx of preservationists united in opposition to the city's plan.

In Defense of Hetch Hetchy

Prior to 1905, advocates of Yosemite National Park had exhibited little awareness of San Francisco's plans for Hetch Hetchy. And once Secretary Hitchcock had denied the city's application for a second time and city supervisors had passed a resolution disavowing pursuit of a Hetch Hetchy dam, there seemed little reason to be worried about the valley's preservation. But following the earthquake of April 1906, and after the Schmitz/Ruef administration had collapsed under charges of graft and corruption, the possibility of needing a defense of Hetch Hetchy drew the attention of Muir and many other Sierra Club members. Fears intensified once James Garfield held a hearing in San Francisco in July 1907 to discuss reactivation of the city's dormant Hetch Hetchy application. Neither Muir nor any other park defenders of note attended Garfield's hearing, but they were well aware that it had taken place and understood its implications.[85]

A few weeks after Garfield's visit to San Francisco, Muir called together the Sierra Club's board of directors to draft a report affirming Hetch Hetchy to be a "great and wonderful feature of the park, and [one that stands] next to Yosemite [Valley] in beauty, grandeur, and importance." The report also raised the alarm that "if dammed and submerged . . . Hetch Hetchy would be utterly inaccessible to travel[;] . . . campgrounds would be destroyed and access to other important places . . . interfered with." Recognizing that many people in San Francisco supported the city's scheme, the board of directors made a broad appeal opposing local parochial interests: "Whereas we do not believe that the vital interests of the nation at large should be sacrificed and so important a part of its National Park destroyed to save a few dollars for local interests . . . We are opposed to the use of Hetch Hetchy as a reservoir site."[86] Initially drafted for Garfield's benefit, the report was soon distributed to the Sierra Club as a whole. And, while finding a generally supportive audience, it did engender some criticism because of how the club's position could be interpreted as abetting the monopolistic interests of the Spring Valley company. As one member complained, "I most earnestly protest against the Sierra Club being used as a cat's paw to pull chestnuts from the fire for the Spring Valley Water Company."[87]

Despite such objections, most club members embraced the defense of Hetch Hetchy that Muir spearheaded, especially when he drew upon florid rhetoric to chastise the "temple destroyers, devotees of ravaging commercialism, [who] seem to have a perfect contempt for Nature." In early 1908, Muir published his famous essay "The Hetch Hetchy Valley" in the *Sierra Club Bulletin,* where, alongside photographs extolling the valley's beauty, he penned perhaps his most memorable attack on the infidels who wished to defile one of "God's best gifts": "Dam Hetch Hetchy! As well dam for water tanks the people's cathedrals and churches, for no holier temple has ever been consecrated by the heart of man."[88]

After the Garfield Permit had been issued, and in anticipation of House and Senate committee hearings scheduled to begin in December 1908, Muir and Colby apparently took the lead in crafting an illustrated pamphlet filled with Muir's writings on the virtues of Hetch Hetchy as a scenic wonder.[89] The pamphlet—which was not formally attributed to the Sierra Club as a publisher—further espoused the availability of alternative water supply sources and referenced numerous newspaper articles and editorials decrying the assault on the scenic valley. Politically driven, the pamphlet implored people and civic groups all across the United States to write their congressmen and senators and urge them to oppose legislation authorizing San Francisco to take ownership of the entire Hetch Hetchy Valley. In a letter signed by Muir and other Sierra Club members that was printed at the beginning of the pamphlet, the plea went out:

> To all lovers of nature and scenery. . . . We do not believe that a great national property preserved for the enjoyment of the people of the entire nation should be thus unnecessarily sacrificed. . . for the pecuniary benefit of a local interest. . . .We urge you to assist in preventing this incalculable loss to the entire nation by writing, and getting all your friends to write at once to your Congressmen and Senators, [in] Washington D.C. protesting in your own language against this desecration. . . . Now is the critical time. We urge you to write or wire, if only a few words.[90]

In many ways, the public relations campaign of late 1908 and 1909 marked the high-water mark of the anti-dam movement. Thousands of letters and petitions from across the nation flooded congressional offices on Capitol Hill, demanding that the park invaders be stopped. Manson and other city officials tried to make the city's case in testimony before the House and Senate committees, but they failed to present a compelling argument delineating exactly why Congress need to act quickly in allowing the city to take legal control over Hetch Hetchy Valley.[91]

Making the point that no great rush was needed to stave off an impending water famine, the lawyer for Spring Valley sharply criticized the city's motives and claimed that the Hetch Hetchy scheme was simply a ploy to drive down the value of the private company. Complaining that San Francisco's interest in seeking congressional support was little more than a subterfuge designed to provide "a big stick in order that [the city] may wield it over the Spring Valley company and make it come to our terms." While Spring Valley's lawyer may have proffered a fanciful argument in opposing the city's plans for Hetch Hetchy, he did complicate the political scaffolding of the anti-dam coalition. Complaints had already been expressed that Muir and his followers were simply a "cat's paw" for the private water company. Going forward, anti-dam preservationists would not find it easy to draw a clear and convincing line separating their efforts from those of Spring Valley.[92]

Muir himself made no appearance at the congressional committee hearings—in fact, at no point in the Hetch Hetchy saga would he ever travel to the nation's capital and engage with political leaders. However, other anti-dam preservationists, including Robert Underwood Johnson, Horace McFarland of the American Civic Association, and Boston lawyer Edmund Whitman representing the Appalachian Mountain Club, pushed hard against the city in their testimony. In the end, the massive letter-writing campaign and the persuasive objections raised by the city's opponents, proved effective. No legislation advancing San Francisco's cause came to either the House or the Senate floor. By March, the legislative session had ended, and any further action on the city's proposed "land swap" would need to start afresh with a new Congress. As of the spring of 1909, the preservationists' cause was in ascendance.[93]

After Congress failed to authorize San Francisco's initiative to take control of all land in the Hetch Hetchy reservoir zone, the city was on the defensive. The Garfield Permit may have remained in effect, but anti-dam forces had made a strong showing on Capitol Hill, arguing that there was no compelling need for the city to dam Hetch Hetchy Dam in the foreseeable future. A new Congress came into office in March 1909, and the city intended to revive its legislative effort to assume ownership of the valley. Preservationist forces, however, saw momentum on their side and wanted to push for a formal revocation of the Garfield Permit. As Horace McFarland advised at the start of the new legislative session, "[W]e [should] at once get ready for a straightforward, well-organized campaign to have [Garfield's Hetch Hetchy] grant revoked. . . . [We must] plan for a dignified but persistent onslaught on Secretary Ballinger [whom President Taft had appointed to replace Garfield], definitely asking him to revoke the permit. . . . for the reason the people do not want it granted."[94]

The feasibility of alternative supply sources held a prominent place in preservationist objections to the damming of Hetch Hetchy and, in the spring of 1909, Pinchot reached out to Manson for help in countering this crucial line of attack: "I see a great many statements made that there are other and more available supplies than the Hetch Hetchy for San Francisco. Will you kindly give me such info as you can conveniently on this point? It seems to be the one argument upon which the opponents to the San Francisco water supply from the Yosemite Park place their greatest reliance."[95] In response, Manson did not present any hard data or cost analysis justifying the city's position. Instead, in a few pages he offered Pinchot a generalized argument, claiming that "from many of these [alternative sources], it is possible from a physical standpoint disregarding cost and present uses, to obtain a supply, but from a practical and financial standpoint it is not possible." Manson saw pumping and filtering Sacramento River water as "too costly and undesirable," while the Cosumnes River "has no adequate drainage basin and is devoid of suitable storage" and the "Mokelumne River is entirely taken up and all its reservoirs are owned by California Gas and Electric corporations." In summation, Manson simply assured Pinchot, "There is no adequate or available source other than the upper Tuolumne, and Eleanor Creek, from which San Francisco can draw a water supply without depriving others of the use of their water nor is there any supply whatever that is as free from contamination and from complicated opposing interests."[96] Previously, Secretary Garfield had accepted such assertions at face value, without demanding that the city justify in any detail why it needed to dam Hetch Hetchy. In the ensuing year, however, Muir and his supporters had tenaciously pursued the issue of alternative water supply sources, and Manson struggled to conjure a comprehensive and persuasive counterargument that would justify construction of the Hetch Hetchy Dam.

Rather than focusing on why alternative sources were deficient, Manson and Pinchot preferred to raise the specter that private interests (including Spring Valley and investor-owned electric companies) were manipulating Sierra Club members to do their bidding. This was a contentious point and, while not easy to disprove, aroused the indignation of anti-dam advocates. Horace McFarland addressed the issue in a letter to Pinchot: "I have just had opportunity to rigorously cross question Prof. William F. Bade [a member of the Sierra Club board of directors]. . . . I have asked him direct and searching questions with respect to the sources of support for the campaign against the contentions of San Francisco with respect to Hetch Hetchy valley. . . . As to whether or not there was a[ny] water company support, Prof. Bade made the most earnest answer that there never had been the least suggestion of support." This seemed to be an unequivocal denial, but McFarland did acknowledge a significant caveat: "The President of the Spring Valley Water Company, it is said, is a member of the Sierra Club, and he contributes in that way by paying his dues only.[97]

While Sierra Club members protested that they were not beholden to the interests of any private companies, it was undeniable that some Sierra Club

YOSEMITE NATIONAL PARK.

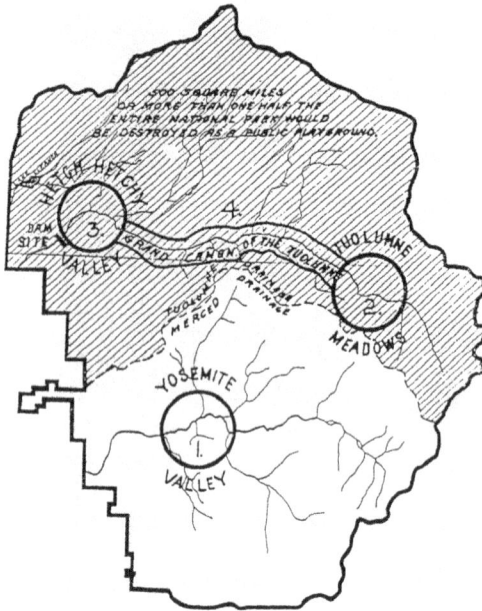

Map publicized by opponents of the Hetch Hetchy Dam in 1909 illustrating their contention that operation of a municipal water supply reservoir at Hetch Hetchy would remove "500 square miles or more than one half of the entire [Yosemite] national park" from public use by campers and destroy this land "as a public playground." (National Archives, Record Group 95, Entry 22, box 4)

members did have ties to private utilities and also that some members of the Sierra Club were supportive of the city's plans to dam Hetch Hetchy. This led, in the spring of 1909, to the formation of a separate anti-dam alliance named the Society for the Preservation of National Parks (SPNP) that, while drawing from the Sierra Club ranks, was intended to include all people who were opposed to the Hetch Hetchy Dam on the grounds of its impact upon the beauties of a national park.[98]

Although the Sierra Club may have been supplanted by the SPNP as the leading anti-dam collective, the effort to defend Hetch Hetchy did not diminish. As the end of the year approached, preservationists again aroused the faithful and published another pamphlet that included a letter from Muir "To the American Public" decrying the city's assault on Hetch Hetchy and accusing the city of "defrauding ninety millions of people [throughout the United States] for the sake of saving San Francisco [a few] dollars"; a listing of possible alternative water supply sources; and a warning (illustrated by a map) that the proposed reservoir at Hetch Hetchy could be used to block all campers and tourists from visiting (and enjoying) the entire upper Tuolumne watershed, including Tuolumne Meadows. Motivated by the presumed renewal of legislation authorizing the city's takeover of Hetch Hetchy, anti-dam proponents stood ready to

LET EVERYONE HELP TO SAVE THE FAMOUS HETCH-HETCHY VALLEY
AND
STOP THE COMMERCIAL DESTRUCTION WHICH THREATENS OUR NATIONAL PARKS

To the American Public:

The famous Hetch-Hetchy Valley, next to Yosemite the most wonderful and important feature of our Yosemite National Park, is again in danger of being destroyed. Year after year attacks have been made on this Park under the guise of development of natural resources. At the last regular session of Congress the most determined attack of all was made by the City of San Francisco to get possession of the Hetch-Hetchy Valley as a reservoir site, thus defrauding ninety millions of people for the sake of saving San Francisco dollars.

As soon as this scheme became manifest, public-spirited citizens all over the country poured a storm of protest on Congress. Before the session was over, the Park invaders saw that they were defeated and permitted the bill to die without bringing it to a vote, so as to be able to try again

The bill has been re-introduced and will be urged at the coming session of Congress, which convenes in December. Let all those who believe that our great national wonderlands should be preserved unmarred as places of rest and recreation for the use of all the people, now enter their protests. Ask Congress to reject this destructive bill, and also urge that the present Park laws be so amended as to put an end to all such assaults on our system of National Parks.

Faithfully yours,

John Muir

November, 1909.

Printed letter from John Muir to the "American Public" distributed in November 1909. This public epistle appeared in a pamphlet published as part of the campaign to block congressional approval of land transfers that would facilitate San Francisco's compliance with the Garfield Permit. (National Archives, Record Group 95, Entry 22, box 4)

continue the fight in the halls of Congress. In addition, they also sought ways to have the Garfield Permit formally revoked.[99] On this latter point, their efforts were soon joined by those of President William Howard Taft and his interior secretary, Richard Ballinger.

"Show Cause"

On his own volition, Teddy Roosevelt chose not to seek reelection to the presidency in November 1908. Subsequently, William Howard Taft took office as president in March 1909 and quickly appointed Richard Ballinger as secretary of the interior in place of James Garfield. Born in 1858 and a lawyer by profession, Ballinger had served as mayor of Seattle, Washington, before being appointed a commissioner of the General Land Office (GLO) in 1907. Ballinger had not played a role in the Hetch Hetchy saga during his time with the GLO, and it appears he did not arrive in his new post with any agenda either pro- or anti-dam.

Because Pinchot remained a prominent part of the public face of the city's Hetch Hetchy initiative, a battle brewing over actions that Ballinger had taken as a GLO commissioner soon began to have an impact on possible changes to the Garfield Permit. Commonly known to historians as the "Ballinger-Pinchot Affair" (or "controversy"), the dispute between the two bureaucrats involved leases of coal lands in Alaska that, in Pinchot's view, were overly generous to J. P. Morgan/Guggenheim industrial interests. The particulars of the controversy are not germane to the Hetch Hetchy story, but in brief, Pinchot saw the leases as evidence that the Taft administration was backtracking on the conservation ideals that Teddy Roosevelt (and his acolyte Pinchot) had championed during his time as president. Not content to keep his criticism of Ballinger out of the public limelight, in August 1909 Pinchot publicly accused Ballinger of being too willing to grant lucrative water-power permits to private syndicates. A month later, he arranged for Ballinger's chief accuser in the Alaska coal leases controversy to have an audience with Taft in which he could make his case to the president. To Pinchot's dismay, Taft quickly came to the defense of Ballinger, declaring that the controversy was overblown and of little import.[100]

With this as context, in early October President Taft visited Yosemite National Park and, like Roosevelt six years prior, was greeted by an amiable John Muir. Over the course of three days, Muir escorted the president through the Yosemite Valley. By all accounts, the two men greatly enjoyed each other's company, with Taft expressing admiration for the Sierra landscape and Muir calling him "the merriest man I ever saw & he made all his company merry." Although Taft's itinerary did not include a visit to Hetch Hetchy, later that month Muir and Ballinger packed into the Tuolumne watershed and inspected the site of San Francisco's proposed reservoir. By that time, Ballinger had come to seriously question whether San Francisco needed to inundate Hetch Hetchy for a municipal water supply, with Muir noting in a letter describing his interactions with the secretary, "Ballinger I also liked & the H.H. scheme seems doomed."[101]

The Garfield Permit had been issued in line with powers accorded the secretary of the interior through the 1901 Right-of-Way Act, and this same act empowered the secretary to revoke any existing permits. After his visit to Hetch Hetchy with Muir, and while his coal lease dispute with Pinchot continued to fester, Ballinger began to pursue a plan to remove the most problematic component of the Garfield Permit. He assigned George Smith, director of the US

Geological Survey, and the engineers Louis Hill and E. A. Hopson of the US Reclamation Service to investigate whether San Francisco truly needed to dam Hetch Hetchy to ensure a reliable, long-term water supply. Manson had done little to document precisely and in detail why other sources were inadequate, and Ballinger was skeptical of the city's broad claim. But he wanted confirmation of his suspicion before amending the Garfield Permit.[102]

Given that the Grunsky/Manson schemes for developing the Tuolumne were limited to a supply of 60 million gallons per day, it was hardly unreasonable to think that other sources might also provide this relatively modest flow. Perhaps alternatives would cost a bit more, and perhaps irrigation developments and hydroelectric power projects would complicate the operation of alternative systems. But it was not particularly difficult to postulate how the city could thrive without relying upon a reservoir at Hetch Hetchy. The possibility also remained that development of the Eleanor and Cherry Creek basins alone in the upper Tuolumne watershed could meet the city's needs for decades to come—and do so in ways that did not unduly interfere with other uses of the park. In his report to Ballinger, USGS Director Smith made exactly this point, describing the Lake Eleanor region as encompassing a "relatively unattractive and inaccessible portion of the High Sierra, being as a consequence much less liable to intrusion by tourist and campers," that could be readily "devoted to the purposes of a municipal water supply, and subjected to the necessarily stringent sanitary control incident thereto with a minimum of interference with the rights of the public." In conclusion, Smith advised the secretary that "the Lake Eleanor project is amply sufficient to meet the present and prospective needs of the city and that it is not necessary that the Hetch Hetchy Valley should be available to San Francisco for the purposes of a municipal water supply."[103]

Relying upon the work of Smith, Hill, and Hopson, on February 25 Ballinger informed San Francisco officials that they needed to "show why the Hetch Hetchy Valley and reservoir site should not be eliminated from said [Garfield] permit." In all subsequent references to this request/demand from the secretary, it would be referred to as a "show cause" (not "show why") order, but the intent remained the same. Ballinger planned to revoke the terms of the Garfield Permit as they pertained to the damming of Hetch Hetchy. But first he would allow city leaders to make an argument in their defense. Should they falter, he would then excise use of Hetch Hetchy from the Garfield Permit.[104]

In the weeks prior to Ballinger's issuance of his "show cause" order, two important events occurred that had bearing on the Hetch Hetchy controversy. The first involved President Taft's dismissal of Gifford Pinchot from his position as chief forester of the US Forest Service on January 7, 1910. The grounds for this firing did not concern Hetch Hetchy per se but instead involved Pinchot's persistent criticism of Ballinger for his role in leasing Alaska coal tracts. For Pinchot, being fired in this circumstance was akin to a badge of honor, evidence of his dedication to protecting the nation's natural resources from wasteful exploitation by private capital.[105]

After leaving the Forest Service, Pinchot continued to fan the flames of the Ballinger-Pinchot Affair, prompting contentious hearings on Capitol Hill during the spring and summer of 1910. Pinchot remained in the public eye as conservationists began to critique Taft's commitment to the Progressive cause, but after January 1910 he detached himself from San Francisco's effort to dam Hetch Hetchy. Three and a half years later he would briefly participate in a House committee hearing relating to the city's project. However, for over forty months after being fired by Taft in January 1910, the former chief forester played no role in the city's struggle to build a dam in Yosemite National Park.

The second event involved two referendums presented to the San Francisco electorate for approval on January 10, 1910. The first requested a $45 million bond authorization to pay for the planned Lake Eleanor/Tuolumne system. On this question the electorate vote was overwhelmingly positive: 32,866 yeas versus 1,609 nays. A two-thirds majority was clearly met, reflecting city residents' enchantment with a mountain water supply. The second referendum involved an agreement negotiated between the city and Spring Valley for sale of the private water company at a price of $35 million. More than three decades had passed since such a transaction was first proposed, and the city finally seemed ready to close the deal. But some taxpayers believed the price tag to be too steep, and the Building Trades and San Francisco Labor Councils passed a joint resolution opposing the proposition because "the properties were not worth the money." When the votes were tallied, they came to 22,068 yeas versus 11,722 nays. A simple majority was easily met, but the yeas fell 454 votes short of the required two-thirds threshold. Fewer than 500 votes constituted a slim margin of defeat, but it was a defeat nonetheless. Spring Valley remained a private company with all its assets intact.[106]

In retrospect, the story of Hetch Hetchy would no doubt have played out differently if a few hundred people had switched their votes in January 1910 and approved the Spring Valley purchase. It would not have quenched the city's desire for a Tuolumne water supply, but it would have removed Spring Valley's leadership as a potential source of opposition to the city's plans going forward. And it would also have obviated the contention that, in some nefarious way, preservationists seeking to protect Hetch Hetchy were acting as pawns for the private water company.

With notice of Secretary Ballinger's "show cause" demand in late February, however, Manson and the city's leadership faced a bigger problem than losing a bond referendum authorizing purchase of the Spring Valley system. In the face of Ballinger's antagonism, they needed help in their battle to keep the Hetch Hetchy component of the Garfield Permit alive. In mid-March the city reached out to a prominent consulting engineer in Providence, Rhode Island, someone experienced in the technology and politics of municipal water supply and wise to the ways of legislative lobbying. In ways impossible to predict at the time, the Hetch Hetchy saga—and whether the valley was to be park or reservoir—was poised to evolve in new and unexpected ways.

CHAPTER 2

Origins and Evolution of a Consulting Engineer

By the time he began championing Hetch Hetchy for use as a municipal reservoir in the spring of 1910, John R. Freeman reigned as one of America's premier hydraulic engineers. Sustained by lucrative consulting fees, he was among the nation's most highly compensated engineers of any specialty (from 1910 through the end of 1913, San Francisco alone paid him more than $50,000).[1] While financially prosperous in later life, Freeman did not benefit from inherited wealth or conspicuous social pedigree in attaining this success. His place among the nation's technological elites accrued through years of hard work, starting in 1872 as a $1.25-a-day engineering assistant in the industrial city of Lawrence, Massachusetts.

Many leaders of the Progressive movement, including Teddy Roosevelt and Gifford Pinchot, were born into prominent families, a providence that helped pave their way as cultural and political leaders. In contrast, Freeman's upbringing aligned more with John Muir, with roots in rural farm life that obligated him to find professional and economic success through sustained, conscientious labor.[2] Decades of rigorous toil—centered on hydraulic engineering, industrial-scale fire suppression, hydroelectric power generation, and municipal water supply—undergirded Freeman's efforts on behalf of San Francisco. While he was not inevitably destined or preordained to become the great advocate for damming Hetch Hetchy, one can nonetheless see in his family heritage and work ethic, in his education, and in his professional endeavors the foundation of his engagement with the "progressive" problem of ensuring a long-term municipal water supply for San Francisco.

As a challenge to would-be biographers, Freeman's career did not follow a simple, single course, especially after he became president of the Providence-based Manufacturers Mutual Fire Insurance Company in 1896. Had he wished, his service as a leading insurance executive could easily have engaged his professional attention full time. But he chose a different path, finding the means to both meet his responsibilities in the insurance world and carry out an extensive practice as a consulting engineer. To characterize him as a "workaholic" might seem strong, but Freeman excelled at multitasking, and when the need arose, he could work seventeen-hour days for weeks at a time. Although we will never know exactly why he drove himself so hard, his work ethic proved crucial to San Francisco's success in winning approval for the Hetch Hetchy Dam. This chapter illuminates noteworthy facets of Freeman's life prior to his engagement with San Francisco; as the venerable adage proclaims, "the child is father of the man." Any hope of comprehending the skills, perspective, and adamancy that Freeman brought to his work on Hetch Hetchy requires knowledge of his upbringing, his education, and key aspects of his far-ranging career.

Born into the farming community of West Bridgton, Maine, on July 27, 1855, John Freeman came from a family of old Yankee stock. Although spared a childhood of deprivation and want, he was hardly a scion of New England's patrician elites. Located west of Portland and some forty miles from the Atlantic coast, the family farm of Nathaniel and Mary Elizabeth Freeman encompassed about one hundred acres of woodland and laboriously cleared fields.[3] In a good year, the tract could supply food for the family's sustenance and perhaps produce an income of a few hundred dollars from logging and pastoral agriculture. Freeman's early years were akin to those of many other rural youths in nineteenth-century America, trapping woodchucks and joining his grandmother on "visits to woods and meadow" in search of herbs. At age eleven his father gave him a small shotgun, whereupon he had "great sport in the neighboring woods." And, in prelude to his life path, his father also showed him "how to dam the brook in the nearby pasture with stones and sod, making a small pond for [his] flock of geese." Together, they built into this dam "a sluiceway, a flume, and several overshot water-wheels."[4]

The rural landscape had its attractions, but young John was also enthralled by a treasured gift from his mother, a "large toy locomotive run by clock springs . . . which [he] took apart and put together repeatedly." This fascination

John R. Freeman with parents Nathaniel and Mary Elizabeth, circa 1867. Two siblings died in infancy, and John was raised as an only child. (John Ripley Freeman Papers [MC 51, box 1], MIT Libraries, Cambridge, Massachusetts)

with machinery later led to investigations of how sewing machines worked, and, while in high school, he earned "a little handy cash . . . repair[ing] discarded machines of various makes." He also recalled his "thrills" when his father took him to "the Atlantic Mills [in Lawrence] to see its big Corliss Steam Engine and the wonderful big gearing of its turbine room."[5]

Freeman may have experienced an idyllic early childhood, one that sparked memories of "cheerful pictures of winter evenings by [the] fireside."[6] Nonetheless his mother, Eliza, chafed at life in the Maine backcountry. Seeking broader prospects for herself and her young son (two siblings did not survive infancy), she left West Bridgton in the early 1860s.[7] Upon moving to the industrial city of Lawrence, Massachusetts, she ran a boardinghouse for the Atlantic Cotton Mills; later, she opened a "fancy goods shop" with her sister in Portland, Maine. Aside from some urban sojourns as a mechanic, sawyer, and mill watchman, John's father remained on the farm in West Bridgton, requiring the young Freeman to shuttle seasonally between urban and rural domiciles. As he explained in the autobiography he was writing at the time of his death, in her first years of married life his mother "rejoiced in having a farm of her own, but soon its narrowness and monotony became apparent . . . [and] she sought a broader experience at about the time I became of school age." She also resolutely determined that her son "should have better opportunities than the scope of the farm and country school offered."[8]

While father Nathan possessed the mechanical skills of a Yankee farmer, mother Eliza constituted the driving force behind John's engagement in the world of modern industry. In his teenage years, she returned to Lawrence, enabling her son to attend high school and obtain a firm foundation in the sciences and math. Freeman described his mother as the impetus underlying his desire to attend college, lauding her determination "that I should have opportunity for a good education." She recognized her son's technical aptitude, and "upon learning about the Massachusetts Institute of Technology recently established . . . she visited it alone, made many inquiries of its secretary, and determined that [her son] should be prepared to enter in due course."[9]

By the spring of 1872, Freeman was ready to act upon his mother's dream and, forgoing his senior year at Lawrence High School, take the MIT entrance examination. The test, however, would not be administered until October. During the interceding summer, he took his first steps toward becoming a hydraulic engineer, hiring on as an assistant with the Essex Company in Lawrence. In nineteenth-century America, there was no better place to learn the art and science of water power than in the heart of the Merrimack River valley.

Water power was vital to the industrialization of America, and in the early nineteenth century, no watercourse proved more important than the Merrimack River, a major stream flowing south from New Hampshire's White Mountains into northeastern Massachusetts.[10] After the end of the War of 1812, Francis Cabot Lowell and fellow Boston-based investors began planning for large-scale textile factories patterned on British industry and in 1821 purchased farmland adjoining the Merrimack River's Pawtucket Falls.[11] There they created the city of Lowell, Massachusetts, a huge manufacturing center featuring a dense assemblage of textile mills driven by the power of falling water. In the 1830s, famed hydraulic engineer James B. Francis took charge of the canals, waterwheels, and (later) turbines that provided Lowell's mills with a generating capacity of more than 10,000 horsepower. Through Francis's work at Lowell, the Merrimack River became world famous in the realms of water power and hydraulic engineering.[12]

Following Lowell's success, investors quickly sought out other power sites along the Merrimack. Downstream from Lowell lay Bodwell Falls, the river's last major drop before the current was absorbed by tidal flow. Much power was available at these falls, but a huge investment of capital—estimated at one million dollars—was required to finance a massive "Great Stone Dam" (over thirty feet high and nine hundred feet long) that could supply power to an array of textile mills rivaling Lowell. In 1845, Abbot Lawrence obtained a state charter for this new industrial leviathan, soon named in his honor. Engineering for the enterprise fell to Charles Storrow, a compatriot of Francis's who took the reins as treasurer and chief engineer of the newly formed Essex Company.[13]

An 1829 Harvard graduate, Storrow studied hydraulics in Paris and, in 1835, published *A Treatise on Water-Works for Conveying and Distributing Supplies of Water*.[14] Under his supervision, work on the Great Stone Dam and the Essex Company's main power canal required three years, reaching completion in the fall of 1848. Once the canal system became operational, the core business of "SX" (as Freeman referred to the company) was to lease water diverted by the dam to textile merchants who would use it to power their looms, spindles, and other machinery. By 1849 the city's first mills were in place, with almost a dozen operational by 1860.[15] The scale of these enterprises is manifest in the city's rapid residential growth. In 1845 the population of the farmland that would become Lawrence was about 200; in 1850 it exceeded 8,000; by 1870, almost 29,000 residents, including Eliza Freeman and her son, John, called the city home. In little more than twenty years, Lawrence ("The Queen City") had become one of America's largest industrial centers.[16]

Stretching 900 feet across the Merrimack River at Lawrence, Massachusetts, the 35-foot-high Great Stone Dam was completed by the Essex Company in 1848. By 1870 Lawrence was one of New England's major centers of textile manufacture, drawing thousands of workers to its water-powered mills. Author's collection.

Lower Pacific Mill, Lawrence, Mass.

Postcard view of the Pacific Mills in Lawrence, circa 1907. This scene features workers crossing the Essex Company's mile-long power canal that runs along the northern bank of the Merrimack River. John Freeman first began working for the Essex Company as an engineering assistant in the summer of 1872. Author's collection.

During his first summer with the Essex Company, Freeman worked as a rod-man on surveying teams in Lawrence and also helped "in setting the penstocks and turbines for the Lawrence Duck Company." By that time Storrow had hired Hiram Mills, an 1856 graduate of Rensselaer Polytechnic Institute in Troy, New York, to assume the duties of chief engineer (the older engineer remained as corporate treasurer and chief executive of the enterprise until 1889, making frequent visits to Lawrence).[17] Under Mills's supervision, the young Freeman undertook "water measurements and some experimental work," tasks that served as harbingers of his later career.[18] After three months' tutelage, Freeman headed off to MIT in October 1872. But while a student he continued to spend summers working for "SX," with his starting wage of $1.25 per day doubling in 1874. In total, during his student years he earned more than $850.[19] Over the course of four years, the MIT tuition plus board came to $1,800, requiring Freeman to take a $400 loan from the school and to borrow money from his parents and Aunt Ellen. He kept careful track of his family's contribution and by 1880 had repaid it all, along with his debt to MIT.[20]

From his early years Freeman valued the monetary rewards that came from hard work, and he also appreciated the sacrifices that his parents made to advance his professional aspirations. For Nathan and Eliza Freeman, a college education lay beyond what was possible for a West Bridgton farmer and his wife. For their son it was a different story. He was to become an early graduate of a new and innovative institution of higher education, one dedicated to educating students in a range of scientific and engineering disciplines, as well as the liberal arts.

Boston Tech

In 1861 the Commonwealth of Massachusetts formally chartered the Massachusetts Institute of Technology (MIT), creating a school dedicated to "the advancement of the Mechanic Arts, Manufactures, Commerce, Agriculture, and the applied sciences generally."[21] Under the leadership of William Barton Rogers, the institute held its first classes in 1865 and the next year opened an imposing four-story academic building on Boylston Street in Boston's Back Bay, soon gaining the appellation "Boston Tech." Two years later the school's first class graduated, with thirteen students receiving degrees in mechanical engineering, civil and topographical engineering, mining and metallurgy, and science and literature.[22] Befitting the school's core mission, most of Freeman's courses at MIT focused on mathematics, graphic arts and mechanical drawing, and the sciences (physics, chemistry, geology). But the school's curriculum was broad, and he also devoted significant time to studying English, French, German, and logic and philosophy—in essence, the liberal arts.[23] In his autobiography he mentions that the school's educational curriculum extended beyond technical disciplines, specifically praising George Howison, his "Professor of the Philosophy of Science, also of Formal Logic" as a formidable mentor.[24]

Notably, surviving lecture notes from a Howison-taught class reveal that as a student Freeman was introduced to a range of concepts lying beyond what might be thought appropriate for a young engineer.[25] These notes include references to psychology, logic (as a "science of the mind"), and the nature of language, making the point that "logic is important to language . . . [and] language may be described in loose terms as a sort of garment in which the mind clothes itself in order that an individual may communicate his thoughts to others." In his notes, Freeman further recorded how language had two attributes: one "as an expression of the individual" and the second "as an instrument to influence the mind."[26]

Freeman's attempts to keep up with an erudite nineteenth-century academic offer evidence of his early exposure to notions of "psychology" and "the mind" and the use of language as an "instrument to influence the mind." It might seem odd to focus on an engineer's education as it relates to such concepts. Freeman, however, was not a typical engineer, at least not in the way he could forcefully engage with nontechnical issues. Although he was no great wordsmith comparable to someone like John Muir, Freeman had a different gift: he could craft clear and direct prose that was not clotted with verbose and cloying wordplay. As evident in his 1912 report on Hetch Hetchy, and in a multitude of other reports, memos, and letters produced during his lifetime, he could communicate ideas and arguments in clear and direct language that was accessible to a diverse, nontechnical readership, one that included the general public as well as businessmen, bureaucrats, and politicians.

There may be no way to directly connect Howison's MIT lectures from the 1870s to Freeman's *Hetch Hetchy Report*. Nonetheless, Freeman's early introduction to psychology as a social concept, and to the power of language as a social force, deserves our attention. Freeman drew upon such basic ideas when, in the phrasing of Professor Howison he wielded them as "an instrument to influence the mind" in promoting his plans to dam Hetch Hetchy.[27] And it is significant that after completing his *Hetch Hetchy Report* in September 1912, he requested that San Francisco's city attorney "mail copies of [his] Hetch Hetchy report to the names given below . . . [who] are largely men who have helped in the preparation." There, tenth on the list, is Professor George H. Howison, then teaching at the University of California, Berkeley.[28]

Freeman was a diligent student, and in 1876 he graduated from "Boston Tech" with a Bachelor of Science degree in civil and topographical engineering. Thereafter, he maintained a robust relationship with his alma mater, staying in contact with classmates, serving as class secretary, and returning to consult with professors and to offer lectures on hydraulic engineering. Most significantly, in 1893 he became a Life Member of the MIT Corporation, a position he held for almost forty years until his death in 1932. As discussed in chapter 5, in 1912–13 he also devoted much time and energy to planning for what is now the school's world-renowned Cambridge campus along the north shore of the Charles River.[29] During Freeman's lifetime, MIT's standing went from a regional technical school, albeit one with great ambitions, to a major university with an international reputation in science and the engineering arts. His reputation as an engineering expert was no doubt enhanced because of ties to the school that, some forty years prior to his work on Hetch Hetchy, had opened its doors to an ambitious young man hailing from Lawrence High School and the rural environs of West Bridgton, Maine.

"SX" and Water Power

Upon graduating in June 1876, Freeman and a cadre of MIT classmates decamped to Philadelphia to explore the Centennial Exposition and its wondrous exhibits celebrating America's industrial accomplishments. This trip—apparently his first outside of New England—proved a fitting capstone to Freeman's hard-earned college education. However, upon returning home he faced the economic reality of the lingering business depression. To his good fortune the Essex Company was willing to bring him on as a full-time employee, but the pay remained at $2.50 per day, and he would not get a raise to $3.00 per day for another two years.[30] Nonetheless, he had a professionally respectable job and was positioned to learn much from the engineer who hired him: Hiram F. Mills.

Born in Bangor, Maine, in 1836, Hiram Mills attended Rensselaer Polytechnic Institute in Troy, New York, and graduated with a Bachelor of Science degree in 1856. Over the next decade he worked on a variety of engineering projects including the Brooklyn, New York, water works and the dam across the Deerfield River in western Massachusetts that provided power for building the Hoosac Tunnel. This latter project connected him with Charles Storrow (an engineering consultant for the tunnel), and when, in 1869, Storrow wanted to delegate some of his Essex Company responsibilities, he appointed Mills the firm's chief engineer.[31] Mills had hired Freeman for his summer engagements while a student at MIT and in 1876 became his full-time boss. Over the next ten years and beyond, he would be an invaluable mentor and counselor.

The heart of the Essex Company's business lay in leasing what were termed "mill powers" to firms that relied upon flow diverted by the Great Stone Dam to energize their looms, spindles, and other machinery. "Mill power" was a concept used by New England industrialists to quantify the amount of water necessary to operate large factories and provide a standard for annual water-power leases.[32] By the 1880s, companies served by the Essex Company could contract for a flow of 26.66 cubic feet per second dropping 28 feet for a period of sixteen hours each day, with this flow defined as one mill power. Lawrence-based companies such as the Pacific Mills and the Atlantic Cotton Mills paid $1,200 annually for every leased mill power provided by "SX." In addition, extra flow was often available; this was marketed as "surplus water" to the company's customers.[33]

To ensure proper payments, the company needed accurate measurements of the water delivered to each mill. This task fell to Freeman and other assistants charged with accounting for "the water drawn day by day by each of the 75 turbines along the canals." As Freeman explained, this work "necessitated measurements of the controlling apertures in every turbine . . . [and] carefully measuring the velocity and area of the current of water in the penstock or tail race." Freeman often spent evenings "at the office planning [his] work for the next day," and upon rising early, he would get in "about an hour of study, or reading technical journals, before office work began . . . determined to be ready for a better job." This diligence did not pass unnoticed; he soon rose to be Mills's principal assistant engineer.[34]

Soon after Freeman joined "SX" full time, he witnessed a major legal battle fought over the firm's economic interest in the flow of the Merrimack. The dispute centered on the expansion of Boston's municipal water supply system westward into the Concord River watershed. In the 1840s, Boston had constructed a fourteen-mile-long aqueduct to draw water from Lake Cochituate.[35] In the 1870s, the city extended this system by building a canal connecting Lake Cochituate with the Sudbury River about 10 miles farther west. On its face, this was a perfectly logical way to increase the city's water supply. But the Sudbury River flowed north into the Concord River, which, in turn, entered the Merrimack just below Lowell. Consequently, any Sudbury flow diverted to Boston was water that would never reach the power canals at Lawrence. In other words, the new Sudbury diversion necessarily reduced the power capacity of the lower Merrimack. And less water meant less revenue for the Essex Company.[36]

The dispute between the Essex Company and Boston's Cochituate Water Board boiled down to a simple question: what financial damages were suffered by the Essex Company because of Boston's Sudbury diversion? As a junior engineer Freeman did not help postulate the company's legal strategy, but he was involved in assisting Mills's analysis of water flow records collected by the company beginning in 1840s. As Freeman explained, the "preparation of evidence for this case" involved documenting "the daily run-off of the Merrimack at Lawrence, which . . . soon became a more elaborate and precise record of the daily discharge of a large river over a period of many years." Such work provided Freeman with an invaluable understanding of streamflow analysis that he drew upon in his later consulting work.[37]

For Freeman, the importance of the Sudbury water case extended beyond hydrological matters, as he also participated in the legal tribunal where the dispute was heard. When the case came to trial, the company's lawyers required ready access to the accumulated hydrological data, and Freeman was appointed "custodian of the records" responsible for their care. Positioned in the courtroom, Freeman had an "opportunity to observe the play of wit between the eminent counsel," a remarkable assignment that offered insight into the legal and political character of American society, something far removed from the typical labors of most young engineers.[38]

By 1880 Freeman was earning $4.00 per day at "SX." In addition, he was aiding Mills on private consulting projects throughout New England, earning a $5.00 per diem for these special assignments, which included extensive work on water power development at Manchester, New Hampshire; Sewell's Falls, New Hampshire; and Bellows Falls, Vermont.[39] He also became "occupied with the details of many hydraulic researches by Mr. Mills" and was "particularly occupied with some of Mr. Mills' experiments on the laws of the flow of water in pipes and smooth open channels, all of which was highly instructive." During his time with "SX," Freeman's work went beyond the day-to-day responsibility of measuring water flow delivered to each of the company's leases and also encompassed more theoretical investigations that, in his later career, helped enhance his professional reputation.[40]

During his later years with "SX," Freeman began handling small assignments on his own account, starting in 1880 with "improving a small water power on the Contoocook River at Fisherville [New Hampshire]." This work involved excavating a small canal and "the devising of various improvements in the setting of turbines at the Amsden-Whitaker Furniture Factory." He later recalled how he "slept very little that first night" because of "trying to make sure of the correctness of [his] answers to the owners."[41] The next year, he further advanced his work as a consultant, helping a riparian landowner in Turners Falls, Massachusetts, determine the value of his property in relation to a proposed new dam and power canal. After a site visit, Freeman earned $20.00 for a four-page analysis of his client's asset. Early on he charged only a modest fee for such reports, but they served as an important portent of his future consulting work.[42]

Freeman was a hard worker who prided himself on his loyalty to the manufacturing elites who dominated the Merrimack River valley. But after almost ten years labor he wearied of his position in the corporate hierarchy. Yes, he could stay in place and, some twenty years down the road, presumably ascend to the position of chief engineer of the Essex Company. But he wanted more from his professional life than waiting for his boss to retire or die. By the mid-1880s, his income reached about $2,200 per year, a considerable sum for a one-time farm boy from West Bridgton but not dramatically more than when he started working full time in 1876. Although Freeman did not lead an ostentatious life, he enjoyed building a library of engineering books and acquiring sophisticated surveying equipment. In addition, in the summer of 1885 he met his future wife, Bessie

Clark, and a desire for a home and family entered into his life plans. Perceiving that "the future of the field at Lawrence was narrow," he decided to move on.[43]

In early 1886, Freeman considered myriad career moves, with an offer from the Boston Manufacturers Mutual Fire Insurance Company to be an "engineering inspector and water supply specialist" seeming the most attractive. To his mind, "a year or two of this experience would give me a lot of contacts with factory owners throughout New England . . . [before opening] my own engineering office."[44] When Freeman informed Mills of his plans to leave "SX," their meeting was "at first a bit stormy." But soon the clouds cleared and the two men reached an amiable accord, with Freeman later fondly recalling how "through the remainder of [Mills's] life he was one of my very best friends." Nine years after Freeman left the Essex Company, this professional friendship would be evidenced in Mills's recommendation that his one-time assistant be appointed a member of the politically powerful Massachusetts Metropolitan Water Board, charged with overseeing a major expansion of Boston's water supply system.[45]

In retrospect, it seems easy to situate Freeman's Essex Company career within a context of corporate capitalism, where water was harnessed as a commodity for the benefit of a commercial elite.[46] Under the tutelage of Hiram Mills, Freeman learned the business of water power with an enterprise that, through technological expertise and political savvy, sought every means possible to squeeze out profitable mill powers from the Merrimack River. But to focus exclusively on the capitalistic character of the Essex Company's interest in water resources is to miss a significant part of the story. Soon after Freeman's departure from "SX," Hiram Mills embraced another role, one that highlighted "public service" as a more broad-minded professional aspiration. In 1886, the fifty-year-old Mills became a member of the Massachusetts State Board of Health, a position he held for the next twenty-eight years (during this time he also remained chief engineer of the Essex Company). As chair of the board's Committee on Water Supply and Sewerage, he was instrumental in improving water quality for residents throughout the state and helping plan new water supply systems. He also organized the Lawrence Experiment Station, where the technology of large-scale water filtration was pioneered within the growing field of sanitation engineering.[47]

Mills and his fellow travelers in the public health movement transformed the nation's water supply in the early twentieth century and helped bring relatively "pure" (or at least much less polluted) water to millions of Americans.[48] This represented one of the great technological successes of the Progressive Era, and it is important to appreciate that this transformation was, to no minor degree, guided by an engineer like Mills whose professional roots were inextricably tied to private industry. This was also the case for his protégé John Freeman, who would work simultaneously for the interests of capital (both in the insurance world and as a consultant for investor-owned electric power companies) and for municipally owned water supply systems. It might seem easy to separate the realms of "private" and "public" within the domain of Progressive Era engineering, but in truth they were closely intertwined. Nowhere would this be more apparent than within the dynamics of Freeman's rapidly evolving career after he left "SX."

Mutual Fire Insurance: From Inspector to President

On March 1, 1886, Freeman began work as an inspector for the Boston Manufacturers Mutual Fire Insurance Company, one of several interrelated enterprises offering fire insurance for mill owners. The business of factory mutual insurance began in 1835, when Providence-based Zachariah Allen organized a group of textile entrepreneurs to pool their resources and provide "mutual" fire insurance for their mills. The underlying motive was simple: general fire insurance companies (or "stock" insurance companies) refused to reduce policy premiums for factories that invested in fire suppression technologies (such as "slow burn" construction and reliable water pumps).[49] To Allen and his compatriots, this seemed disingenuously exploitative. If mill owners were willing to operate safer factories, why shouldn't they benefit from reduced insurance rates? And if stock insurance executives resisted this logic, then why shouldn't financially shrewd industrialists seek out other, less costly, alternatives?

In essence, the "factory mutual" enterprise offered a way for businessmen to lower premiums by including factories in an insured "risk pool" only if their owners acted to reduce the prospect of fire damage. To further distribute liability, these industrialists created a web of interrelated companies that often shared coverage for individual mills; by 1875, factory mutuals in toto had carved out a substantial market, underwriting $375 million in risk.[50] By the early 1880s, eighteen factory mutual companies were operating in New England. To a degree, they competed with one another. However, they also shared a desire to reduce fire losses and develop technologies and inspection regimens that limited the likelihood of conflagration. Recognizing their common interests, these enterprises convened a "conference" (or the "Associated Mutual Fire Insurance Companies") to administer an Inspection Department charged with developing—and enforcing—design and operating standards.[51] Organization of the Inspection Department was the brainchild of Edward Atkinson, a cotton factory entrepreneur and president of the Boston Manufacturers Mutual Fire Insurance Company.[52] Atkinson hired Freeman as an inspector for both his company and the larger conference of associated mutual companies.[53] Freeman held his new boss in great esteem, later describing how "Mr. Atkinson's vision and faith in modern scientific training . . . [brought] the factory mutuals into lines based on scientific study and real fire prevention engineering."[54]

Freeman's new job was based in Boston, but his work entailed frequent trips to factories throughout New England. Every mill differed in its particulars, but to qualify for the insured risk pool, each needed to adhere to basic performance standards. As an inspector, Freeman would arrive at a factory unannounced and then scrutinize "from the top to bottom of every building . . . looking first of all for dirt or lack of prompt removal of waste. . . . All extinguishing appliances were examined in due course, and finally a test was made of the fire pumps in full operation." Upon completing his inspection, Freeman drafted a "Report Card" delineating "substantially all that the insurance executives needed for judging of the desirability of continuing the insurance on the risk."[55]

Reveling in his new job, Freeman exhibited a skill and enthusiasm that quickly impressed Atkinson. Within a few months, Freeman was upgraded to "Special Inspector and Engineer" charged with both inspection work and taking on research experiments with automatic sprinklers, hydrants, hoses, pipes, and other fire protection technology.[56] Over the course of the next few years, he developed an automatic sprinkler (which he patented and sold the rights to for $1,500) and, at the Washington Mills in Lawrence, supervised extensive testing of hoses, nozzles, and pipes to ascertain how "fire streams" could achieve maximum performance in extinguishing conflagrations. These investigations proved enormously productive and spawned two papers published in the *Transactions of the American Society of Civil Engineers*: "Experiments Relating to the Hydraulics of Fire Streams" (1890) and "The Nozzle as an Accurate Water Meter" (1891). Remarkably, both papers won the society's prestigious Norman Medal in recognition of their contribution to engineering science.[57]

Freeman's reputation was on the rise, and the two Norman Medals reflected his growing professional standing.[58] In 1889 he joined a delegation of the American Society of Civil Engineers (ASCE) and American Society of Mechanical Engineers (ASME) that held a joint meeting with the British Institute of Civil Engineers in London (this was the first of many European trips made during his lifetime). He maintained his standing in the ASCE and in 1904 served as the organization's vice president; the next year he held the presidency of the ASME. Much later (1922) he was selected president of the ASCE. Within the world of factory mutuals, Freeman's stature also continued to grow. In 1890 he took charge over the Associated Factory Mutuals' Inspection Department, a position through which Atkinson introduced him to "the great industries of the United States—textile mills, paper mills, great machine shops, locomotive works, and watch factories." Of equal import, this work also brought him "into intimate contact with many great captains of industry."[59]

Freeman's original intent had been to work as an inspector for a year or two and then set up a private consulting practice. This plan, however, quickly fell by the wayside as he became engrossed in the business of fire insurance. In early 1896 his prominence in the industry was formally recognized when, upon the death of Henry Ormsbee, longtime president of the Providence-based Manufacturers Mutual Fire Insurance Company, he was awarded this prestigious post at an annual salary of $12,000. In a mere ten years he had risen from the staff position of inspector to the presidency of a major insurance enterprise. He was only forty years old.[60] After taking control of Manufacturers Mutual, Freeman and his family (he had married Bessie Clark in December 1887 and they had four children within eight years) moved from suburban Boston to Providence, where, when not traveling as a consultant, he resided for the rest of his life.[61] By 1903 he was president of five other mutual companies located in Providence, and he remained head of the six firms until his death; under his leadership these companies' insured risk rose from less than $200 million in 1896 to over $3 billion in 1930.[62]

Freeman's new executive position encompassed many responsibilities, but the most politically significant entailed protecting Manufacturers Mutual from the predatory grasp of the tax man. The details of such work need not be described here, but issues as to what constituted "profits" as opposed to "premium deposits" returned to constituent members were of great consequence in terms of tax liability. One of Freeman's preeminent tasks was to lobby for favorable regulations and rulings by politicians and government officials at the city, state, and federal level.[63] This responsibility became especially important after passage of the 1909 Corporation Tax Act, prompting him to visit Washington, DC, to meet with Treasury Department officials, legislators on Capitol Hill, and even the president of the United States. In these conferences he would explain how factory mutuals differed from traditional stock insurance companies and lobby for advantageous tax treatment under the existing law or in proposed amendments. When the battle over Hetch Hetchy enflamed Washington, DC, in

1912–13, Freeman's experience in navigating political currents in the nation's capital and elsewhere proved especially helpful to San Francisco.

Freeman's stature in the fire insurance industry was not limited to corporate boardrooms and the offices of politicians and tax regulators. His work also positioned him at the forefront of a burgeoning movement to protect cities from massive urban fires. In the twenty-first century, there are still dangers of rural and exurban wildfires wiping out towns and outlying communities. However, the prospect that a raging inferno might destroy a major city center has been severely mitigated by fire codes spearheaded in the Progressive Era. This did not occur by happenstance but was the result of professionals like Freeman who worked to encode fire safety into the heart of urban governance.

As a leader in fire protection engineering, Freeman attracted nationwide acclaim for his investigation of Chicago's Iroquois Theatre fire, a horrific tragedy that killed almost 600 people on December 30, 1903. During a matinee performance of the play *Mr. Blue Beard,* sparks from electric "spot lighting" equipment ignited scenery backdrops stored above the stage. Waves of acrid smoke quickly billowed into the main auditorium. More than 1,800 frantic patrons, including many children, raced to escape the choking cloud, but blocked exits and locked doors trapped hundreds in the theatre. Some were trampled to death, but most died from suffocation induced by smoke inhalation.[64]

Within hours after the disaster. Chicago businessman Charles Crane (whose two nieces died in the blaze) reached out to Freeman, imploring him to undertake an independent investigation to determine "how such fearful disasters can be prevented."[65] Freeman agreed and spent several months analyzing how the deaths of hundreds could have been averted. Blocked exits, doors that opened inward, and poorly designed routes of egress did not escape Freeman's criticism. But for him, the key culprits lay in the absence of automatic sprinklers above the scenery loft and the lack of automatically opening roof vents that could have easily allowed the accumulating smoke to escape skyward.[66] Freeman devoted his 1905 ASME presidential address to the Iroquois fire, and soon his study "The Safeguarding of Life in Theaters" was distributed to city officials across the United States.[67] Freeman suffered disappointment that many municipal codes were not quickly updated to standards outlined in his Iroquois report, but this did nothing to diminish his national reputation. As historian Scott Gabriel Knowles notes in his book *The Disaster Experts: Mastering Risk in Modern America,* Freeman stood in the pantheon of Progressive Age fire suppression technologists, heralded as "one of the nation's foremost authorities on fireproof construction and fire safety engineering."[68]

Public Water Supply: Boston

When Freeman signed on as an insurance factory inspector in 1886, he put his work as a consulting engineer on hold. But his desire to engage in projects on his own account never disappeared. In July 1895 his first major step as an independent consultant came when he accepted a prestigious appointment from the Commonwealth of Massachusetts. The assignment: help lead the newly authorized Metropolitan Water Board in expanding Boston's regional water supply system. Not long after, he also joined two court-administered appraisal boards charged with setting a fair price for the municipal purchase of the privately owned Gloucester Water Supply Company and the Newburyport Water Company.[69] Although an unabashed proponent of free enterprise, Freeman did not confine his consulting work to investor-owned initiatives. He recognized the role of government in fostering sanitary water supplies, and here he drew a line separating domestic water supply from privately financed hydropower projects.

Underlying Freeman's appointment to a politically sensitive post on the Metropolitan Water Board was the endorsement of his mentor/colleague Hiram Mills. By 1894 the Essex Company's chief engineer had chaired the Massachusetts State Board of Health's committee on water supply and sewerage for eight years. During that time, Mills had become a major figure in the public health and sanitation movement, and his opinions mattered to the state's political hierarchy. Boston's plans for expanding its water supply system were closely aligned with the interests of the State Board of Health; in supporting Freeman's involvement in these plans, Mills signaled that his protégé shared his progressive ideals regarding the sanctity of municipal water supply.[70]

Authorized by state law in 1895, the Metropolitan Water Board oversaw an ambitious project that eventually would deliver over 100 million gallons of water per day from the upper Nashua River to greater Boston. Included as part of the scheme was a large storage reservoir (impounded by the 200-foot-high Wachusett Dam) that would inundate almost 5,000 acres in the towns of Boylston, West Boylston, and Clinton. As an alternative to the Wachusett Dam, the State Board of Health considered building a water filtration plant drawing water directly from the Merrimack River. But filtration technology for such a large system was deemed too risky, especially if the Nashua River diversion system proved politically viable.[71]

At its core, the Nashua/Wachusett initiative represented a classic Progressive Era public works project, justified because it would provide the greatest good for the greatest number. But almost two thousand people would lose their

Wachusett Dam nearing completion in 1905. Freeman served as a commissioner for the Metropolitan Water Board during planning for this 205-foot-high water supply dam serving greater Boston. Author's collection.

businesses and domiciles to provide the city of Boston and surrounding communities with a bountiful reserve of high-quality water. So why should some rural and small-town folk be forcibly displaced (albeit with modest monetary compensation) so that a rising swell of suburban homeowners and city dwellers could flourish with access to safe and abundant water supplies?

Herein lay a key dilemma of the Progressive Age, and one bearing relevance to the conflict over Hetch Hetchy. "Progress" was (and is) in the eye of the beholder, and the benefits of proposed public works were/are rarely distributed to universal acclaim. To the dismay of citizens in the Nashua Valley who were destined to lose their homes, farms, and businesses, the Massachusetts state legislature enacted legislation empowering the Metropolitan Water Board to take control of several square miles of privately owned land needed to deploy a massive storage reservoir. Progress requires trade-offs and a balancing of societal costs and benefits. Not everyone would be happy with the Nashua River scheme, but it was determined by the body politic that, in aggregate, the citizens of the commonwealth would benefit from this far-reaching system.

For Freeman, the existential question posed by the Nashua/Wachusett project was easy to resolve. The parochial interests of the few needed to make way for the legitimate desires of the many, especially if the goal was to propagate a high-quality urban water supply. Perhaps if the rural landscape to be flooded were in West Bridgton, Maine, and family friends and loved ones were to be forced from their homes, he might have thought differently. But, in fact, he expressed no qualms about the costs exacted upon those whose property lay within the reservoir "take" zone. They were simply collateral damage, an unfortunate minority left bereft by a public works project designed to serve a larger—and legitimate—public good.

In his autobiography, Freeman highlights a major innovation in water supply technology that he proposed to the Metropolitan Water Board a few months after becoming a commissioner. As originally planned, the Wachusett project was to carry water from the Nashua watershed via an open-flow (that is, non-pressurized) conduit feeding into the Chestnut Hill reservoir. At Chestnut Hill, large steam-powered pumping engines would then provide high-pressure service to consumers throughout greater Boston. For Freeman, extensive reliance upon steam-powered pumps represented a long-term waste of money (coal was costly and the pumps required the constant attention of skilled mechanics/operators). As an alternative, he proposed building a nine-foot-diameter, pressurized steel pipeline connecting the Wachusett Reservoir to high-level distribution reservoirs near Boston and—by relegating the expensive steam pumps at Chestnut Hill to emergency service—dramatically increase the system's long-term economic efficiency.

Inspired by his experience in fire suppression, Freeman prepared estimates for an aqueduct providing "20 or 25 lbs. per sq. inch higher pressure by gravity for the entire city of Boston without pumping . . . afford[ing] upward of 100 lbs. per sq. inch working pressure on almost every fire hydrant."[72] Orally briefing his fellow commissioners in October 1895, he quickly followed up with a sixty-page written plan detailing his desire to take advantage of "the great elevation of the Nashua Reservoir [385 feet above sea level]" and deliver "high pressure water by natural gravity flow without pumping."[73] Obsessed with the possibilities of this innovation, he confided to his diary that his new scheme entailed "the hardest that I have ever worked in my life. Go to work commonly at 4 A.M. and work until 11 or 12 P.M. Sundays and all." What drove him was a progressive desire to both improve fire safety and save money, estimating that "this scheme will save $250,000 per year in pumping expenses, $150,000 in insurance costs, [and] $50,000 in Fire Department expenses." For himself, construction of the high-pressure system promised to be "the chief engineering triumph of my life."[74]

Freeman recognized that impending contracts would soon "settle the course of development for all future time," and he implored his fellow commissioners that "it is *now or never* with a *high pressure gravity water supply*."[75] But despite his hopes, resistance came quickly, with legal advisors arguing that the radical scope of his plan would necessitate new—and politically treacherous— enabling legislation. He also received pushback from the Water Board's chief engineer, Frederic Stearns, who, in his capacity as chief engineer for the State Board of Health, had designed the open-flow, low-pressure Wachusett system that Freeman castigated as needlessly wasteful.[76] Although he considered Stearns a "good friend" and maintained a respectful relationship with him in later life, Freeman ruefully noted, "Stearns [was] quietly doing all he can to bring my scheme into disfavor."[77] Fighting for his plan, Freeman enlisted support from other engineers, including Hiram Mills, who he claimed "was very strongly inclined to support my views." Despite such lobbying, the naysayers remained intransigent. In the face of legal concerns that "if we went back to the legislature, we could not get away with another bill so extremely favorable," Freeman backed down. He reached this decision reluctantly, later bemoaning that "never for a moment [have I] failed to believe that [my plan] was much the better. As I have looked back in later years, I doubt if I did right in yielding."[78]

By the end of December 1895, Freeman realized that the Metropolitan Water Board would never approve his plan for a high-pressure aqueduct. Facing an unappealing future in which his technological talents would go unappreciated (or at least underappreciated) by the Water Board, a new professional opportunity suddenly arose. On January 3, 1896, Henry Ormsbee, the long-serving president of the Manufacturers Mutual Factory Insurance Company in Providence, died at age seventy-five. For the first time in a generation this prestigious position became open, and within a matter of days, the company's directors asked Freeman to serve as Ormsbee's successor. He quickly accepted. On February 4 he presided over the first of many meetings of the company's board of directors, and a new phase of his life and career opened before him.

Perhaps if leaders in Massachusetts had embraced his high-pressure scheme he would have declined the Manufacturers Mutual's presidency. But that is not how things worked out, and he soon began relocating his family and professional base to Providence. In April he formally resigned from the Metropolitan Water Board, but not before experiencing firsthand the politics undergirding large municipal infrastructure projects. Equally important, the deployment of high-pressure conduits became ingrained in his vision of state-of-the-art urban water supply.[79]

Public Water Supply: New York City

Upon becoming president of Manufacturers Mutual, Freeman fully embraced his new executive responsibilities. Nonetheless, he recalled, "I had no intention of settling down contentedly like my predecessor Mr. Ormsbee, with insurance affairs bordering my horizons. . . . I therefore made the change, but with the clear and firm understanding that problems of engineering would nonetheless continue to be the chief interest of my business life."[80] Following this path, he maintained a vibrant consulting practice that, in the summer of 1899, brought him into the cauldron of New York City water supply politics.[81]

Freeman's relationship with Boston's water supply plans lasted but a year before he resigned as Water Board commissioner. In contrast, his service as a consulting engineer for New York City encompassed more than two decades of service and survived the administrative regimes of seven mayors. In 1902, city officials sought to hire him full time as chief engineer of the Department of Water Supply, Gas and Electricity, but he fended off such entreaties.[82] While he prized his role in helping plan (and realize) what became the city's Catskill Aqueduct, he always wanted the freedom to maintain his presidency of Manufacturers Mutual and also pursue other consulting opportunities. In this he established a model that guided his later relationship with San Francisco.

Efforts to guarantee a reliable water supply for New York City dated back to the turn of the nineteenth century, and by the 1830s, public discontent with private ownership of the city water supply was rife. At that time, voters approved a municipally owned, forty-one-mile-long aqueduct drawing water from the Croton River in Westchester County. Completed in 1842, the Croton Aqueduct and fifty-foot-high Croton Dam proved a tremendous success, impelling the city's growth through midcentury. In the 1880s, work started on a "Second Croton Aqueduct"; once the second aqueduct became operational in 1891, the city authorized construction of the New Croton Dam to expand storage to almost 20 billion gallons.[83]

Presumably, the enlarged Croton system would have been enough to ensure New York City with a reliable water supply for many years to come. However, the city's relentless population growth, combined with rising per capita use of water, raised concerns of an impending water famine.[84] Such fears only intensified when, in 1898, the five boroughs consolidated into a civic jurisdiction covering more than three hundred square miles. To help meet—or take advantage of—the city's future needs, a group of politically connected businessmen took control of the privately owned Ramapo Water Company and, armed with a commanding charter from the state legislature, proposed a private water supply scheme to complement the municipally owned Croton system. In his history of New York City's water supply, David Soll explains the contentious nature of these plans to sell water to the city from sources in the Catskill Mountains, emphasizing that "the company had the ear of Tammany Hall, the notoriously corrupt Democratic Party machine that had regained control of municipal government." And with Tammany influence undergirding its efforts, "the company sought to exploit its exclusive control of potential water sources to egregiously overcharge New York City for access to these waters."[85] The extent of these impending "overcharges" was substantial, calculated at "200 million gallons a day for 40 years at $70 per million gallons . . . amount[ing] to $14,000 per day or $5,110,000 a year."[86]

Freeman entered New York's public works arena in the summer of 1899 when fears arose that the Tammany-linked Ramapo proposal would become a costly, long-term burden for city businesses and taxpayers. At the behest of F. C. Moore, president of the New York–based Continental Insurance Company (note the insurance industry connection), Freeman was hired by City Comptroller Bird S. Coler to study the water sources available to the city as a counter to the Ramapo scheme.[87] With this entrée, Freeman initiated an in-depth study of potential water sources that covered hundreds of square miles, extending from groundwater tracts on Long Island to the Housatonic River along the New York/Connecticut border and to myriad sources within the Hudson River valley (including the Hudson River itself).[88]

To Freeman, the most advantageous way to expand New York City's water supply did not involve sources west of the Hudson River such as the Ramapo interests proposed. Instead, he favored building a large storage dam across the Housatonic River about seventy-five miles north of the city and then drawing water (which he estimated could exceed 700 million gallons per day) through an easy-to-build conduit into the Croton watershed and thence on to the metropolis. The technological/economic merits of the Housatonic project were undeniable. However, it failed on political grounds because the damsite and much of the proposed reservoir lay about a mile east of the Connecticut/New York state line. Despite Freeman's hope that "in a neighborly spirit the State of Connecticut will grant any power needed for New York to complete and perfect its title without hindrance," state authorities in Hartford refused to endorse the plan.[89] To Freeman's dismay, political considerations—and not technological merit—would govern the issue and preclude New York City from tapping into the Housatonic's voluminous flow.[90]

Published in the spring of 1900, Freeman's five-hundred-page report failed to spark approval for his favored Housatonic aqueduct, but it did help untrack the Ramapo initiative and ignite municipally sponsored efforts to seek out new water sources.[91] Freeman remained intensely involved in these efforts; after declining an offer to become chief engineer of the city's Department of Water Supply, Gas and Electricity he was appointed to the city's newly formed "Commission on Additional Water Supply." This commission expanded upon his 1900 report and focused on sources lying within the state of New York. After weighing myriad possibilities, including filtering water pumped from the Hudson, the commission recommended Esopus Creek (flowing eastward out of the Catskills near Kingston) as the preferred alternative. Of course, this was the same source proposed by the Ramapo Company for its system, but the commission's plan called for a radically different means of water storage; whereas Ramapo called for seven relatively small reservoirs with a combined storage capacity of about 12 billion gallons, the city's plan called for a large dam and reservoir with a capacity over 210 billion gallons.[92]

The Commission on Additional Water Supply completed its report in late 1903, and their investigation became the focus of the city's campaign to obtain approval from the state legislature. In June 1905, Mayor George McClellan won legal authorization for the city to tap into Esopus Creek for "an additional supply of pure and wholesome water." After the law's passage, the city established a board of water supply to oversee construction of the new system; later that summer, Freeman became consulting engineer to the new board.[93]

The particulars of Freeman's extensive engagement with the board of water supply lie beyond the scope of this book, but the culturally complex process necessary to bring the $184 million Catskill project to completion represents one of the most ambitious infrastructure ventures of the Progressive Era: more than two thousand people were removed from villages and farmsteads in the Ashokan reservoir take zone; by 1908 some 3,900 laborers were employed in construction; and by the end of 1909 over $64 million in contracts had been let. Acknowledged as a key technical visionary of the Catskill Aqueduct, Freeman helped oversee the design and construction of the 126-mile-long conduit (with a minimum design capacity of 500 million gallons per day) that featured the 200-foot-high Ashokan (Olive Bridge) Dam, the 307-foot-high Kensico Dam, and, harkening back to his rejected plan for Boston's Wachusett Aqueduct, a 3,022-foot-long, 14-foot diameter, pressurized tunnel running under the Hudson River.[94]

Freeman's work for New York City enhanced his stature as a proponent of municipally owned public works and also brought home to him the value of

New York City's Catskill Aqueduct under construction, circa 1910. Freeman became consulting engineer for the city's Board of Water Supply in 1905, a position he held until his death in 1932. Author's collection.

Completed in 1915, the 240-foot-high Ashokan (or Olive Bridge) Dam near Brown's Station, New York, is a key component of New York City's Catskill water supply system. Author's collection.

preparing large, authoritative publications justifying the need for, and value of, major public works projects. This experience played an important role in his later engagement with San Francisco. And his work for New York City also attracted the attention of civic leaders in Los Angeles who needed skilled engineers to assess a bold plan for a new 233-mile-long aqueduct.

Public Water Supply: Los Angeles

Built between 1907 and 1913, the Los Angeles Aqueduct stands as one of the most famous (or infamous) urban water projects of the Progressive Era.[95] Depending upon one's point of view, this aqueduct is either reviled as an audacious urban water grab that stole water from a bucolic rural valley or celebrated as a farsighted construct that created the hydraulic foundation for modern-day southern California. Of these two perspectives, Freeman definitely favored the latter, seeing the aqueduct as a welcome enterprise fostering growth of a burgeoning metropolis. By 1906 William Mulholland, chief engineer of Los Angeles's water supply system, had devised plans for an aqueduct carrying water from the Owens River along the eastern slope of the Sierra Nevada down to the coastal plain. But civic leaders and prospective purchasers of the city-issued construction bonds needed assurance verifying the project's viability. A board of consulting engineers experienced in dam building and municipal water supply would be called upon to assess the city's audacious scheme. With his municipal experience at Boston and New York City, Freeman was a logical choice to take on this task.

As William Kahrl explains in *Water and Power: The Conflict over Los Angeles' Water Supply in the Owens Valley,* this consulting board was key to moving the aqueduct project forward and "represented fulfillment of the city's pledge during the bond election of 1905 not to embark too deeply upon the project until the wisdom of the enterprise had been independently and reliably confirmed."[96] During November and December 1906, Freeman and his colleagues on the board—including Frederic Stearns (chief engineer of Boston's recently completed Wachusett Dam) and James D. Schuyler (former California assistant state engineer)—carried out a review of Mulholland's plans including a field investigation of the proposed right-of-way.[97] In its final report, the board found "the project in every respect feasible . . . and full of promise." However, they did recommend a significant realignment of the aqueduct's right-of-way down San Francisquito Creek, allowing for two major hydroelectric power plants capable of "supply[ing] 49,000 horse power continuously, 24 hours per day,

View of the Jawbone Canyon Siphon carrying the Los Angeles Aqueduct across the western Mohave Desert, 1913. Freeman would use features of this aqueduct as a template for his Hetch Hetchy design. (Department of Public Service of the City of Los Angeles, *Complete Report on Construction of the Los Angeles Aqueduct* [Los Angeles, 1916], 194)

and every day of the year." For Freeman, the integration of hydroelectric power into a municipal water supply system represented something new. Neither Boston's nor New York City's new aqueducts featured any capacity to generate large quantities of hydropower. But conditions in California were different.[98]

The Los Angeles consulting board took no part in overseeing construction of the city's aqueduct. Nonetheless, Freeman's experience evaluating plans for the system proved of lasting import. First, it introduced him to issues affecting public land and municipal water supply in California. Second, he came to appreciate how vast quantities of hydroelectric power could be generated along pipelines descending from the Sierra Nevada. Third, it brought him into professional contact with the aqueduct's chief engineer, William Mulholland, a relationship that would be rekindled in unexpected ways during Freeman's work for San Francisco.[99]

Charles River Dam and Mystic River Improvement

Two other large-scale municipal improvement projects in Freeman's pre–Hetch Hetchy résumé warrant special mention, particularly because they reflected his belief that hydraulic engineering interventions could enhance environmental quality. For the first of these, he served as chief engineer for the Committee on Charles River Dam in 1902–3 charged with investigating problems posed by the unsightly and malodorous tidal mud flats bordering the Charles River between Boston and Cambridge.[100] Of course this was a landscape familiar to Freeman, as it lay but a short distance from MIT's campus in Boston's Back Bay. The second involved improvements to the Mystic River, a smaller stream lying immediately north of the lower Charles.

The issue of the Charles River was integrally tied to problems involving the navigability of Boston Harbor as well as the tracts of tidal marshland that, since the eighteenth century, had been drained and filled with rocks and soil to further fuel the city's urban growth. As Michael Rawson recounts in *Eden on the Charles,* in 1894 the Massachusetts State Board of Health and the newly formed Metropolitan Park Commission jointly proposed "that the state dam the Charles River for sanitary and recreational purposes. A high stable water level [for the lower Charles basin] would not only permanently cover the offensive areas along the river's banks, but also encourage recreational boating and park construction."[101] Although many residents in Boston and Cambridge supported the plan, the harbor commissioners were opposed because they believed the river "to be an important source of scour for the harbor channels" and thus vital in maintaining harbor depth sufficient for deep-draft ships. For the moment, the Harbor Commission prevailed.[102]

Despite the protestations of navigation interests, a civic desire to ameliorate conditions on the lower Charles did not abate, and in 1901 the state legislature revisited the controversy, authorizing a "committee to consider the advisability of constructing a dam across the Charles River between the cities of Boston and Cambridge."[103] The newly formed Committee on Charles River Dam soon recruited Freeman as their chief engineer, and during 1902, he and a team of assistants analyzed the effect of a dam at Craigie Bridge (the modern-day site of the Science Museum). The resulting study engaged issues involving freshwater river flow, saltwater tidal infusion, silt and sedimentation, sewerage and pollution, health concerns (including malaria and mosquito breeding), navigation, and the effects of storm flooding.[104]

A detailed parsing of Freeman's investigation is unnecessary here, but one aspect of Freeman's justification of the desirability of building the dam and inundating the Charles River tidal flats is important. The committee's final report, largely consisting of a voluminous investigation by Freeman, includes more than twenty photographs documenting existing conditions within the estuary. In addition, it features four photographs illustrating harbor scenes at

View of the Charles River tidal flats abutting Boston's Back Bay, circa 1902. This photograph was published as part of Freeman's report advocating construction of the Charles River Dam. (*Report of the Committee on Charles River Dam* [Boston, 1903], 33)

Hamburg, Germany, that in Freeman's view graphically illustrate how a dam could improve the beauty of a tidal/riparian landscape. Freeman's report as chief engineer certainly offered a sanitary justification for the proposed improvement, but he also advanced an aesthetic rationale whereby the Charles River Dam offered a "magnificent opportunity at comparatively small expense for replacing unsightly mud flats and unclean muddy shores now having indifferent surroundings by a great water park." To support this argument, he praised "the Alster basin at Hamburg, Germany," emphasizing how "the advantages of wholesome recreation" could be created "near to the great centres of population and convenient of access to people of moderate means and limited leisure" and stating that "a more beautiful landscape will be made possible by the construction of this dam."[105]

Soon after recommending the Charles River Dam, Freeman assisted the Metropolitan Park Commission in examining "improvements on Mystic River which, as recommended by the Landscape Architect [Olmsted Brothers], will consist of various drives and walks, and, if possible, the placing of a tidal dam in the neighborhood of Cradock Bridge [near Medford City Hall]."[106] The Mystic

Freeman used this view of Hamburg's Alster Basin, circa 1900, to illustrate how a dam across the lower Charles River Dam could inundate, and beautify, the tidal flats separating Boston and Cambridge. (*Report of the Committee on Charles River Dam* [Boston, 1903], 65)

River enters Boston Harbor just north of the Charles River at Charlestown, running through Medford and the lowlands surrounding Mystic Lake. Specifically, the legislature charged the commission with alleviating problems attending the shallow ponds, fetid wetlands, and standing pools of water in the Mystic River/ Alewife Brook watershed, conditions that plagued a growing suburban community. In accepting this assignment, Freeman told the commission chair that "the problem is in the main, a sanitary problem, and sanitary considerations such as the prevention of malaria and the discontinuance of stagnant or polluted pools that might become breeding places for the ordinary mosquito or the malarial mosquito, must largely control the engineering design."[107]

During the spring and summer of 1904, Freeman pushed his enquiry forward, engaging his mentor, Hiram Mills, to formulate maximum flood flows for the Mystic basin and also drawing upon the skills of MIT biologist William Lyman Underwood, who was a "naturalist of exceptional information in all that pertains to the breeding of mosquitos and the causes of malaria."[108] In his final report (September 1904), Freeman included photos of pools and marshland with captions highlighting "Anopheles breeding here" and emphasized how interviews with "twelve prominent practicing physicians" revealed that

Postcard view circa 1950 looking across the Charles River basin toward the MIT campus in Cambridge. Fifty years earlier, Freeman had envisaged how a dam impounding the lower Charles River could create "a more beautiful landscape." Author's collection.

"malarial disease was already prevalent and that it was . . . spreading from the cheaper houses near the marshes and claypits to the more expensive residential districts on the upland." What largely started out as a project focused on landscape park design and controlling tidal flow evolved into one where, for Freeman, "questions of public health are paramount."[109]

Freeman's reports on both the Charles River Dam and the Mystic River affirmed a belief that humankind can, through technological means, both enhance nature and protect public health. The scale and locale of these two Massachusetts river basin projects and San Francisco's proposed Yosemite reservoir were quite different, but to his mind they collectively represented how a broad-based public good could be attained through an engineered alteration of the hydraulic landscape. In the Mystic River project, he became an advocate for public health, especially in regard to mosquitoes and mosquito-borne disease. Later, when heralding the benefits of damming Hetch Hetchy, he would decry how swarms of mosquitoes infesting the valley's marshy bottomland were well positioned to feast upon summer visitors.[110] Along the lower Charles, Freeman avowed that a dam across the river's mouth would create "a more beautiful landscape" and offer "wholesome recreation" for people of "moderate means and limited leisure," not simply the well-heeled elites. Significantly, he would later argue that a dam at Hetch Hetchy could also serve such progressive ideals, with city-built roads opening up the northern reaches of Yosemite National Park to a multitude of less-than-affluent visitors. The transformation of the Charles River basin from a muddy tidal estuary into an urban lake and recreational attraction engaged a multitude of civic actors, and crediting Freeman as its sole progenitor would be brash.[111] But as chief engineer for the Committee on Charles River Dam he played a vital role in creating the sailboat-speckled vista that is now—open to one and all—a beloved part of Boston's cultural geography.

Hydroelectric Power

Freeman's insurance work and his participation in municipal water supply and civic improvement projects consumed much of his professional resources, but he held strong in his desire to serve private sector clients. He never left the worlds of factory mutuals or municipal water supply, but drawn by the lure of generous consulting fees, he found a way to propagate a vibrant private practice focused on the nascent field of hydroelectric power.

The 1890s brought electricity into the heart of American urban life, and by the start of the new century, long-distance, high-voltage alternating current (AC) transmission technology allowed for the profitable development of once-remote water-power sites and eventually helped foster the creation of expansive regional power grids.[112] The precepts of water-power economics that Freeman honed with the Essex Company may have focused on mechanically driven power transmission systems, but the principles he learned as a young engineer provided an excellent foundation for evaluating the viability of hydroelectric power schemes. While hydroelectricity was in many ways a novel technology promising new possibilities of economic development, the basic principles of water power still held meaning as the world adopted long-distance AC transmission networks.

Freeman was always proud that his expertise had value for hydropower investors seeking to make "dollarable" the exploitation of natural resources.[113] For example, by the early 1890s, Freeman's friendship with metallurgist Alfred E. Hunt, an MIT classmate who had business ties to Charles Martin Hall (patentee of electrolytic aluminum smelting), brought him into the orbit of the Pittsburgh Reduction Company (which in 1907 became the Aluminum Company of America—Alcoa).[114] The electrolytic production of aluminum from bauxite consumes huge quantities of electricity, and for more than twenty years as a consultant for Hall's Alcoa enterprise, Freeman helped guide the hydropower development of the St. Lawrence River at Massena, New York, power plants at Niagara Falls, and eventually the construction of hydropower dams in the Smoky Mountains of North Carolina and Tennessee.[115]

For himself, he wanted nothing more than to be properly compensated for his technical skills and the hard-won knowledge that guided his advice to corporate leaders. In this, he succeeded. By 1910 prospective clients were expected to pay an annual retainer of $2,500 simply for the privilege of his working on their project. This retainer would be supplemented by $100 per day for work undertaken in Providence and $200 per day for fieldwork. And even this fee structure could be ratcheted up if the market was willing to bear the freight. For example, in 1909 he was brought in to review plans for the hydroelectric dam being built across the Mississippi River at Keokuk, Iowa. For this project he demanded—and received—an annual retainer of $3,000 and a fee of $200 per day for both fieldwork and time spent in his Providence office.[116]

Freeman's financial interest in hydroelectric power extended beyond retainers and consulting fees; it also encompassed equity in investor-owned enterprises.[117] Over the years his investments grew, spurred by both his financial savvy and connections made within the investment community. For example, by 1910 the Boston-based engineering firm Stone & Webster had taken control of the Keokuk hydro project and was marketing shares in the newly formed Mississippi River Power Company. Because of his close relationship to Stone & Webster (both Charles Stone and Edwin Webster were MIT graduates) and his ongoing work as a consultant for the Keokuk project, he was given an opportunity to buy into the company on a "confidential" basis before shares were publicly offered. The deal was a good one: for $105,000 (to be paid in seven installments of $15,000), Freeman received 1,500 shares of preferred stock with par value of $100 (total $150,000) paying 6 percent dividends. In addition, he received "a bonus [of] $60,000 of common stock."[118]

Although there was some risk attached to this investment, Freeman had insider knowledge of the enterprise and presumably would have advance warning of potential difficulties should they threaten the project's financial health. One might wonder why the Mississippi River Power Company would offer Freeman such an attractive discount on its shares, but evidently they wanted to

Interior view of the Keokuk Dam powerhouse on the upper Mississippi River. Freeman was both a consultant for and an investor in the Keokuk project, and he visited the damsite frequently during the time he worked for San Francisco on Hetch Hetchy. Upon completion in 1913, Keokuk ranked among the largest hydroelectric power projects in the world. Author's collection.

keep him on board as an advisor. And having such a prominent and respected engineer as an investor would also help bolster the company's standing when shares were offered on the public market. Freeman's name possessed cachet within the financial world; as a consequence, he reaped a shareholder reward for his association with the Keokuk project.

In relation to Hetch Hetchy, the hydroelectric power project of most immediate importance was the Great Western Power Company's scheme to exploit the power potential of the Feather River north of Sacramento. Freeman's first professional engagement in the Golden State came in 1905 when New York investors asked him to evaluate the company's plans for a large dam at Big Meadows that would supply water to a series of downstream power plants.[119] Subsequently, Freeman developed a close relationship with Great Western corporate leaders through his role as consulting engineer; he also became an investor, soon controlling "600 shares each of the [company's] common and preferred stocks."[120]

In 1909 he began work designing Great Western's Big Bend Dam and Las Plumas power plant on the lower Feather River below Big Meadows, and in April 1910, he journeyed west to California to assess construction problems afflicting the project. It was during this trip that he first met with San Francisco city officials and agreed to assist their efforts to dam Hetch Hetchy.[121]

Money and Pride

By 1910, Freeman may not have been in the ranks of America's multimillionaire financial elite, but he was among the most highly compensated engineers in America. He was also among the most respected, by the public at large (for his work in fire protection and urban water supply), by leaders in America's financial/investment community, and by numerous politicians at the local, state, and federal levels. Notably, his reputation had attracted the attention of President Teddy Roosevelt, who in March 1907 and December 1908 appointed him as a consulting engineer to accompany Secretary of War (and later President) William Taft on inspection tours of the Panama Canal.[122]

However, despite his professional standing and experience, Freeman was not infallible in terms of engineering judgment or his ability to sway decision makers to his point of view. A case in point is the Big Bend Dam project for the Great Western Power Company, the very project that took him to California in April 1910 and provided San Francisco officials with an opportunity to approach him regarding the Hetch Hetchy project. As described in my book *Building the Ultimate Dam*, in 1909 Freeman had proposed a 160-foot-high dam to divert water from the Feather River into the company's newly planned Big Bend/Las Plumas power plant, but Great Western's recently hired California-based vice president H. H. Sinclair believed construction of the high dam at the site to be impractical. Sinclair argued for a less expensive, and less problematic, low dam in its place. In the face of his objections, and much to Freeman's dismay, in late May 1910 corporate management in New York sided with Sinclair and approved the low dam. Freeman may have held 1,200 shares of Great Western stock, but his influence only went so far. Challenged by an experienced utility manager with knowledge of the California power market, he could not overcome concerns that his plans were ill-advised and too costly.[123]

Freeman was hardly happy about the decision to abort his plan for a high dam at Big Bend. Nonetheless he kept his anger in check, or at least he did not publicly protest. When an unanticipated death in early 1912 brought changes to Great Western's corporate leadership, he reacted quickly. Through adroit lobbying with board members, he found a way to reinsert himself into the Feather River project. And as Sinclair weakened under the debilitating effects of tuberculosis in the fall of 1912, Freeman gained control over engineering plans for the Big Meadows Dam and spurred the company's abandonment of the multiple-arch dam design that Sinclair had favored.[124]

Viewed more broadly, Freeman's reintegration into the affairs of the Great Western Power Company was as much a matter of pride and professional status as it was about consulting fees, and here we can gain important insight into his relationship with San Francisco vis-à-vis Hetch Hetchy. Unquestionably money played a role in his work for the city, but pride in a design of his own making—one that featured an aqueduct akin to his proposal for Boston's Nashua/Wachusett project—and a desire to prevail in the face of adversity also served as compelling motivators. Like many people, Freeman did not enjoy losing once he had set his mind on a goal. However, he was unusual in the prodigious amount of time and energy he was willing to expend in pursuit of professional achievement, especially if he thought an adversary might prevail at his expense. Freeman was not an instigator of the Hetch Hetchy controversy, but once he fully embraced San Francisco's plans for a Yosemite water supply it became *his* crusade and it would be a blot on *his* prestige if the city faltered. Money certainly mattered to Freeman, but losing—especially to opponents deemed less than worthy—was something he took very personally.

By the spring of 1910, Freeman's experiences in the realms of hydraulic engineering and municipal water supply made it eminently reasonable for San Francisco officials to seek him out as a counselor on the Hetch Hetchy project. However, he did not straight away take charge of planning for the municipal dam and aqueduct and immediately mold it to his vision. His relationship with the city, and with its engineer Marsden Manson, would evolve dramatically over the following two years, and only then would his expansive plans for the city's water supply begin to reach a wider audience, one stretching from California and points west to the East Coast and eventually to the centers of political power in Washington, DC.

CHAPTER 3

Marsden Manson

A Failure to "Show Cause"

Threatened with revocation of the Garfield Permit, City Engineer Marsden Manson wired Freeman on March 19, 1910, asking for help. Explaining that Secretary of the Interior Richard Ballinger had "cited [the] city to show cause why Hetch Hetchy permit should not be revoked," Manson inquired if the Providence engineer would be willing to appear in Washington and "testify as to possibilities of this source [Hetch Hetchy] and probable future growth and needs of these [Bay Area] cities."[1] Exactly what inspired Manson to reach out to Freeman—someone he had never met or previously corresponded with—is uncertain.[2] But given Freeman's experience with major municipal water supply systems in Boston, New York City, and Los Angeles, and the fact that he possessed some familiarity with the California waterscape through work with the Great Western Power Company, he was certainly qualified to assess San Francisco's plans. And as someone who had served both as vice president of the American Society of Civil Engineers and president of the American Society of Mechanical Engineers, he possessed sufficient professional gravitas to bolster his authority on questions of civic infrastructure. In the world of hydraulic engineering and municipal water supply, Freeman's opinion mattered.

In the spring of 1910, Freeman had no compelling need—financial or otherwise—to jump headlong at the chance to assist San Francisco in its time of bureaucratic peril. If the city was willing to accept his terms, however, then perhaps he could be of help. On March 21, Freeman wired Manson a tempered

89

response ("[I] don't know enough about case to help you, also my preconceived notions somewhat adverse to your view"), but he was willing to confer with the city engineer as an offshoot of a mid-April trip to California on Great Western Power Company business. Meetings in the Bay Area with Manson and other San Francisco officials on April 16–18 won him over to the city's cause.[3]

Freeman's willingness to engage with the city was far from altruistic—as a paid consultant he required an annual retainer of $2,500 plus a per diem of at least $100 (he charged a $200 per diem for time spent away from Providence). If the city could bear the freight, however, he was happy to help represent San Francisco at the upcoming hearing.[4] As he explained to a business colleague in Providence, "While in San Francisco I enlisted under the banner of those who are willing to spoil that 'paradise for campers' on the floor of the Hetch Hetchy, for the benefit of the future water supply of San Francisco."[5]

Upon returning from California, Freeman started drafting a "brief" (or memo) that offers insight into how he first approached the Hetch Hetchy project. Unquestionably he was a quick study, but given that he had been introduced to the subject only a few weeks prior, it is still remarkable how arguments supporting the city's position found ready expression in his own voice. In this brief he worked through the city's case, devising a narrative of advocacy and support that, in various ways, later appeared in his 1912 report on Hetch Hetchy.[6]

To Freeman, urban development represented the highest and best use of a region's water resources. Other, less important uses, such as those championed by John Muir and his followers, would have to be deferred or abandoned. In Freeman's view, "the greatest good to the greatest number of people will not be secured by reserving the valley floor at Hetch Hetchy for camp purposes, but will be secured by . . . [using] Hetch Hetchy for a great reservoir." Reacting to complaints from anti-dam proponents, he further professed that damming Hetch Hetchy would require no drastic exclusion of visitors from the park, going so far as to say that a mountain reservoir and associated roads would increase visitation and enhance the beauty of the landscape and stressing that "perhaps a hundred-fold more people would thus come to the valley to see the beauties of the region. . . . The Hetch Hetchy valley will have its beauty increased by the addition of this large lake."[7]

Freeman also appreciated that a storage reservoir at Hetch Hetchy could best be justified if the Bay Area sustained a large and growing population. To him, a truly progressive twentieth-century city could not stint on providing a bountiful water supply for both present and future residents. Accordingly, he declared that "it is reasonable, prudent and indeed obligatory [that] municipal governments secure a generous future supply of the best and purest water obtainable. . . . [L]ike Los Angeles, Boston, New York, and Baltimore, [San Francisco should be able to plan] its future water supply in a comprehensive way." Beyond issues of water quantity, Freeman also worried about quality, embracing public health as a great progressive ideal, proclaiming that "the health of a community depends upon the purity of its water supply," and counseling that "the detention of the water which enters a great storage reservoir provides the greatest of all safeguards against disease germs and the results of pollution for in this period of detention disease germs perish."[8]

In a broader political context, Freeman appreciated that other Californians held an interest in the Tuolumne River, but he contended that San Joaquin Valley farmers would not suffer by the city's storage of floodwaters at Hetch Hetchy. And storage in the Sierra would also work to enhance hydroelectric power generation along the lower Tuolumne—thus serving the "fundamental provisions of conservation."[9]

In drafting his brief, Freeman refrained from addressing a key question confronting the city: specifically, were any other water sources capable of meeting the Bay Area's long-term need in place of Hetch Hetchy? This was an issue frequently raised by anti-dam preservationists, and it lay at the heart of Ballinger's "show cause" demand. But, having no specific data on alternative sources, Freeman simply ignored the question, instead boldly asserting, "Hetch Hetchy and Lake Eleanor will soon be needed in the case of San Francisco, to supplement the present sources of supply now owned by the Spring Valley Company." Although referencing the possibility of pumping water from the Sacramento River, he otherwise made the easy—but unverified—assumption that only the Hetch Hetchy/Lake Eleanor system could meet the future needs of the greater Bay Area.[10]

Overall, Freeman trumpeted the progressive social benefits of a bountiful mountain water supply. Offering no specific rationale for why San Francisco *needed* to dam Hetch Hetchy, the brief instead makes a broad-based argument supporting the city's desire for a mountain reservoir; it also advocates for the city in terms that required no specialized understanding of hydraulic engineering or water supply technology.

The Ballinger Hearing

After leaving Providence on Sunday night, May 15, Freeman arrived in Washington, DC, the next morning. Meeting with Manson, the newly elected mayor

P. H. McCarthy (of the Union Labor Party), and City Attorney Percy Long, he soon learned that Interior Secretary Ballinger had pushed the "show cause" hearing back a week until May 25. Taking advantage of his presence in the nation's capital, Freeman made a point of arranging for the San Francisco party to meet with President Taft in the White House. Although largely a courtesy call, this conference did attest to Freeman's ability to access the highest levels of political power. On Thursday, Freeman returned to Providence for the weekend; three days later he again boarded the overnight train for Washington.[11]

As it turned out, during this second trip Freeman spent but a single day in the capital focused on Hetch Hetchy affairs.[12] Late Monday, he headed back north to prepare for a major corporate confrontation over the future of the Great Western Power Company's Big Bend Dam.[13] As a result, he was not present when the Ballinger hearing finally convened two days later. On the surface his absence might appear a bit strange, but by the time he had left the nation's capital, Manson and the San Francisco entourage believed that no great edict on Hetch Hetchy would be forthcoming that week. As documented by historian Kendrick Clements, Manson had learned that the Hopson-Hill report presented to Ballinger the previous fall contained data generated by a consulting engineer—Phillip Harroun—who had previously worked for the Spring Valley Water Company. And Harroun had also been hired by the Society for the Preservation of National Parks to help represent their cause at the Ballinger hearing.[14]

None of this was public knowledge at the time, but Clements makes a convincing argument that San Francisco's leadership privately advised Ballinger of the political imbroglio that would ensue if word of the perhaps biased data was leaked to the press. It is also likely that President Taft had become aware that studies underlying Ballinger's "show cause" request of February 25 may have been tainted (or at least suffered from the appearance of possible tainting). This would explain Taft's action on May 12, when he authorized appointment of an "advisory board" of three officers from the Army Corps of Engineers to assist the secretary. In Ballinger's words, these army officers were to "prepare such findings as seem to them the evidence warrants, advisory to the Secretary of the Interior." The president's appointment of this board proved to be no minor event. The Hetch Hetchy controversy now featured a new and important focal point, as the Army Board was empowered to evaluate all engineering issues involving San Francisco's plans and publicly report upon their findings.[15]

At 10 a.m. on May 25, Ballinger, accompanied by the formally titled Advisory Board of Army Engineers, convened his hearing. A contingent of anti-dam activists, including Berkeley professor William Bade representing the Sierra Club, Boston attorney Edmund Whitman representing the Society for the Preservation of National Parks, and Horace McFarland, president of the American Civic Association, argued for immediate revocation of the Garfield Permit. In rejoinder, San Francisco officials urged its continuance; however, the city offered no evidence for exactly why its needs could not be met by other available sources. In fact, City Attorney Long readily acknowledged that "at this time [we are] not prepared or able to take a proper showing or to show cause . . . on account of a lack of information and data on which a proper showing should be based."[16] If Freeman had attended the hearing, he certainly would have supported the city, but his presence would have had no major impact given Long's admission regarding San Francisco's inability to "show cause."[17]

At noon, Ballinger excused himself, leaving the Army Board to preside over an afternoon session. The next morning the secretary brought the proceedings to a halt, proclaiming that the American people wanted to know "whether it is a matter of absolute necessity for the people of [San Francisco] to have this source of supply [Hetch Hetchy]; otherwise it belongs to the people for the purpose for which it has been set aside [as part of Yosemite National Park]."[18] To answer this question, Ballinger ruled that the city would be given a year to investigate and "determine whether or not the Lake Eleanor basin . . . together with all other sources of supply available to the city will be adequate for all present and reasonably prospective needs of said city . . . without the inclusion of the Hetch Hetchy Valley."[19] In other words, Ballinger decreed that the city would need to affirmatively defend its claim to Hetch Hetchy and demonstrate why all other possible water sources, including the Lake Eleanor watershed, were inadequate to meet its future needs. The city was granted one year to prepare a report making its case. On June 1, 1911, Ballinger was to reconvene the "show cause" hearing and, predicated on the city's submissions and an evaluation by the Army Board, render judgment on possible revocation of the Garfield Permit. One year, and then the fate of Hetch Hetchy would be decided. At least that was the plan.[20]

Ballinger's decision to punt the "show cause" hearing back a year was not a defeat for the anti-dam collective, yet it fell far short of their highest hopes. For the time being, the Garfield Permit remained in place—albeit in bureaucratic limbo—and all that Muir and his followers could do was await completion of the city's investigation. Perhaps such a report would never appear and Ballinger would bring their troubles to a welcome end a year hence. But if San Francisco did submit a completed report, then they would be forced anew to defend Hetch Hetchy.

For the city, Ballinger's decision came as welcome news, especially for Manson. Granted a cushion of twelve months, the city engineer felt no compunction to take rapid action on fulfilling the "show cause" directive. As events played out—and the city was given an extension to "show cause" until the summer of 1912—Manson would prove remarkably ineffective in addressing the complex tasks entwined with the city's claim on Hetch Hetchy. But despite his limitations he had no desire to relinquish control over the project, and for two years he sought to maintain appearances that everything was fine with the city's water plan. In the end, his efforts proved mere chimera.

In a July 1912 letter to the Army Board published in Freeman's report on Hetch Hetchy, San Francisco's mayor and city attorney state that the report was "the result of a two year's personal study on the part of Mr. Freeman, working with a staff of skilled engineers," and praise the "intelligent zeal and loyalty which [Freeman] has brought to his task."[21] No mention is made of Manson or his responsibility for the city's Hetch Hetchy studies until May 1912. By all appearances, this letter seems to imply that Freeman was given charge over the city's efforts immediately following Ballinger's decision to delay the "show cause" hearing. In turn, historians have taken a cue from the mayor and city attorney's missive and largely assumed that Freeman began managing the city's investigations in June 1910. Most notably, in his 1965 history of Yosemite National Park, Holway Jones avers that in the wake of Ballinger's truncated hearing, Freeman "was now contracted to do an exhaustive investigation by the city."[22] However, while it is true that Freeman never broke off his relationship with the city after May 1910, he hardly had control over an "exhaustive investigation" during the time that Manson actively served as city engineer. In fact, at times prior to the summer of 1912 Freeman would express a willingness—almost a desire—to excuse himself from the city's Hetch Hetchy initiative so he could pursue other professional opportunities. Through the spring of 1912, Manson oversaw the city's faltering effort to "show cause," while Freeman served as a city consultant, albeit one who gradually exerted ever more influence over the project.

Summer in Europe

Freeman hastily left Washington, DC, on May 23, 1910, and during the late spring and early summer, he devoted most of his professional attention not to Hetch Hetchy but to Great Western Power Company matters and other consulting projects. Manson updated him on early correspondence with the Army Board, and Freeman offered feedback on issues such as filtration of the Sacramento River.[23] But his most direct work related to Hetch Hetchy came in submitting a bill to the city for services rendered during April and May—the tab came to $4,000 plus $115.15 for expenses.[24] This was no small sum, but Freeman had made his fees clear when the city engaged his services.

On July 16, Freeman steamed out of New York City on the RMS *Celtic* for a six-week sojourn in Europe—one paid for on his own account and without support from San Francisco. Upon disembarking, he spent time in Birmingham and London on business affairs and then joined family members vacationing in Germany. Although his travels were largely personal, he was alert to what might be of value to his professional clients. Most notably, he kept an eye out for precedents and possibilities that, down the road, he might draw upon when advocating for the Hetch Hetchy Dam. Thus, while traversing southern Norway with his wife in August, he took note of mountain highways and wagon roads rimming steep-cliffed fjords; these latter vistas offered spectacular views that he perceived might correlate with roadways passing above the Hetch Hetchy reservoir. With striking foresight, he collected photographs of these scenes, several of which later appeared in his published *Hetch Hetchy Report*.[25]

Freeman spent much of the summer of 1910 in Europe. When visiting Norway, he observed how roads built along steep-cliffed fjords might serve as a model for a road circling the Hetch Hetchy reservoir. This view of Lake Oifjords was included in his 1912 *Hetch Hetchy Report* (p. 18).

From Bergen he crossed the North Sea to Britain and set out on an extended tour of water supply systems serving Glasgow, Manchester, Liverpool, and Birmingham.[26] As in Norway, he looked for ways he might draw upon European precedents when advocating for San Francisco's mountain water supply. In a letter to Manson, Freeman described how he had "hired an automobile and inspected the Thirlmere watershed and reservoir of the Manchester [municipal water] supply, which, as you know, is in the very heart of the nature lover's stamping ground in England. The roads which were built by the Manchester Waterworks now form the great highway for tourists' coaches, automobiles and bicycles through this beautiful region of the English lakes."[27] Freeman recognized the political significance of Manchester's success in building the Thirlmere system in the face of environmental opposition, advising Manson that the secretary of Manchester's water board had provided him with a report describing how "bishops and other prominent citizens protested by all that was sacred against defiling and debasing this beautiful region for purposes of municipal water supply. But the Secretary told me that everyone now admitted that the

In the late nineteenth century, Manchester's plan to develop Lake Thirlmere as a municipal water supply reservoir engendered much opposition because of its location in the picturesque English Lake District. But once the project was built, it did not impede broader use of the rural watershed. Freeman visited the region in August 1910 and, with the caption "on the public highway in the catchment of Lake Thirlmere," used this travel image in his *Hetch Hetchy Report* (p. 47).

While touring Britain in August 1910, Freeman visited Birmingham's newly completed Craig Goch Dam in Wales, describing it as "the most beautiful dam in the world." This image is from Freeman's *Hetch Hetchy Report*; his caption draws attention to the "inviting roads" around the reservoir built "chiefly for the pleasure of the public" (p. 51).

work of the Board had been helpful to the nature lover, particularly in their building of a magnificent carriage road along the west shore of the lake."[28]

Farther to the south, Freeman visited Liverpool's Vyrnwy Dam in North Wales, "which has a great highway, patronized by the nature lovers, along its shores"; he also inspected "the chain of reservoirs recently built in the mountains of Central Wales for the additional water supply of the city of Birmingham." As part of this latter system, Birmingham had recently completed the Craig Goch Dam, a massive storage dam that Freeman greatly admired. To his mind, Craig Goch was "the most beautiful water works dam in the world. . . . A dam precisely like this would certainly add to the beauties of Hetch Hetchy."[29]

Freeman's European trip had been planned before Hetch Hetchy became a part of his professional portfolio. However, during his travels he kept his San Francisco client in mind, paying attention to rural roads in Norway and the dams and watersheds that served major British cities. He appreciated that if the city were to prevail in its Hetch Hetchy plans, it would need to confront the objections of the "nature lovers" in ways that transcended mere technical arguments. On his own initiative he began formulating a visual narrative in which the damming of Hetch Hetchy could be tied to a larger international movement to construct attractive urban water supply systems and, simultaneously, make the rural landscape more accessible to vacationing travelers.

Freeman and Manson

On August 27, 1910, Freeman left Liverpool on the RMS *Lusitania* (the same ship later torpedoed by a German U-boat), arriving in New York six days later.[30] Upon reaching his Providence office he discovered that, to his dismay, the consulting bill he had presented to San Francisco the prior June languished unpaid. Manson blamed the delay on difficulty in selling Hetch Hetchy bonds, but Freeman was unimpressed.[31] He expected clients to treat him with respect and not allow accounts to linger in arrears for months on end. Angered, he went so far as to threaten abandoning the city as a client, admonishing Manson and pointing out, "I am well aware that [my fees] . . . are higher than most engineers charge . . . but I seem to have plenty of clients who are ready to meet these figures."[32] Freeman further reinforced the point that San Francisco held no special claim on his time, warning that "in all sincerity, I would today be willingly excused from further attention to the Hetch Hetchy problem, because of some particularly interesting and important work nearer home." Exactly what this "work nearer home" may have encompassed is unclear. But Freeman was putting Manson on notice: his association with a slow-to-pay San Francisco was not necessarily a high priority.

Despite frustrations, Freeman was nonetheless willing to explore the Hetch Hetchy backcountry as part of an early fall trip to the Pacific Coast.[33] After leaving Providence on September 20, he made stops in Chicago and Minneapolis before spending five days inspecting hydropower dams near Helena, Montana. He reached San Francisco on September 30 and spent the next day focused on Great Western Power Company business before joining Manson aboard the night boat to Stockton enroute to Hetch Hetchy. After daybreak, the two engineers "started by automobile at 7 AM via Knights Ferry, Chinese Camp, Stevens Bar, and Oak Flat Road . . . [and] took photos and studied maps for Aqueduct routes all along."[34] The next day they traveled by wagon to Hog Ranch (the city-owned tract lying just outside the boundaries of the national park that is now known as Mather), where they switched to "saddle horses and pack horses [and set off] to Hetch Hetchy Valley—arrived just about sunset on October 4th."[35] Over the next few days, Freeman and Manson explored Hetch Hetchy and trekked up to Lake Eleanor and Cherry Creek, all the while discussing damsites and aqueduct routes. After almost a week on the road, Freeman made it back to San Francisco on Saturday morning, October 8.[36] Quite famously (or infamously), Gifford Pinchot never visited Hetch Hetchy Valley, but after October 4 John R. Freeman could never be criticized on such grounds. As a result of his trip with Manson, he gained direct knowledge of Hetch Hetchy, the Tuolumne watershed, and the terrain separating San Francisco from Yosemite National Park.

Freeman and Manson had spent almost a week together in the backcountry, but they appear to have developed little rapport. Freeman had yet to be paid for his work in the spring, and when a check was finally issued on October 13, 1910—after he had left San Francisco for the trip home—it only covered his travel expenses and the $2,500 retainer. Another $1,500 remained in abeyance, something that did little to endear the city engineer to him.[37]

In late October, Manson got word from the Army Board that at least four possible alternative sources of supply (including Lake Tahoe and the San Joaquin River) could be eliminated from the city's evaluation. But others, including the Eel River, the Sacramento River, and the "East Bay Shore Gravels" (associated with the Spring Valley Water Company's Alameda Creek holdings), "should be given further consideration." In particular, the board wanted the Sacramento River to be carefully studied because "filtration has been so thoroughly successful in furnishing pure water . . . [that there] is not a sufficient reason for rejecting the process for San Francisco." On this last point, Manson asked Freeman for suggestions on who might undertake a study focused on filtering the Sacramento.[38] Later in the fall, Manson expressed frustration over his interactions with the Army Board and complained to Freeman that "the way the problem is framed up to the so-called impartial Board of Federal [Army] Engineers the more iniquitous and tyrannical becomes the action of the government." Seemingly hoping for political relief from the demands of the Army Board, he broached the idea that Freeman might "find time to visit Washington in the near future and have a conference with the President."[39]

Soon after, Manson requested a copy of Freeman's 1900 New York City water supply report to serve as a guide in analyzing alternative sources.[40] While happy to oblige, Freeman nonetheless pointedly opined, "My belief is strong that you will fail to secure the desired concessions from Uncle Sam unless you are able to present a deeper and broader study of the whole subject [of the need to dam Hetch Hetchy]." And sensing that Manson was perhaps hoping that he might become more engaged in helping the city, Freeman preemptively demurred, wondering "how much time I am justified in putting into detailed studies myself."[41]

Soon after the New Year, Manson again asked if Freeman might have time for "a consultation with the President" and expressed his hope that the East Coast engineer would "be able to visit Washington in connection with this."[42] In January Freeman did visit the nation's capital before setting off to reconnoiter an irrigation project in Texas and visit the Keokuk Dam across the Mississippi River north of St. Louis.[43] Only upon his return east at the end of the month did he respond to Manson's entreaty. Voicing displeasure, Freeman explained that during his recent visit to Washington he "did not call on the President" because "[I] did not feel that I had the data or the problems in shape to answer some of the questions he might ask." Further separating himself from the city engineer, he described how an associate had "asked me in Boston a few weeks ago if I had ever investigated the real necessity of [San Francisco] going to the mountains in the near future for a supply. . . . I told him that I had not and that was not my problem, but yours."[44]

In early February, 1911, Manson offered a muted response to Freeman's grumblings but, more importantly, advised his erstwhile colleague that he had "applied for . . . an extension of the time in which to report, from June 1st to December 1st next." He added, "This will give me time to work up a report on the basis of your suggestion."[45] As it turned out, a month later the "show cause" hearing did undergo a major adjustment, with Ballinger resigning as interior secretary, ostensibly due to health issues. As a consequence of Ballinger's departure, Manson would receive a six-month reprieve on the June 1 deadline. In addition, Freeman would have an opportunity to engage with the new secretary and begin advocating an expansive plan for Hetch Hetchy, one of his own making that was not confined by the strictures of the Garfield Permit. Before I describe these changes, however, another aspect of Freeman's involvement with San Francisco's water supply infrastructure requires attention. Significantly, this involves Freeman's relationship with the Spring Valley Water Company, activities unexplored in any previous histories of the Hetch Hetchy controversy.

Spring Valley Water Company

After his tramp into the Sierra with Manson, Freeman might well have relaxed for a few days, but that was not his style. Upon reaching San Francisco the morning of October 8, 1910, he first "went to Hotel St. Francis & cleaned up."[46] Refreshed, he then met with the vice president of a new prospective client: the Spring Valley Water Company.

In stories about San Francisco's quest for a municipally owned water supply system, this company almost always appears as the great villain. To its many detractors, Spring Valley constituted a rapacious private enterprise that, thanks to a charter dating to the 1850s, dominated the region's water resources and forced city residents to pay dearly for a product often considered to be of mediocre quality. Above all, it was the Spring Valley monopoly that animated the city's drive for a public water system and gave many progressive conservationists a rationale to support the damming of Hetch Hetchy. In the memorable phrasing of Mayor James Phelan, the battle over Hetch Hetchy was fought to save the city from "monopoly and microb[e]s," with Spring Valley standing in the way of a better civic future.[47] So what did the company want from Freeman?

In the 1870s the Spring Valley interests purchased control over the Alameda Creek watershed east of Palo Alto. This basin fed the Sunol aquifer, and the company soon connected this subsurface supply to its peninsula system feeding San Francisco from the south. Calaveras Creek was a prime tributary of Alameda Creek with a reservoir site capable of storing billions of gallons of floodwater. Although the Spring Valley leadership recognized that at some point the city would likely purchase the firm's water supply system (in January 1910, the company had almost sold its assets to the city for $35 million—see chapter 2), they also realized that they could enhance the value of their holdings by continuing to develop the resources under their control. So, in the summer of 1910, Spring Valley began planning a large storage dam across Calaveras Creek. Under the direction of Vice President S. P. Eastman, the company engaged a Los Angeles–based engineer to develop an earth embankment dam design featuring a concrete upstream face.[48] Eastman deemed it prudent to seek out an independent review of the proposed design, and this led him to John R. Freeman.

In August 1910, Eastman had asked Freeman for his "opinion . . . on the type and sufficiency" of plans for the Calaveras design and requested "what will your charge be?"[49] From Liverpool, Freeman wired Eastman of his interest in the assignment "unless incompatible [with] my engagement [with] city [regarding] Hetch Hetchy."[50] With that possible conflict clearly stated, he allowed, "I expect to be in San Francisco within three or four weeks." At that time, he could conceivably review Spring Valley's plans for Calaveras Dam, but only "after first seeing the site."[51]

After their initial meeting on October 8, Freeman and Eastman spent the next day on a field trip to Sunol and the Calaveras Dam site. They quickly reached an understanding of how the Providence engineer might be of service to the private utility. With the acquiescence of City Attorney Percy Long, Freeman agreed to review the company's plans for an earth embankment dam.[52] As phrased by Eastman in a follow-up letter, Freeman had "gone over the possible sites in the Calaveras Gorge," and he was authorized to "approve, modify, or disapprove of the design and specifications" of the proposed dam.[53] On October 11, just as Freeman was leaving California, Eastman codified their financial arrangement: "[I] am enclosing a check for $2500, the amount of retainer. Actual time spent on [Calaveras Dam] work . . . will be [billed] at the rate of $100 to $200 per day."[54]

In Eastman and Spring Valley, Freeman found what he liked: a corporate client with deep pockets who was willing to pay him promptly—and well—for his services. Whereas he had yet to receive any recompense from San Francisco (the city's first partial payment would not be forthcoming for another two days), by the time he departed from the Bay Area he carried with him a $2,500 check from the often-demonized Spring Valley Water Company.[55]

As originally conceived, Freeman's work for Spring Valley was to "approve, modify, or disapprove of the design and specifications" for a proposed earth embankment dam, but the word "modify" could be construed rather broadly and encompass an entirely new design. Upon his return to Providence, he began to evaluate the earth dam proposal, soon apprising Eastman, "I am inclined to seriously doubt permanence and wisdom of [the] earth design notwithstanding its great economy." Claiming that he would "keep an open mind," he nonetheless averred, "[The] present data inclines me strongly toward . . . [an] arched concrete dam."[56]

Perhaps he did keep an open mind regarding the proposed earthen design, but most of his interest in the subject over the coming weeks involved gathering data about recent problems with earth embankment structures; as part of this inquiry he corresponded with both Arthur P. Davis (chief engineer of the US Reclamation Service) and J. B. Lippincott (assistant chief engineer of the Los Angeles Aqueduct).[57] At the start of the new year, he reiterated his interest in a massive "arched concrete dam," and Eastman encouraged this approach: "I look with much favor upon your suggestion to include in your report a structure of massive concrete rubble."[58] A month later, Eastman told Freeman to stop any further consideration of the earth dam proposal and instead focus on "your alternative design" so that it could be "submitted to us before April first."[59]

Over the next several weeks, Freeman and his staff developed plans for a concrete gravity dam across Calaveras Creek. Forwarded to Eastman on April 15, 1911, these plans outlined a massive design approximately 250 feet high with a "cross section safe by gravity if standing alone, regardless of arch." The imposing structure would require 390,000 cubic yards of concrete and "pudding stones" and cost an estimated $2,369,000. This was considerably more than estimates for the rejected earth dam design, but Freeman assured Eastman, "[P]ersonally, I believe that nothing materially less substantial or less expensive will satisfy the proper demands for safety." And he also assured Eastman that "another argument" favoring the massive concrete arch "is found in the psychology of having a dam that will impress every dweller in the valley and every citizen who sees it as being secure beyond the shadow of a doubt."[60]

Perhaps vexed at the hefty price tag, Eastman did not immediately accept Freeman's proposal. But neither did he reject it. Instead, he acknowledged, "[T]he tentative design which you have submitted is, I believe . . . best suited . . . to serve our purposes" and then, anticipating that Freeman would be in the Bay Area over the summer, advised that "certain details of your design . . . I will take up with you upon your arrival." In the meantime, the company would continue work clearing the foundations and investigating the site geology. For Freeman, this was far from bad news as his relationship with Spring Valley entered a new phase. Although his design for Calaveras Dam was yet to be approved for construction, his consultancy remained in place and he would continue work on the project during his upcoming trip to California.[61]

Freeman was well compensated by Spring Valley, and this constituted a powerful incentive for him to cultivate the company as a client. But it is also important to understand that in working for the private utility, Freeman did not place himself in opposition to the city's plans for a municipally owned water supply. Instead, the Calaveras Dam could be envisaged as a likely—ultimately inevitable—component of publicly owned infrastructure system in the Bay Area that would complement the Hetch Hetchy/Lake Eleanor system. And if was possible to get paid by two clients in pursuit of such a vision, then why should he complain?

Given the seemingly "anti-progressive" character of the Spring Valley water monopoly, Freeman's ability to operate in the politically contested zone separating public and private interests represents one of the most astonishing aspects of his engagement in San Francisco's civic affairs. Later that summer, his relationship with public and private interests in the region became further complicated

During the summer of 1911, Freeman advised the Pacific Gas and Electric Company on the design of Spaulding Dam. This 1913 view shows the privately financed hydro-electric power dam under construction across the Yuba River in northern California. In his *Hetch Hetchy Report*, Freeman referred to his Spaulding/Yuba design as "much the same" as his proposed Hetch Hetchy Dam. Author's collection.

when he began advising the Pacific Gas and Electric Company (PG&E) on its plans to build Spaulding Dam in the upper reaches of the Yuba River.[62] The fact that he still maintained a financial relationship with Great Western Power Company, PG&E's corporate rival, only added to the complexity of his place within the tempest swirling around Hetch Hetchy. In retrospect, Freeman's ability to straddle the politically charged boundary separating public works from private corporate interests is truly remarkable and speaks to his distinctive, almost singular, status as a consulting engineer.

Ballinger Out, Fisher In

Manson's concern about meeting Ballinger's "show cause" deadline quickly dissipated when the interior secretary resigned in March 1911 and Taft replaced him with Walter Fisher, a former president of the Conservation League of America and a Taft-appointed member of the Federal Railroad Securities Commission.[63] Ballinger's exodus from the president's cabinet had nothing to do with Hetch Hetchy but instead related to health problems and, perhaps more importantly, to the political imbroglio attending the "Ballinger-Pinchot Affair." Gifford Pinchot believed that Alaska coal leases issued by the General Land Office, which Ballinger supported, represented a betrayal of the ideals of progressive conservation. Pinchot questioned the legality of the leases and brought the issue to Capitol Hill. Widely publicized congressional hearings proved embarrassing to Ballinger (although he was exonerated of wrongdoing) and ultimately spurred his resignation as interior secretary.[64]

For Manson, Ballinger's departure came at an opportune time, and he gloated to Freeman that "matters have considerably brightened by the recent change in the head of the Department of the Interior." City Attorney Long was planning to head east to press the city's case in the nation's capital. As part of this initiative, Manson advised Freeman that it would be "highly desirable, if not essential, that you should see the president and fully acquaint him with the situation and prepare his mind for such presentations as you and Mr. Long may subsequently make."[65] Two months earlier, Freeman had brusquely rebuffed Manson's entreaty that he intercede with the president. However, the departure of Ballinger dramatically altered both the political landscape surrounding Hetch Hetchy and Freeman's own orientation toward the city's plans. Now he was eager to engage with Taft and the new interior secretary.

Walter Fisher was a Chicago lawyer whom Freeman had come to know during his work in the Windy City dating back to the Iroquois Theatre Fire investigation in 1904. As he pointedly informed Manson, "It so happens that the new Secretary of the Interior is one of my good personal friends of some seven years standing, and in fact his son, who is a classmate of my own son at Harvard, spent last Sunday at our home. You can rest assured, therefore[,] that Mr. Fisher will accept whatever I tell him about the Hetch Hetchy as my candid and sincere belief . . . [but you] can be equally certain that I will not violate his confidence by advocating any project that I do not believe in or concerning which I am not fully informed."[66]

With Ballinger gone, Freeman perceived an opportunity to advance a new scheme that, in radically diverging from constraints proscribed by the Garfield Permit, opened up the possibility that the city might totally reconceptualize its plans for a mountain water supply. Joining with City Attorney Long (but not Manson), on April 6 Freeman met with Taft and Fisher. In a follow up letter to Fisher, Freeman made clear that he had no compelling ties or commitments to the city's prior proposals: "I am afraid that I am not an entirely subservient counselor, for against the wishes of certain of the San Francisco people I have been recommending the policy of opening this case wide-open, wiping all the old agreements off the slate in the belief that something better for the city than the Garfield Permit was obtainable. . . . After leaving your office Mr. Long told me he should rely on my advice."[67] This letter—and the meeting it references—marks a key inflection point in the history of the Hetch Hetchy controversy. First, it indicates that Long and Freeman were beginning to collaborate in ways that did not include City Engineer Manson. Second, in advising Fisher that he wanted to wipe "all the old agreements off the slate," Freeman began actively campaigning for a new plan that positioned the damming of Hetch Hetchy as the centerpiece of a capacious mountain water supply and hydropower system.

In essence, Freeman believed the Garfield Permit to be grossly deficient because it required the combined Lake Eleanor and Cherry Creek watersheds be fully exploited before the city could impound a reservoir at Hetch Hetchy. To Freeman, this constituted an unacceptable inefficiency because first, the watershed above Hetch Hetchy was much larger than those feeding Lake Eleanor and Cherry Creek, and second, the narrow Hetch Hetchy dam site allowed for greater storage with a shorter, less expensive dam than would be required at both the Lake Eleanor and Cherry Creek dam sites. While recognizing the city's need to address the "show cause" demand inherited by Fisher, Freeman believed that the city should propose a completely new initiative, one centered around the immediate construction of a large storage reservoir at Hetch Hetchy and a high-pressure pipeline/aqueduct that—instead of the 60 million gallons of water per day (mgd) proposed by Grunsky and Manson—could deliver 400 mgd to San Francisco without pumping (i.e., it would be a fully gravity-flow aqueduct akin to what he had proposed for Boston in 1895). As later detailed in Freeman's 1912 *Hetch Hetchy Report,* this conduit would also allow for the generation of over 150,000 horsepower of electricity at city-owned hydropower plants.

In the short term, Freeman recognized the political need to defend the Garfield Permit in terms of "show cause." But starting in the spring of 1911, his ultimate goal was something far greater than what had been offered in 1908.[68] No longer would Freeman simply be advising San Francisco on how Manson's efforts to "show cause" might succeed. Now he had a project of his own to promote, one that, to his mind, could provide "something better for the city" on a scale dwarfing all prior city plans.

1911: Summer and Fall

As a result of the Ballinger-to-Fisher transition, Manson was blessed with much-needed breathing space. However, getting an extra six months to "show cause" did not alter the basic task confronting the city (as requested by the city engineer, Fisher pushed the deadline back to December 1, 1911). At Freeman's urging—and with Long's support—Manson began to seek out engineers to assist in his investigations.[69] Most notably, this led to the engagement of the New York–based engineer Allen Hazen to analyze the Sacramento and San Joaquin Rivers as possible sources of pumped/filtered water.[70] The Hazen Report proved to be the most comprehensive of all the studies carried out on possible "alternative" sources to Lake Eleanor/Hetch Hetchy. For his part, Freeman began to focus on plans for a newly envisaged Hetch Hetchy Dam and Aqueduct.

During the late spring, there was little Freeman could do in pursuit of his Hetch Hetchy and Calaveras Dam initiatives until his summer trip to the West Coast. A review of his diary and other files for May and the first half of June reveals that he spent almost no time working for the city and only a modest amount of time on Calaveras Dam. The bulk of his consulting work focused on Keokuk Dam and a proposed hydroelectric dam on the Cheat River south of Pittsburgh.[71] The affairs of Spring Valley and San Francisco would be addressed when he decamped for the West Coast in late June.

On June 23, Freeman sailed down Narragansett Bay from Providence and spent the evening socializing with a vacationing President Taft.[72] The next day he headed west on the 20th Century Limited, reaching San Francisco four days later. Never one to dawdle, within half an hour of his arrival he was in Manson's office, and the next morning he set off on a "tour of inspection to Spring Valley sources of supply," followed by an examination of the Calaveras Dam site.[73] On Saturday, July 1, he joined Manson, Long, former city engineer C. E. Grunsky, and sanitation engineer Allen Hazen in a conference convened by the Army Board. There, Colonels Biddle, Spencer, and Cosby along with their assistant H. H. Wadsworth were updated on the status of Manson's efforts. But more importantly for Freeman, the gathering offered him an opportunity to promote his proposed restructuring of the city's plans. As he recorded in his diary, "I made a long and strong plea [to the Army Board] for beginning with building the Hetch Hetchy [Dam] first and a wagon road to be built by the city to make it available [for visitors]."[74] The vision that he had shared with Secretary Fisher in April to get "something better for the city" thus received a wider airing.

The Army Board had scheduled a trip to Hetch Hetchy to start a week thence; in the meantime, Freeman visited the Muir Woods and took a "steamboat up river to Sacramento looking up filtration problem with Hazen. . . . Good site for intake at Rio Vista."[75] He also began to focus on helping the city with the valuation of Spring Valley properties, a complex procedure that would provide a foundation for any future sale; over the course of the next year he would continue to be involved in the politically contentious process of trying to set a "fair" valuation of the water company's assets.[76] Later in the week he held conferences with Long, Grunsky, and J. H. Dockweiler, an experienced California-based hydraulic engineer hired to help advance the city's work. In a letter to Long, Freeman asserted, "[I]t will greatly help in my preparation of the City's case if Mr. Grunsky and Mr. Dockweiler can immediately begin their studies on the special problems discussed in our interview." Specifically, Dockweiler was to investigate the use of water by the Turlock and Modesto Irrigation Districts and determine the amount of irrigable land that could be reasonably watered by the districts' diversions from the Tuolumne River. On a broader level, Grunsky was to consider the totality of water available in streams tributary to the Sacramento and San Joaquin Rivers and contrast this with the amount of irrigable land in the Central Valley. The goal of both investigations was to demonstrate that San Francisco's claims to flow from the Sierra Nevada could be met without negatively impacting the state's agricultural interests.[77]

On July 10, Freeman, Manson, and Hazen headed to the Sierra, joining up with the Army Board at Hog Ranch. Supported by "52 horses and mules," the party entered Hetch Hetchy two days later to explore the damsite and reservoir zone. Unfortunately, Freeman did not write detailed diary entries covering his time in the mountains, but he did make note of "great sport in watching swimming horses and mules across the Tuolumne River." Two more days were spent visiting Lake Eleanor and Cherry Creek before he split off from the group and returned to San Francisco on July 16. He then "got mail at Hotel and left for Vancouver at 11:20 am."[78] No matter that he had been in the Sierra backcountry for a week of arduous travel. Other clients beckoned and his time on the West Coast was limited.

Upon leaving San Francisco, Freeman first inspected hydropower projects in British Columbia and then headed to Boise, Idaho, where the US Reclamation Service was planning the massive concrete gravity Arrowrock Dam.[79] He returned to San Francisco on July 30 and, during the following three weeks, divided his time between Hetch Hetchy (in early August he asked William Mulholland, chief engineer for the Los Angeles Aqueduct, for cost data on tunnel construction), Calaveras Dam work for Spring Valley, and Pacific Gas and Electric's plans for a storage dam across the Yuba River (on August 15–16 he visited the Yuba/Spaulding Dam site preliminary to preparing a concrete dam design for PG&E). Taking time for some socializing (and politicking), he also attended the Bohemian Grove retreat on August 12–13.[80]

Secretary Fisher was scheduled to visit Hetch Hetchy Valley in the early fall, but Freeman could not delay his departure from California and accompany the secretary's party into the Sierra; he had other obligations to meet. After departing San Francisco on August 23, Freeman spent a week in Manitoba, Canada, studying the power potential of the Winnipeg River. He then inspected Keokuk Dam before arriving back in Providence in early September.[81] Despite an absence of over two months, he would remain in New England for only a few weeks before heading off on yet another business trip, this time to Central America, where he would consult on Mexico City's water supply and on a hydroelectric project in northern Mexico.

Prior to this extended journey out of the country, however, Freeman took time to counsel Manson and Long as to how they could advance the city's claim on Hetch Hetchy during Secretary Fisher's upcoming visit.[82] Of special importance is Freeman's advice that Manson should make Fisher acutely aware of how few people had ever made the journey to Hetch Hetchy and how this state of affairs could be rectified by roads built as part of the dam project. Specifically, he urged Manson, "While at Yosemite obtain for me complete copy [of] Hetchy visitors' register since first established, including dates, names and residence. . . . Show Secretary [Fisher] this explaining extent of recent advertising and emphasize the benefits of making valley accessible to ten or hundredfold this number, instead of selfish private camping reserve for a few solitude-loving cranks."[83]

In highlighting how few people had ever actually visited the valley, Freeman perceived an opportunity to portray anti-dam proponents as "solitude-loving cranks," disdainful of the common person who, absent a modern road system, could never hope to make the laborious and expensive trip to Hetch Hetchy. In his open letter to the American public in 1909, John Muir had decried "the determined attack . . . made by the City of San Francisco to get possession of the Hetch Hetchy Valley as a reservoir site, thus defrauding the ninety millions of people for the sake of saving San Francisco dollars."[84] Freeman saw such exuberant language as an opportunity to stress how few of the nation's "ninety millions" had actually been making the trip to the valley. And by culling data from the national park's visitor's logs, the city's argument could be given a seemingly objective foundation.

For City Attorney Long, Freeman had a different assignment, one that would reinforce his plan to place the development of Hetch Hetchy as a centerpiece of the city's mountain water supply: "Don't let Secretary [Fisher] get away with idea that confirmation of the Garfield Permit is all the City needs. I am stronger today in the hope than ever that the City get something much broader and better."[85] Of course, this plan had been broached with Fisher back in April and Freeman had also presented it to the Army Board in July, but he wanted to make sure that it remained fresh in the secretary's mind. Significantly, Long was hesitant to embrace Freeman's vision of building the Hetch Hetchy Dam first, counseling him in early October, "I am somewhat uncertain about the wisdom of asking the Secretary for a broadening of the permit. . . . Manson and [Stanford Professor] Marx [also] seem to think that we should stand upon the Garfield Permit and not antagonize the [irrigation] districts."[86] From a historical perspective, Long's reluctance in the fall of 1911 to unconditionally endorse Freeman's plan for "something broader and better" than what was codified in the 1908 Garfield Permit is important, especially in highlighting Freeman's singular role in gestating and promoting the new initiative.

By early October, Long acknowledged to Freeman that, with the December 1 "show cause" deadline rapidly approaching, "Manson told me yesterday that he would be unable to secure all of his reports by December 1st and suggested a continuance until April 1st [1912]." [87] At Manson's request, Fisher soon agreed to a further three-month extension to March 1, 1912.[88] But the question remained: Would Manson ever be able to craft a comprehensive, persuasive report? Freeman was skeptical. Yes, Allen Hazen was doing a good job investigating the Sacramento River, but Freeman was frustrated by Grunsky's ineffectiveness, admonishing Manson, "[Y]ou must earnestly get your hooks and spurs into Grunsky and inject more energy into his collection of data" on alternative sources.[89] The clock was ticking, and Manson seemed incapable of organizing a convincing response to the Interior Department's "show cause" demand. Seventeen months had passed since the Ballinger hearing in May 1910, and Manson was seeking yet more time to complete his report.

At the beginning of October, Freeman and his wife headed south to Mexico, but prior to leaving, he advised Manson that he "had been pushing this Mexico matter . . . forward [as] rapidly as possible so as to have [his] time free for some strenuous work on . . . Hetch Hetchy matters . . . [and] planned to devote the month of November until the first part of December very earnestly to getting the city's case ready."[90] However, Freeman did not return until mid-November. As he then told Manson, "after five weeks absence in Mexico . . . I find my desk piled so high with important matters requiring attention that several days will elapse before I can get back to my studies of San Francisco['s] problems. . . . I take it that no one will suffer by a little delay."[91]

In early December he reviewed Allen Hazen's recently completed analysis of a "Filtered Water Supply for San Francisco" and congratulated the sanitation engineer on "the admirable way in which you have worked up this report."[92] But any hope that Freeman could get back to San Francisco's "problems" in a timely manner was dashed by personal heartbreak. On Monday, December 11, he tersely informed Manson, "Our youngest son [Nat] is lying between life and death from infant paralysis. . . . For this reason there will be some delay in my attending to your work."[93] Later that day his son died. Over the coming weeks and into January, Freeman and his wife struggled to overcome their family's trauma, but the emotional impact lingered. Two months after Nat's death he confided to Secretary Fisher, "In December our youngest son, a boy in his tenth year, in perfect health, was attacked by infantile paralysis and survived only four days. . . . This terrible blow no[t] only impaired my working efficiency (insomnia, etc.) but left Mrs. Freeman on the verge of a nervous breakdown from which she is now slowly recovering."[94]

By February Freeman was regaining his professional equilibrium, but amid all the other demands on his time there was no way that he could rapidly complete his "Hetch Hetchy studies." And to Freeman's dismay, the complexity of these studies had assumed a new and demanding political dimension.

Spring Valley Water Company II

Through the end of 1911 and into the New Year, Freeman remained involved with plans for Spring Valley's Calaveras Dam.[95] However, his ability to work for both the city and the private utility on matters related to municipal water supply was becoming ever more untenable. It was one thing for Freeman to serve as a consultant on the design of Calaveras Dam, as this structure would presumably one day become part of a municipally owned system. But the task of placing a "valuation" on Spring Valley's assets for a possible future sale was a matter in which the city and the utility had very different interests. As objective as he might wish to be, Freeman could not serve both masters in advising what a fair purchase price might be.[96]

It was one thing to place a value on the existing infrastructure (e.g., pipelines, water mains, the Crystal Springs dam, etc.), but what about the future value of water resources controlled by Spring Valley? In particular, how much water could be reliably drawn from the Alameda Creek watershed, including untapped groundwater in Livermore Valley? Determining the magnitude of the Alameda system's ultimate capacity was an important issue for two reasons: first, if more water could be reliably drawn from the Alameda system, this would increase the value of the company's overall holdings and justify a higher price when sold to the city; and second, the more water that the Alameda watershed could provide to meet the city's needs, the longer it would take for the city to require a reservoir at Hetch Hetchy. In fact, if the potential of the Alameda system was large enough, it could mean that the city might never need to dam Hetch Hetchy to foster the Bay Area's long-term growth.

Realizing that Freeman's increased involvement with the city's "valuation" initiative made it impossible for him to advise the private utility on the supply capacity of the Alameda watershed, Eastman sought out another source of expert advice. Highlighting yet again the complexity of the public works/private interest dynamic, Spring Valley engaged the consulting services of William Mulholland and J. B. Lippincott for an assessment of the Alameda system. In early February 1912, these two Los Angeles–based engineers submitted a report that was later supplemented by many others commissioned by the company.[97] At that time, Mulholland's primary job was chief engineer for the municipally owned Los Angeles Aqueduct with Lippincott serving as his chief assistant. A few years earlier, Freeman had worked with both of them while serving on the aqueduct's board of consulting engineers (see chapter 2). The fact that, in early 1912, Mulholland and Lippincott began a relationship with Spring Valley placing them in opposition to the city's interest represents a little-known aspect of the Hetch Hetchy story, one that passes unmentioned in recent histories of the controversy.[98] But Freeman understood the significance of their work, quickly realizing that their reputations as experts in California water supply and their optimistic yield projections for the Alameda system could prove troublesome for San Francisco.

On February 20, Freeman acknowledged to city officials, "With regard to the Spring Valley purchase and the water from Alameda Creek watershed, I have . . . word from Mr. Eastman that a report by Messrs. Mulholland & Lippincott . . . is on its way to me. . . . [P]lease bear in mind that it is a pretty big job to place values on some of these rights and expectations."[99] After receiving the report a few days later, he bewailed, "I am almost appalled by the mass of material I have laid out on these [Hetch Hetchy–related] subjects at home on my study table."[100] A week later Freeman again took note of the "mass of data"

Ca. 1910 view of the Spring Valley Water Company's underground Sunol filtration gallery in the lower reaches of the Alameda Creek watershed. Freeman would eventually question the company's estimate as to how much water could be drawn from such subsurface sources. Author's collection.

that Spring Valley had provided to him relative to Alameda Creek sources.[101] A convincing rebuttal of Spring Valley's arguments regarding the Alameda watershed's potential yield became yet one more obligation that the city—relying upon Freeman's expertise—had to address if it hoped to make a persuasive case for needing Hetch Hetchy Dam.

Through early 1912, Freeman had much to worry about other than Hetch Hetchy/Spring Valley. He needed to complete his reports for Mexico City and the Mexican Northern Power Company. He continued to be engaged with the Keokuk Dam project and with New York City's Board of Water Supply. He had also been invited by MIT's President Maclaurin to "take charge of designs for New Technology," as the school began planning for a new campus in Cambridge.[102] And Freeman was intensively engaged in lobbying for favorable tax treatment of mutual fire insurance companies, meeting twice with President Taft—on February 16 he "called on President of U.S. [and] [h]ad very pleasant interview for 20 minutes on Insurance Tax and political matters" and on March 22 he was "very busy re-drafting statement on tax for submission to President," which, the next day, led to an "[i]nterview with President Taft at 12:30."[103]

In the context of Hetch Hetchy and the supposed antagonism of northern California's investor-owned electric power companies, what is striking about Freeman's consulting work in the first part of 1912 is that he also remained highly engaged with both the Great Western Power Company and Pacific Gas and Electric. He had been largely separated from the former company since his plans for Big Bend Dam had been abandoned in May 1910. But with the death of Great Western's president, Edwin Hawley, on February 1, questions about Big Meadows Dam (scheduled to start construction in the spring) arose among board members, and Freeman was offered an entrée back into the company's affairs.[104] With PG&E, he had started consulting on Spaulding Dam the previous year, developing plans for a concrete gravity dam that were submitted in March 1912; construction would commence by late summer. Given how the private power companies serving San Francisco were often castigated as great opponents of the municipally owned Hetch Hetchy system, it is notable how Freeman was engaged with both firms during the pivotal year of 1912.

The Baton Passes

While Manson dawdled in making San Francisco's case to Secretary Fisher and the Army Board, the political life of San Francisco kept evolving. After winning the mayoral election in September 1911 over incumbent P. H. McCarthy, on January 8 James "Sunny Jim" Rolph (a Republican reformer) took charge of the city—a position he would hold for the next nineteen years. While Rolph was firmly committed to damming Hetch Hetchy and had no desire to backtrack on the long-delayed project, the change of administration nonetheless placed stress on Manson to meet his responsibilities as city engineer.

After Rolph had been in office for only a few weeks, Manson expressed his frustrations to Freeman regarding the political nature of his position: "The change in administration has imposed a great deal of detail work in the making up of reports. . . . As this involves the renewal of political and personal animosities . . . it is an unwelcome and impossible duty."[105] Later harkening back to this theme—and implicitly offering an excuse for his dalliance on Hetch Hetchy—in late March Manson sought to explain to Freeman his inability to address his "show cause" responsibilities: "I have practically neglected the entire [Hetch Hetchy] matter for three months owing to the remarkable press of work in connection with a very large number of old projects which have been revived."[106]

— WHY —
Civic Center.
Hetch Hetchy.
Municipal Railways.
Twin Peaks and Stock-
    ton St. Tunnels.
$12,000,000 School Pro-
    gram.
$2,000,000 Relief Home.
$7,000,000 Sewer System.
$20,000,000 in Streets
    and Boulevards.
Aquatic Park & Marina.
Parks and Playgrounds.
San Francisco Hospital.
**We need ROLPH to
complete these projects
and to protect them.**

ROLPH

FOR MAYOR
BY PUBLIC DEMAND

James "Sunny Jim" Rolph became mayor of San Francisco in January 1912 and held the office for the next nineteen years. This election campaign card trumpeted his broad support for civic improvements, among them the city's Hetch Hetchy project. Author's collection.

Recognizing how pressure was building upon Manson, and in acknowledgment of how San Francisco had come to see Freeman as a key figure in its plans for Hetch Hetchy, on January 20, City Attorney Long informed Fisher that San Francisco "has engaged Mr. John R. Freeman of Providence, Rhode Island, as Consulting Engineer in this entire matter. . . . This City expects to be governed by Mr. Freeman's conclusions and judgment as to the presentation of this case."[107] Upon hearing this news, Manson wrote Freeman directly, noting that the city attorney "has suggested that instead of following the usual procedure, all reports, data, etc. intended for the Secretary of the Interior and the Board of Engineers, shall be forwarded through you rather than through this office." But Manson was unwilling to concede administrative preeminence to Freeman, pointing out that "this is a little awkward as the City Engineer is authorized by both general and specific provisions of the law to act for the City in the Department at Washington."[108] Awkward indeed, and at least for a few months, leadership of San Francisco's Hetch Hetchy initiative languished in a bureaucratic limbo.

In accord with Long's directive, Freeman met with Secretary Fisher in mid-February and again a month later; as a result, he obtained a further extension of the "show cause" deadline to July 1.[109] But as long as Manson remained city engineer, Freeman's ability to ensure that the city could meet any deadline remained at best problematic. Yes, Manson had directed former city engineer Grunsky and other assistant engineers such as Max Bartell to undertake studies of some alternative rivers/watersheds. However, the presentation of data assembled by these investigations remained incomplete and, for the time being, beyond Freeman's purview.[110]

As spring approached, Long apprised Freeman that "the situation in the City Engineer's office is such that a change may take place at almost any time."[111] Concerns were heightened when the Army Board informed Manson in April that he would need to investigate four watersheds that he had previously requested be excised from consideration, making clear to the city engineer that "the Feather, Yuba and Eel Rivers, and the Bay Cities Water Company [i.e., resources encompassing the American and Cosumnes Rivers] ... should be investigated and reported upon."[112]

Complaining to Freeman that the Army Board's recent demands were "absurd," at the end of April Manson assured him that "in the near future" he would send to the Providence engineer reports on the capacities of the "McCloud, Stanislaus, Mokelumne, American and Tuolumne Rivers."[113] Included in this letter was a "list showing dates and data of letters sent to [the] Board of Engineers" by Manson and a separate "list of dates and data sent to Mr. Freeman." Covering a time frame from June 1910 through April 1912, these lists denoted 36 items (7 sent to the Army Board and 29 to Freeman) and referenced a mishmash of letters, reports, maps, photographs, runoff data, proposed bills and legislation, and newspaper clippings. The only substantive and complete report related to alternative water supplies was Hazen's study, "Filtration of Sacramento and San Joaquin Rivers," submitted in December 1911. Even a cursory review of the two lists provided by Manson makes clear that, almost two years after Secretary Ballinger's "show cause" directive, the city engineer had assembled little of substance justifying the need to dam Hetch Hetchy. The moment of truth was at hand.

By that time, confidence in Manson by city leadership had vanished; in early May, City Attorney Long made a cross-country trek to Providence so that he could confer directly with Freeman over the weekend of May 10–12.[114] Viewed in retrospect, the purpose of this meeting was to devise a new plan for advancing the city's Hetch Hetchy project, one premised upon Manson's impending departure. Thus, in the midst of their consultations, Freeman provided Long with a thirty-two-page report titled "Progress on Studies under Supervision of J. R. Freeman, Consulting Engr." In this memo Freeman detailed his work relative to the Hetch

Hetchy project (largely focused on the pressurized gravity-flow aqueduct) and also reviewed the status of work that Manson had been overseeing, sharply noting that he was "much disturbed over the non-receipt of information" relating to alternative sources of supply.[115] In a subsequent telegram to Manson, Freeman reiterated this concern, expressing his "fear [that the] Army Engineers will be disappointed, vexed, and possibly prejudiced after City asking for two years delay for investigating, unless you present something definite . . . they will be dissatisfied with mere rehash of old data."[116] This was hardly a new criticism, but with Secretary Fisher demanding action, it now assumed greater urgency.

Two years had elapsed since Manson had set out to "show cause" and justify the city's need to dam Hetch Hetchy. Despite a series of time extensions, he had failed, and his quixotic journey came to an ignominious end. On May 18, Mayor Rolph wired Freeman with news that "Manson's health has broken and his physician has ordered him to suspend work." Rolph also requested that, as soon as possible, Freeman seek yet one more "show cause" extension from the secretary, this time for an additional sixty days. The hope was that somehow the Hetch Hetchy project might be kept alive under Freeman's direct leadership.[117] On May 22, Freeman made the trip to Washington, DC, and, joining with Percy Long, met with both Fisher and the Army Board to plead for yet one more delay. A terse telegram to Rolph outlined the result: "Two hour strenuous interview today with Secretary Fisher. . . . Had great difficulty obtaining further extension because Secretary felt city not diligent enough during past year. . . . [B]est he would grant was extension until August first for alternative projects and insists Modesto [and Turlock Irrigation Districts] and McCloud [River reports] be filed June thirtieth and Freeman's general Hetch Hetchy Report July fifteenth. . . . [I am] confident these dates can be met by strenuous work."[118]

The baton had passed. Manson was out, and a final schedule to "show cause" was in place. Win or lose, the fate of San Francisco's long-delayed Hetch Hetchy project lay in Freeman's hands.

CHAPTER 4

Strenuous Work

The Summer of 1912

After meeting with Secretary Fisher and the Army Board in late May 1912, Freeman returned to Providence to clean up his affairs before heading west for the summer. During this brief interregnum, Freeman wrote to the incapacitated Marsden Manson, expressing "[how] profoundly sorry [I am] to hear of your insomnia." He then outlined for Manson the work that Fisher expected him to accomplish by August, emphasizing how he intended to focus on "new lines very similar to those on which the Los Angeles Aqueduct is being built." Anticipating his arrival in San Francisco two weeks hence, Freeman hoped Manson might "take part in discussing many of the questions" that would no doubt arise. But while he was willing to "discuss" matters with Manson, he avoided saying that he might "work" with him. Now that the city engineer had relinquished responsibility for the Hetch Hetchy project, there was little chance that Freeman would countenance his return to anything remotely approaching a position of authority.[1]

In contrast to Manson, there was one prominent California engineer whom Freeman did seek to involve in his Hetch Hetchy plans, particularly in regard to the high-pressure aqueduct. Perceiving the Los Angeles Aqueduct as a precedent to emulate, Freeman hoped he might formally engage William Mulholland, the aqueduct's chief engineer, as a paid consultant to review his design for San Francisco's water supply. A week after meeting with Fisher, Freeman queried Mulholland, "[If] you could help me out to the extent of a week's time in San Francisco . . . by going over, mainly by automobile, the general route of the [Hetch Hetchy] aqueduct and tunnel line. . . . What would be your terms for a week or ten days [consulting work]?"[2] Although Mulholland deflected this request to become a paid consultant, over the coming weeks the two engineers interacted, not just on the design of the aqueduct Freeman envisaged but also on how development of Alameda Creek might impact the Spring Valley Water Company's financial valuation.[3] By August, however, Mulholland was professionally estranged from Freeman. And by the time the Providence engineer left San Francisco in September, Manson had been officially relieved of all his duties as city engineer. It would be Freeman's plan—and Freeman's alone— that the city would present to Secretary Fisher to "show cause."

For his summer residency in San Francisco, Freeman corralled a team of assistants from his home office. This core group consisted of Horace Ropes (who previously had worked with him on New York City water supply investigations), his secretary/stenographer David Eastwood, and draftsmen Raymond Cranston and Karl Kennison.[4] Freeman's entourage departed New England on June 4 and arrived in Sacramento on Sunday, June 9. There, Ropes detrained and headed south to begin surveys for the aqueduct. Freeman continued on to San Francisco, where, later that day, he booked a suite at the St. Francis Hotel. Over the next two and half months he used this aerie (Room 1229) as a de facto engineering office, a veritable command center where he created the *Hetch Hetchy Report* and detailed his design for the city's mountain water supply.[5]

Because of Manson's prolonged and ineffective dawdling during the prior two years, the schedule demanded by Secretary Fisher was extremely tight. Freeman needed to submit all reports on alternative water supplies by August 1, with reports on the McCloud River and the needs of the Turlock and Modesto Irrigation Districts due by June 30. In addition, his design for the Hetch Hetchy Dam and Aqueduct was to be presented to the Army Board by July 15. As part of this, he had to validate greater San Francisco's prospective need for a water supply system capable of delivering 400 million gallons a day. And he would be called upon to help negotiate the city's purchase of the Spring Valley water supply system. If this was not enough, he planned to create an elaborate printed report that would bring together all facets of the city's plan and, drawing upon a multitude of illustrations, make a public case for why the damming of Hetch Hetchy would enhance the environment and foster a bounty of social benefits. To help quell the political influence of the anti-dam opposition, he also recognized the importance of engaging with the press in espousing the city's need for a reservoir in Yosemite National Park.

During the summer of 1912, Freeman occupied a penthouse suite—room 1229—in the St. Francis Hotel, which served as both domicile and engineering office. Author's collection.

Immediately upon his arrival in the Bay Area, Freeman met with reporters at an impromptu press conference. In their headlines and coverage, local newspapers offered an approving take on Freeman and his ambitious plans. He was cast not as some meddlesome outsider but rather as an expert and valuable ally in countering the protestations of ill-advised "nature lovers." For example, the *Evening Post* reassured readers, "Noted Engineer Quiets Fears over Hetch Hetchy: Great Inland Lake Will Help, not Mar, Beautiful Mountain Scenery," while the *Examiner* exclaimed, "S. F. Must Have Hetch Hetchy, Says Expert: John R. Freeman, World Famous Engineer, Arrives to Complete Data."[6] And the *Chronicle* quoted him on his inspiration and goals for the project: "My report on the Hetch Hetchy will be based upon my experience with the New York and Los Angeles aqueducts. . . . [T]he plans include an imposing dam, beautiful auto roads around the lake, and other features which will make the site more accessible and even more scenically attractive than it now is."[7]

For its part, the *Call* made clear that Freeman was not interested in simply drafting an apolitical water supply report, describing how "[t]he objections of the 'nature lovers' in regard to destroying the scenic beauty of Hetch Hetchy Valley were branded by Freeman as 'ghost stories'[;] . . . he further designated their talk as 'moonshine' and was ready to prove that less than 300 persons visit the valley yearly."[8] Freeman's work would get further newspaper coverage in July, but for the moment, he had taken a direct shot at the "nature lovers" and started campaigning for his plan.

"Up at 5:15 AM"

Taking no time to rest after his cross-country trek, Freeman embarked upon a grueling schedule that continued for the next eleven weeks. Excerpts from his diary detail how he spent his first days on the job, evincing the "strenuous work" his assignment entailed:

> Up at 5:15 AM [on Monday, June 10] and unpacked and began on plans of days work & draft of Report. At 10 called on C. E. Grunsky and reviewed progress on Feather River [report] etc. Left card for Eastman [of the Spring Valley Water Company]. 2 PM met Mayor and Supervisors. 2:45 sat by Mayor at meeting and addressed Board of Supervisors for 20 minutes on water problems. Apparently very well received. . . . 4 PM long social call on Marsden Manson who I find is a nervous wreck. All hands worked until 11 PM on unit costs of tunnels etc.[9]

The next morning, he was "up at work at 5:30 AM on steel pipe unit costs" and "worked steadily" until 11 PM. The next day he was again working at 5:30 AM and in the evening attended a conference in the mayor's office that went so late that "boys were calling the morning papers as we came home." The weekend offered no respite, and on Saturday he rose at 5:30 a.m. in "preparation for an interview with Mayor, etc." On Sunday he was "up at work at 6am" and reviewed "Ropes' reconnaissance of aqueduct line across [the San Joaquin] Valley" before retiring early at 10 p.m. The following Monday he was back at it at 5:30 a.m. making "final review of pipe costs" for the "San Joaquin line."[10]

And so it went, as June passed into July and then to August, with scarcely a day off for rest or recreation. Charles Whiting Baker, editor of *Engineering*

News, witnessed firsthand Freeman's work regime and later offered this description: "[I] visited Mr. Freeman's suite of rooms on the top floor of the St. Francis Hotel while the work was in full swing. There, desks, drafting tables and apparatus, reference books, maps, photographs, manuscript and printer's proof in all stages crowded the customary hotel furniture into the background."[11]

Before delving into the particulars of Freeman's *Hetch Hetchy Report,* it is worth reflecting on what drove him to pursue this intense work regime— and how he managed to maintain his health and mental acuity during weeks of seventeen-hour workdays. In a letter to his son Roger sent near the end of his second week in San Francisco, he privately complained of "insomnia" and how he had been "awake at 2 and further sleep impossible." In this private message he also bemoaned the difficult situation that he had inherited from Manson and his concern that, if he faltered, he would be left as a scapegoat for the city's poor showing: "I am up against the hardest proposition ever . . . [and will be blamed for] any mistakes made without any consideration for the rush."[12]

Fear that he might be discredited within the public arena is something that, day after day, drove him to succeed. Although he would be well compensated for his service to the city, the dread of failure, and an inherent understanding that he could never publicly shame (or blame) Manson, animated his completion of the *Hetch Hetchy Report.* As the weeks went by and he realized that, despite severe time constraints, he could plausibly meet Fisher's demands, he became empowered by his ability to thrive under pressure. When his birthday came at the end of July—little more than a month after his complaints of insomnia—he confided in his diary that he was in fact reveling in his work: "57 years old today. In best of health. . . . Mentally stronger and clearer than ever before and able to concentrate for long hours in this cool California San Francisco air. Am feeling fine considering the hard work and long hours and small exercise of past 6 weeks."[13]

Over the course of the summer, Freeman's energies did not diminish, and he believed he actually gained mental strength as he pulled together his report. This is particularly significant because his printed report is more than four hundred pages long; includes scores of photos, illustrations, and maps; and addresses a multitude of issues related to the city's water supply plans. Anyone reviewing the published document would likely be astonished that it was essentially created by a single engineer with a small staff of assistants laboring for less than three months. To say that the *Hetch Hetchy Report* was the product of "strenuous work" is unquestionably true. But to explain exactly how Freeman maintained his stamina and health during the summer of 1912 remains an enduring mystery. While Manson collapsed under the pressure of "show cause," Freeman flourished.

Freeman's *Hetch Hetchy Report* is no doubt a remarkable document that transformed San Francisco's plan for a Tuolumne River water supply, but it is hardly flawless; its organization is repetitive at times, and it fails to present a compelling and comprehensive analysis of the water supply alternatives available to the city. Reasons underlying the report's shortcomings in large part relate to the extremely short deadline under which Freeman operated. Some portions of the text—most notably the seventy-page introduction/overview of Freeman's proposed Hetch Hetchy Dam and Aqueduct—are clear and easy for a reader to follow. But the descriptions and analysis of the water supply alternatives are at times less than edifying. As it turned out, Freeman had relatively little to do with analyzing and reporting upon the various alternatives, leaving such work to Manson and then—after Manson's incapacitation in May 1912—largely relying upon former city engineer Grunsky to complete the investigations. But Grunsky proved to be less than effective, and over the course of the summer, Freeman became acutely aware of his shortcomings. Given the time constraints demanded by Secretary Fisher, however, Freeman could do little but work with what Grunsky concocted and, in editing and organizing the *Hetch Hetchy Report,* try to embellish it and make it look more authoritative. In this, he had but limited success.

400 Million Gallons a Day

When Freeman first began lobbying Secretary Fisher and municipal officials in the spring of 1911 of his desire to provide "something better for the city," his vision extended beyond San Francisco proper and included other cities and communities clustered around San Francisco Bay. Drawing upon his knowledge of Boston's Metropolitan Water District, he envisaged a regional system that would depend upon the upper Tuolumne River as its major source of water supply. So the question arose: how many people would such a regional system need to serve at, say, the end of the twentieth century? This was an important figure to postulate, because it would provide justification for why greater San Francisco needed to act decisively in committing to an expansive water supply infrastructure.

In forecasting the region's future population, Freeman looked to work undertaken by Stanford professor Charles D. Marx. Marx believed that the Bay Area population of 773,000 (taken from the 1910 US Census) would rise to more than 3.6 million by the year 2000, entailing a daily water consumption of 441 million gallons. In his own estimation, Freeman chose to be a bit more conservative and projected a population in 2000 of about 3.1 million but still needing a supply of at least 400 million gallons per day (mgd).[14] In the end, both these estimates proved shortsighted, as the 2000 population of San Francisco, Greater Oakland, and the counties constituting modern-day Silicon Valley totaled about 4.5 million. While Freeman's and Marx's prognostications may have undershot the mark a bit, they did not overestimate the capacity of growth possible for the Bay Area.[15]

The original Hetch Hetchy system proposed by City Engineer Grunsky in 1902 and revised by City Engineer Manson in 1908/11 was designed to provide the city with a supply of only 60 million gallons per day. With his Hetch Hetchy system Freeman promised to supply more than six times this amount, and it was crucial that he justify the need for such a huge quantity. As a result, the report describes in considerable detail how the water supply infrastructure for Oakland, Berkeley, Hayward, Albany, Richmond, and other East Bay communities could be integrated into a San Francisco–based metropolitan district served by a capacious Hetch Hetchy system, one capable of supplying much more than Grunsky or Manson ever imagined.[16] Freeman's plan was to build a separate offshoot aqueduct line to serve the East Bay region that would connect directly to the main Hetch Hetchy Aqueduct at the Irvington Gate House (near Alameda Creek); this subsidiary aqueduct would connect by gravity flow into an enlarged Chabot Reservoir located in the hills above Oakland.[17]

In July 1912, Oakland Mayor Frank Mott initiated plans to convene East Bay politicians and community leaders in a "joint conference" to explore how an "inter-municipal water district" might purchase the assets of the local Peoples Water Company that served the Oakland environs.[18] But in the end, and despite Freeman's entreaties, the East Bay communities declined to cast their lot with San Francisco, and no single, comprehensive metropolitan district for the entire Bay Area ever materialized.[19] During the summer of 1912, however, such a regional system seemed plausible enough that Freeman could use it to justify why San Francisco and its neighboring communities required an ultimate supply of 400 mgd.[20]

As a counterpoint to the Bay Area's future water supply needs, Freeman also needed to address how farmers in the Modesto and Turlock Irrigation Districts could be assured that agricultural diversions from the Tuolumne River could be sustained even after the city drew 400 mgd into its aqueduct. Rather than simply acknowledge how the annual measured flow of the Tuolumne—calculated to be an average of 2,200,000-acre feet measured at the La Grange Dam for the years 1896–1911 (or a constant flow of about 3,040 cubic feet per second [cfs])—was sufficient to meet the districts' legal claim to 2,350 cfs, Freeman and his engineering colleague J. H. Dockweiler looked to the practical needs of the districts in watering all of their irrigable lands. In total, the districts had a combined area of 257,353 acres. But of this, he believed only 206,000 acres could be profitably irrigated (Freeman also noted how overwatering had already injured much low-lying land). Taking 2.50 feet of water per year as a sufficient depth to ensure crop productivity without further damaging the soil, he determined that a total of 533,050 acre-feet per year (or a steady flow of about 737 cfs) could meet the real requirements of the districts. But even allowing for a diversion of 2,350 cfs at the La Grange Dam, a constant flow of over 650 cfs would still be available to San Francisco, equating to more than 400 mgd. And this would successfully meet the requirements of Freeman's system centered around a large storage reservoir at Hetch Hetchy.[21]

Freeman's Aqueduct and Dam

In rejecting the viability of the Grunsky/Manson 60 mgd scheme, Freeman drew upon the idea that he had conceived for the Wachusett Aqueduct in 1895 (see chapter 2), proposing a pressurized aqueduct that could carry water by gravity flow from Hetch Hetchy to cities throughout the Bay Area. From a technological perspective, this was the most consequential component of his report: a high-pressure pipeline crossing the width of the San Joaquin Valley—and connecting to a tunnel extending under the Mount Diablo Coastal Range—that, without pumping, could deliver water to the Crystal Springs Reservoir twenty miles south of San Francisco at an elevation almost three hundred feet above sea level.[22] In the previous designs proposed by Grunsky and Manson, power generated along the aqueduct was to be used to pump water over the Mount Diablo Coastal Range escarpment near Altamont. But Freeman's system eliminated this requirement.

Another key change that Freeman proposed to the Grunsky/Manson plan involved substituting an almost twenty-mile-long tunnel in place of a mostly open canal connecting the aqueduct intake below Hetch Hetchy with the forebay of the Moccasin Creek hydroelectric power plant. Excavated through hard rock with a cross-section sufficient to deliver 400 mgd, this tunnel would require an additional initial expense over an open canal, but it would be more reliable for year-round use and eliminate much recurring maintenance.

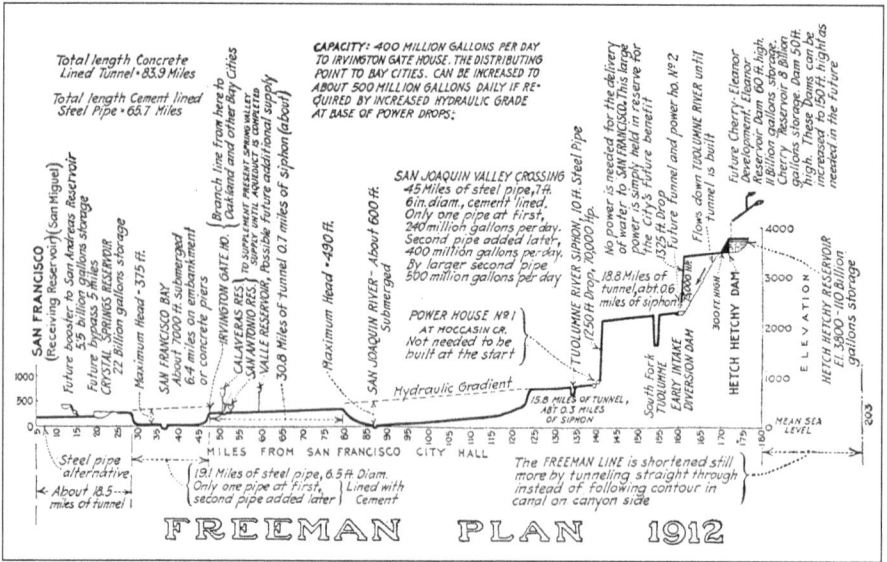

From a technical point of view, Freeman's most important contribution to the Hetch Hetchy project was his proposal for a high-pressure aqueduct that would bring water to San Francisco without the need for pumping. ("A Notable Water Supply Report," *Engineering News*, 68 [December 26, 1912]: 1213)

Freeman's chief assistant Horace Ropes, working with Stanford geologist John Branner, surveyed the aqueduct's right-of-way, providing detailed maps and construction cost estimates covering the 167-mile route from Hetch Hetchy to San Francisco.[23] Freeman also looked to Mulholland's design of the Los Angeles Aqueduct (then under construction) as a template for his Hetch Hetchy proposal. The 233-mile-long aqueduct connecting Los Angeles to the Owens River on the Sierra Nevada's eastern slope included lengthy steel penstocks and hard-rock tunnels; nearing completion in 1912, Mulholland's aqueduct gave Freeman the confidence to advance a similar construct to serve San Francisco.

Although Mulholland never formally consulted on Freeman's aqueduct, the two engineers did meet in San Francisco near the end of June 1912, and previously, during the summer of 1911, Mulholland had provided Freeman with construction data on the Los Angeles Aqueduct.[24] In developing cost estimates for his Hetch Hetchy Aqueduct, Freeman relied upon a figure of $2.25 per day for common laborers that, in his report, he contrasted with the day rate used at Los Angeles: "On the Los Angeles Aqueduct the day laborers . . . receive

TYPICAL HEAVY SECTION
FROM INTAKE TO SAN JOAQUIN VALLEY
To be used thru heavy shattered rock
or soft ground or where subject to
high outside water pressure

TYPICAL SECTION
FROM SAN JOAQUIN TO GATE HOUSE NEAR IRVINGTON

In fairly sound rock
and where outside water
pressure will not be excessive,
amount of excavation may be
lessened, and concrete lining
thinned, at discretion of
Engineer in charge.

Note for all Tunnel Sections
All Lagging and as many
Main Timbers as possible
to be removed before
concreting.

TYPICAL SECTION FOR MINIMUM THICKNESS OF LINING
FROM INTAKE TO SAN JOAQUIN VALLEY
To be used thru sound rock where
the principal use of the lining is for
smoothness and to increase delivery

FEET 0   5   10   15

CONSTRUCTION QUANTITIES PER LINEAL FOOT

	Section A		Section B		Section C	
	Untimbered	Timbered	Untimbered	Timbered	Untimbered	Timbered
Excavation	4.13 cu.yd.	5.04 cu.yd.	3.69 cu.yd.	4.49 cu.yd.	6.40 cu.yd.	7.40 cu.yd.
*Concrete	1.04 "	1.85 "	0.59 "	1.29 "	1.60 "	2.48 "
Timbers		34 B.M.		32 B.M.		43 B.M.
Lagging		60 "		57 "		50 "
Bracing etc.		12 "		12 "		10 "

* Main Timbers assumed left standing

HYDRAULIC ELEMENTS

	Section A&B	Section C
Area	83.6 sq.ft.	130.0 sq.ft.
Wet Perim.	32.8 ft.	40.4 ft.
Hyd. Radius	2.55 "	3.22 "

PROPOSED PRESSURE TUNNELS
HETCH HETCHY AQUEDUCT
John R. Freeman , Consulting Engineer
Providence , R.I.   July 13,'12

Cross-sectional drawings of the "pressure tunnels" that Freeman proposed for his Hetch Hetchy Aqueduct. (Freeman, *Hetch Hetchy Report,* 120)

$2.25 to $2.50 per eight hour day. If the Hetch Hetchy work should be built under [comparable] contracts . . . it is probable . . . that an average of $2.25 per eight-hour day is a fair and liberal assumption."[25] Perhaps it was reasonable for Freeman to pick the low end of the Los Angeles wage scale for common labor when making estimates for the Hetch Hetchy Aqueduct. But this was much lower than the prevailing wage rate in San Francisco, and it was lower than the wage rate used by Allen Hazen estimating the Sacramento River pump/filtration system. For Freeman, this discrepancy in estimated wage rates would later prove a nagging headache.

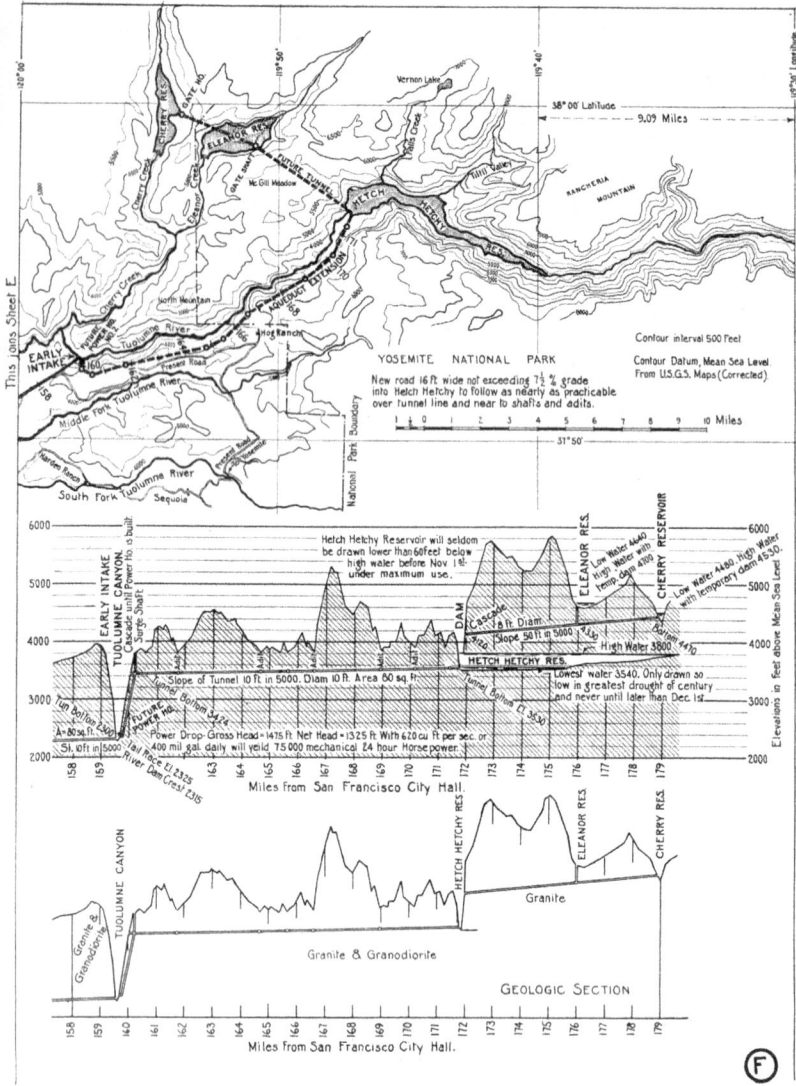

One of several detailed maps showing the cross-sectional profile of Freeman's Hetch Hetchy Aqueduct. This page (117) in the *Hetch Hetchy Report* illustrates how the reservoirs on Cherry and Eleanor Creeks were to connect into the Hetch Hetchy reservoir via a "future tunnel."

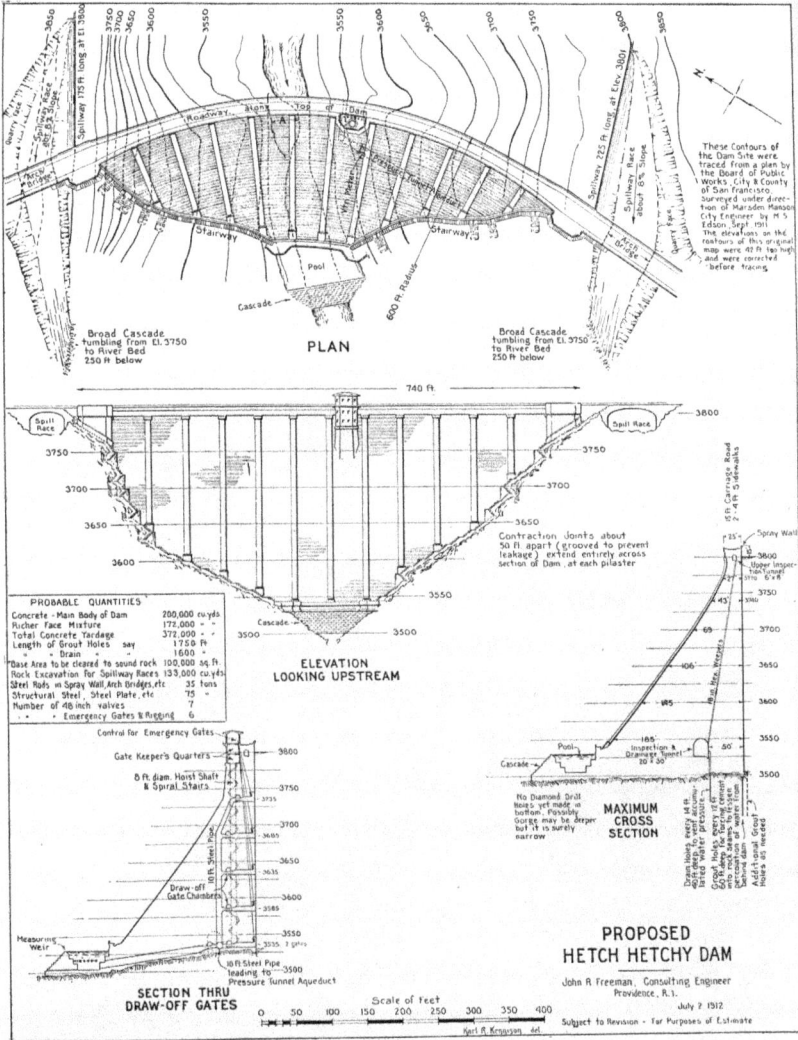

Figure 4.5. Drawing upon his experiences with Boston's and New York's water supply systems, as well as work on hydroelectric power projects, Freeman developed a massive concrete gravity dam design for the Hetch Hetchy site. (Freeman, *Hetch Hetchy Report,* 118)

In contrast to the aqueduct proper, Freeman did not extensively detail the massive concrete curved gravity dam that he proposed for Hetch Hetchy. However, a full page of his report illustrates the dam in plan, elevation, and cross-sectional profile, with some instructions included within the design drawings (such as cement grout to be injected into the foundations and drainage holes for relieving subsurface water pressure). While acknowledging that "no borings have yet been made at the site," he nonetheless assured readers that "from a careful personal examination of the character of the rock formation and the shape of the gorge, I am confident the site involves no great difficulty."[26]

Along with including a photograph of the recently completed Craig Goch Dam (built to supply Birmingham, England), Freeman explained in his report how large dams he had been associated with constituted the inspiration for his Hetch Hetchy Dam: "The design proposed . . . is much the same as the dam that I designed last year for the [Pacific] Gas and Electric Company at the outlet of Lake Spaulding. . . . Twice as tall as the dam proposed at this site ten years ago [by Grunsky], this Hetch Hetchy Dam . . . is of substantially the same type as the two dams [Ashokan and Kensico] that are now under construction by the Board of Water Supply of New York City, for which I have acted as Consulting Engineer."[27] His proposed dam for Hetch Hetchy was also similar to the 240-foot-high Calaveras Dam design he had prepared for the Spring Valley Water Company in the spring of 1911, although that fact was not referenced in the report.[28]

Hydroelectric Power

Beyond promising 400 mgd of water to the Bay Area, a key benefit of Freeman's scheme was that all of the hydroelectricity generated at power plants along the aqueduct could be used for purposes other than pumping or conveying water to city taps. Freeman calculated the ultimate capacity of his system to be a minimum of 157,500 horsepower (hp) and possibly reaching as high as 200,000 hp during periods of high demand. In the context of early twentieth-century California, this represented a tremendous amount of power.[29] Although Freeman relied upon the fieldwork of Horace Ropes (and the consulting expertise of geologist John Branner) to locate the aqueduct's right-of-way, it was his knowledge of the Los Angeles Aqueduct and its planned high-head power plants, combined with his experience with other hydroelectric power initiatives, that gave credibility to his Hetch Hetchy hydropower scheme.

Freeman's Hetch Hetchy system featured three primary power plants, the most prominent taking advantage of a 1,250-foot drop at Moccasin Creek with a proposed capacity of 70,000 hp. This power drop correlates with the steep incline of the "Priest Grade" along the Big Flat Oak Road (or modern-day State Route 120), where, in the space of a few miles, the highway descends over

The Hetch Hetchy Aqueduct's Moccasin Creek power plant upon completion in 1925. The penstocks feeding into the powerhouse are visible on the right. Author's collection.

1,300 feet. The Moccasin Creek power plant came online in 1925 upon completion of the first phase of the Hetch Hetchy's construction.

Freeman's second major power plant was to take advantage of the 1,475-foot drop between the Hetch Hetchy reservoir and the aqueduct's "Early Intake." This development would require excavating a twelve-mile-long tunnel through the canyon wall below Hetch Hetchy Valley; at maximum capacity he projected that it could steadily generate 75,000 hp of hydroelectric power. Not built until the 1960s, this feature constitutes the city's modern-day Kirkwood Powerhouse.

The third component of Freeman's hydropower system was a small powerhouse drawing water from the tunnel that Freeman proposed for delivering water from LakeEleanor and Cherry Creek into the Hetch Hetchy reservoir. Although never built, this facet of Freeman's plan was to take advantage of the fact that the Lake Eleanor and Cherry Creek reservoirs lay at a higher elevation than Hetch Hetchy. A seven-mile-long tunnel conveying the waters of Eleanor and Cherry Creeks to Hetch Hetchy offered the possibility of developing 12,500 hp at a powerhouse operating under a head of about 700 feet. Ultimately the city opted to replace Freeman's plan with a tunnel connecting the Lake Eleanor reservoir to the Cherry Creek reservoir, where a penstock now feeds water to the Holm Powerhouse (completed in 1960) under a head of more than 2,400 feet.

Although the hydropower capacity of Freeman's system was a key feature of the proposed design, in his *Hetch Hetchy Report* he did not recommend immediate development of the aqueduct's hydroelectric power potential. Readers will look in vain for any detailed treatment of how the system would meld into or add to San Francisco's power grid. Instead, they will find a nuanced statement that both trumpets and downplays how his Hetch Hetchy hydropower system could serve the city's future needs:

> The city does not propose in the immediate future to build any plant for the development of hydro-electric power, but it plans to carefully conserve all reasonable opportunities for power development against the time when it may become expedient for the city to undertake all such matters.... [W]ith three large hydroelectric power enterprises [now] bringing electric current to the city and competing actively for business under the oversight of a public service commission, no occasion appears [imminent] for the municipality to go into the power business[;] ... [however,] the time will surely come when this power privilege will be a most valuable asset of the Tuolumne water supply system ... [and it] should be reckoned into the balance sheet when comparing the Tuolumne source with others that present no such potentiality.[30]

Here we see Freeman carefully balancing his competing allegiances to both the private power industry and a large-scale publicly owned water supply (and power) system. The capacity to generate huge quantities of hydropower is what bestowed upon his Hetch Hetchy aqueduct a key economic advantage over all alternative projects (later, the Army Board would value the system's power component at $45 million).[31] Freeman, however, did not recommend that the city use his aqueduct as an impetus for developing a municipally owned power system that might supplant the city's existing investor-owned utilities. Seeking to remain on good terms with both the Great Western Power Company and Pacific Gas and Electric, he walked a politically fraught tightrope on the hydropower issue.

Freeman was well aware of the contentious battleground separating private and public power advocates. Early in the summer he confided to William Mulholland that he recalled the opposition that private power companies had brought against the Los Angeles Aqueduct: "[R]emembering the inverted enthusiasm of the Los Angeles electric companies in a similar case, this part of my program [for Hetch Hetchy] must be discreetly played under the soft pedal."[32] In a related vein, at the end of the summer he wrote to the editor of *Engineering News,* observing, "[In my report,] I have not dared to emphasize the wonderful power opportunities of this [Hetch Hetchy] project for fear of arousing the determined opposition of the three powerful electric power companies supplying San Francisco, with their many banking connections, because of my memories of Los Angeles where the electric interests were believed to be at the bottom of the most bitter opposition."[33]

Of course, what is left out of Freeman's professed reason for presenting the hydropower capacity of the Hetch Hetchy system "under the soft pedal" is that he possessed his own professional motives for not alienating corporate leaders who controlled the private power companies serving San Francisco. This was especially the case with the Great Western Power Company, as he was seeking increased engagement in the company's plans for the Big Meadows Dam. In fact, in early August he took a train up the Feather River from San Francisco and spent a day examining the damsite before returning to complete his *Hetch Hetchy Report.* Freeman was deeply connected to America's privately owned electric power industry, and he wanted to maintain good relations with utility executives, most of whom had little interest in publicly owned power systems.

Tourism and Automobile Roads

In a strictly engineering context, there is no reason why Freeman's Hetch Hetchy plan needed to involve anything other than a technical consideration of what it would take to bring Sierra water to San Francisco. But as early as the summer of 1910, he had begun to envisage how the city's proposed water supply system could purportedly enhance the beauty of the Sierra highlands and also make the rural landscape more accessible to visitors. Travelling through Norway and the British countryside, he sought out scenes and locations that might serve as models for San Francisco to draw upon in promoting a mountain water supply. A few years earlier, Freeman had used European scenes such as Hamburg's Alster Basin to illustrate how a dam across the Charles River could improve Boston's waterscape. For his Hetch Hetchy scheme, he built upon this Charles River precedent in his effort to promote the desirability of building a dam within Yosemite National Park.

Freeman's emphasis on tourism and his belief that the Hetch Hetchy project could both increase the beauty of the valley and make this beauty more accessible is no doubt a remarkable feature of his report. This approach certainly

During Freeman's trip to Europe in 1910, he toured rural Norway. Two years later, his *Hetch Hetchy Report* (p. 14) included this view of a Norwegian mountain road, illustrating how once-remote landscapes could be made accessible through artful engineering.

angered preservationists, who saw it as a deceitful attempt to manipulate public opinion. And to a degree such complaints were justified. Beauty is no doubt in the eye of the beholder, and Freeman was free to argue his belief that the dam and reservoir would add to the beauty of Yosemite National Park. But whether his arguments were always made in good faith is subject to debate.

Freeman's use of photographs showing roads rimming steep-cliffed Norwegian fjords was perhaps fanciful, but it did offer a sense of how a road around the perimeter of the Hetch Hetchy reservoir might appear when the reservoir was filled to the spillway crest. The photograph he used to show how the waterfalls at Hetch Hetchy might appear in a reflection on the reservoir surface was more problematic, because conditions that would foster such a reflection were at best infrequent and always subject to mountain breezes that could easily disrupt any mirror effect.[34]

Another provocative use of photographic images in the *Hetch Hetchy Report* involved views of visitors strolling about the parkscapes surrounding Boston's reservoirs.[35] Freeman wanted to emphasize how use of Hetch Hetchy for a reservoir would not preclude visitors from going to the valley and enjoying the wonders of northern Yosemite National Park. This was driven by preservation arguments averring that *any* use of the upper Tuolumne watershed for municipal

Much of Freeman's *Hetch Hetchy Report* focused on technical issues, but he also argued that the Hetch Hetchy Dam would serve as a scenic improvement to the Tuolumne landscape. For example, Freeman included this image showing how a reflection in the reservoir might make Hetch Hetchy Valley look even more grand. (Freeman, *Hetch Hetchy Report,* 10)

water supply would necessarily lead to draconian restrictions that would forever block visitors from camping in or traversing any part of the park upstream from the reservoir. In response, Freeman sought out images that made such fears appear to be paranoid and overblown.

However, while water supply officials in Boston apparently never publicly objected to Freeman's use of such images, they did privately express concern that it was misleading to make it appear as though the presence of human activity close by Boston's reservoirs posed no sanitary or hygienic issues. In late June, Henry H. Sprague, chairman of the Metropolitan Water and Sewerage Board, specifically apprised Freeman, "The nearness of the trolley line and the highways is bringing people about the [Spot Pond Reservoir] in such a way as to produce a decided menace to the water supply. . . . [W]e should not like you to give the impression that we think that it is desirable to make a watershed attractive to tourists or campers."[36] At the end of the summer, Dexter Brackett,

Freeman used this photograph of visitors to Boston's Middlesex Fells Reservoir to depict how sanitation regulations do not necessarily preclude people from enjoying the environs of urban water supply reservoirs. (Freeman, *Hetch Hetchy Report,* 35)

the Metropolitan Water and Sewerage Board's chief engineer, reinforced this point: "It is the policy of the Board to discourage the increase of population on the watersheds, particularly in proximity to streams, ponds and reservoirs."[37] Freeman did not argue with either Sprague or Brackett after receiving their cautionary missives, but he saw no reason not to use photos of Boston's reservoir system in his *Hetch Hetchy Report.*

Of course, the reason that Freeman wanted to make it appear that construction of Hetch Hetchy would foster an influx of visitors was because of how few people had ever actually visited the valley. He saw this as an opportunity to portray anti-dam advocates as "solitude-loving cranks," elitists interested in their own selfish pastimes and disdainful of the common person. As early as the summer of 1911, Freeman had perceived the public relations value of highlighting Hetch Hetchy's isolation. In anticipation of Secretary Fisher's visit to the valley in September 1911, he had urged City Engineer Manson, "[O]btain for me complete copy [of] Hetchy visitors' register since first established, including dates, names and residences."[38] The data revealed that only a few hundred people per year had been making the trek to Hetch Hetchy, and rarely did anyone residing outside of California visit. The point was accentuated in

Freeman's *Hetch Hetchy Report* under the heading "The Present Inaccessibility of Hetch Hetchy and the Few Who Enjoy It," where Freeman described how in 1909 only 269 people had visited the valley and that this included "cattlemen from adjoining counties and the repeated visits of the city's surveyors."[39] In 1910 only 16 people who were not California residents made the tramp to the mountain valley, while the next year, thanks to a Sierra Club–sponsored excursion, the number jumped to 37. The number of visitors to Hetch Hetchy was quite small compared to the thousands who visited Yosemite Valley every year. And as Freeman's report pointedly observed, "A pair of campers can hardly make the trip if bent on pleasure in less than a week's time nor can they readily do it for a smaller expense than $200 for that portion of the journey beyond the nearest stage line."[40]

Alternative Sources

When Secretary Ballinger first called for the city to "show cause," the intent was to force Manson to consider how alternative sources might meet San Francisco's future needs. Secretary Garfield may have been willing to blindly accept the city's professed need to dam Hetch Hetchy, but Ballinger was not. This meant that the city needed to justify why other, "alternative" sources could not reasonably sustain the city's future growth. Significantly, this issue lay at the heart of objections raised by John Muir and other anti-dam advocates. To their mind, impounding a municipal reservoir at Hetch Hetchy was reasonable only if all other possible sources were, after careful study, deemed unavailable or insufficient to meet the city's future needs.

To be clear, it was not absolutely necessary that the city of San Francisco impound a reservoir at Hetch Hetchy to ensure a long-term municipal water supply; other sources were undeniably available. The question was whether they could be developed in a cost-effective manner and reliably meet a projected future demand of 400 mgd. Also, could other sources provide water comparable in quality to a "mountain water supply" delivered from the upper Tuolumne watershed? Freeman was well aware that, in engineering terms, the most practicable alternative to using Hetch Hetchy involved pumping and filtering water from the Sacramento River. But Freeman believed that filtered water from the Sacramento, even after it was treated to be hygienically safe, was a less desirable product than that provided by the upper Tuolumne. Or at least that was the argument he sought to make.[41]

Leaving aside issues of water quality (or desirability), it is useful to compare the estimated cost of Freeman's Hetch Hetchy plan with the filtered/pumped Sacramento River proposal prepared by Allen Hazen (and submitted to the city in December 1911).[42] This is because Freeman always believed, going back to 1910, that drawing water from the lower Sacramento represented the most viable alternative to Hetch Hetchy in terms of cost. As Hazen's proposal detailed, a system with a supply capacity of 60 mgd and an annual pumping expense of $370,000 was estimated to cost $24 million; doubling the capacity to 120 mgd would incur a total investment of $42 million; and tripling it to 180 mgd would cost $60 million.[43] Although in his report Freeman did not develop a cost estimate for a 400 mgd supply from Hetch Hetchy, he did draw upon his detailed design for the initial aqueduct system to estimate that "the total cost of supply [from Hetch Hetchy], varying from a maximum of 240 million gallons daily to [a] minimum 160 million gallons daily in extreme drought" would be $36.981 million.[44] In other words, an absolute minimum of 160 mgd (and, except for extremely dry periods, close to 240 mgd) from Hetch Hetchy would cost $6 million less than 120 mgd pumped from the Sacramento and over $20 million less than 180 mgd pumped from the Sacramento.

The Sacramento River is the largest waterway in California, with a drainage basin covering more than 26,000 square miles. As illustrated in this ca. 1905 postcard, in its wide, lower reaches the river carries an average flow of about 30,000 cubic feet per second. If pumped and properly filtered, this would be more than enough to meet greater San Francisco's municipal water supply needs. Author's collection.

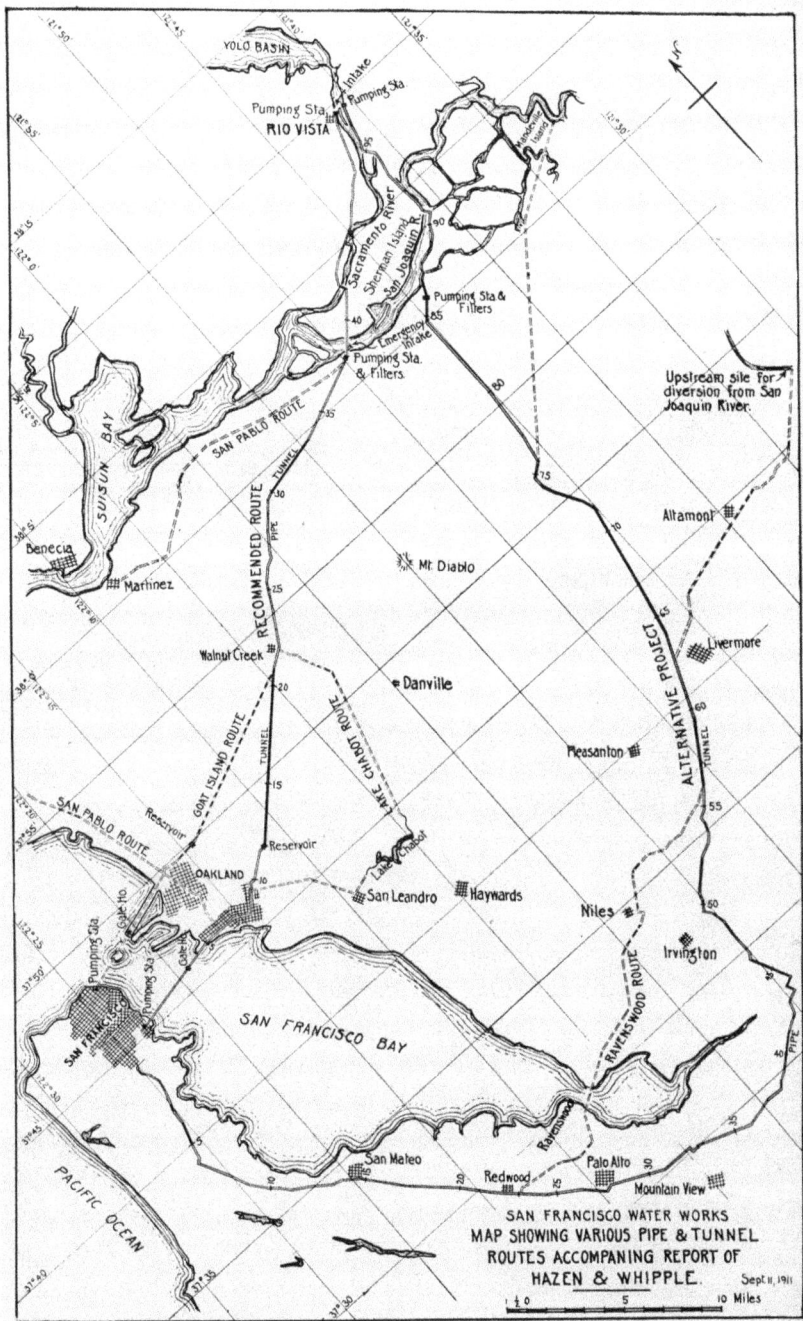

Map showing possible routes of an aqueduct serving San Francisco that, at a site near the river town of Rio Vista, would pump and filter water from the Sacramento River. (Freeman, *Hetch Hetchy Report*, 318)

Was this an overwhelming difference, given the rare and scenic splendor of Hetch Hetchy? Perhaps yes, perhaps no. But to Freeman such a cost differential was more than the city should have been expected to pay in order to preserve Hetch Hetchy. As he opined in his report, "It is a matter of plain common sense that an extra burden of ten or twenty or thirty million dollars should not be placed on the tax payers and other citizens of the cities around San Francisco Bay merely to satisfy the peculiar views of a few solitude lovers."[45]

Moving beyond Hazen's filtered/pumped Sacramento River proposal, Freeman divided the possible alternative sources for the city into two groups. As illustrated in a map featured in his *Hetch Hetchy Report,* these consisted of, first, rivers lying to the north of San Francisco (including the Eel, Yuba, Feather, and McCloud), which would require aqueducts crossing either the northern reach of San Francisco Bay at Carquinez Strait or the Golden Gate directly to the north of the city; and second, rivers lying east of the Bay Area in the central Sierra Nevada (including the Calaveras, Stanislaus, Mokelumne, and combined American-Cosumnes Rivers) whose water could be carried by aqueducts extending south of San Francisco Bay and up the peninsula to San Francisco.[46] Although each potential source in the two groups featured its own particular conditions, Freeman believed there was one important attribute shared by all the alternatives: "This map shows evidence of nothing more direct or with a cheaper aqueduct than the Tuolumne when the extra length for going back to equal altitudes and when cost of pumping from low altitudes are given due weight in the comparison."[47]

To the north, the Eel River was the only stream in the group flowing directly into the Pacific Ocean, but this did not offer any notable advantage. In fact, it was largely a disadvantage: "From the Eel River source, the direct route would be thru a tunnel under the Golden Gate, but the uncertainties and costs attending this particular point of crossing are such as to exclude [it] from serious consideration. . . . [B]y reason of distance, difficulty, cost, pumping, power interference and of presenting only half the quantity ultimately required it quickly may be ruled out."[48]

With the Eel River dismissed, Freeman turned to the other northern sources with a skeptical eye, noting, "The McCloud, the Feather, and the Yuba aqueducts must each cross Carquinez Strait, a mile in width presumably in a deep tunnel the quality of rock which has not been explored. . . . [N]o one of these sources can deliver its water to Lake Chabot [near Oakland], Crystal Springs or the city's reservoirs without pumping."[49] To his mind, there was little reason to devote much attention to these northern sources, because any use of these rivers

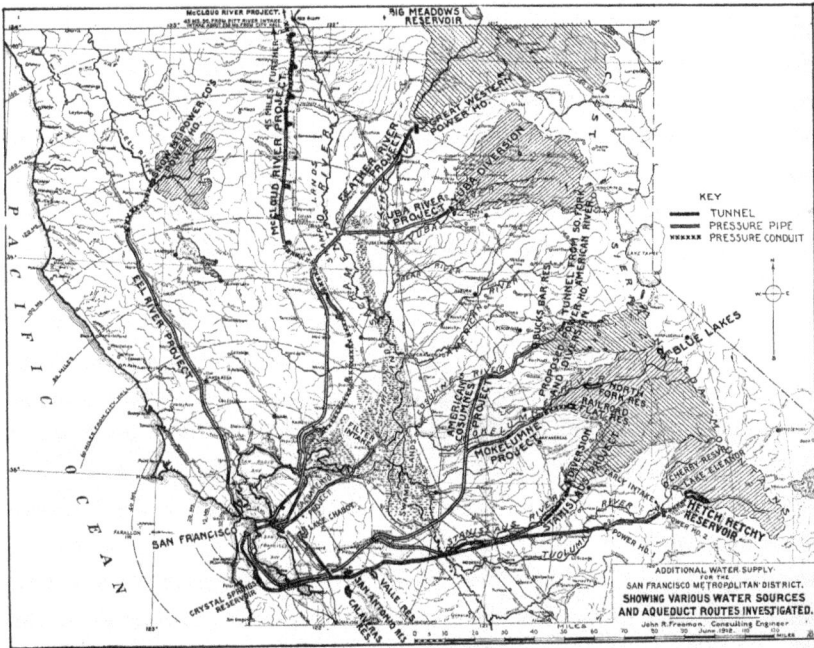

San Francisco's plans to dam Hetch Hetchy for an urban water supply was predicated upon the claim that no other comparable water sources were available to the city. This map was included in Freeman's *Hetch Hetchy Report* (p. 156) to show the locations of other possible water supply sources in northern and central California.

could just as reasonably be taken by pumping and filtering a comparable flow in the lower Sacramento: "If the McCloud, Feather and Yuba Rivers water must be filtered because of taking them from a running stream and pumped because of low altitude and delivery at sea level, one might as well pick up these waters at Rio Vista, on the Sacramento River."[50]

In other words, why build an expensive, lengthy aqueduct when the Sacramento River itself could serve an analogous function by carrying the water into close proximity of the Bay Area? Nonetheless, while Freeman considered it a fool's errand, he still gave the northern streams a modicum of consideration, if only to emphasize their limitations.

The McCloud River is a tributary of the Pit River (and thence the Sacramento) lying along the southeastern slope of Mount Shasta. The stream had engaged the interest of the Army Board of Engineers, and in his *Hetch Hetchy Report* Freeman took time to explain the problems it posed as a water supply source for San Francisco.[51] Along with highlighting the fact that "a simple inspection of the topographic map shows that the McCloud source requires the longest aqueduct of any that are being considered," Freeman observed, "Any practicable point of diversion of the waters of the McCloud River lies downstream from numerous lumber camps and downstream from the town of McCloud which contains a population of over 1500 to 2000 people during the summer months."[52]

In the absence of a large storage reservoir or filtration apparatus, Freeman explained how such human settlement posed a serious risk if the river served as a municipal water supply: "Much professional experience in tracing the relation of typhoid fever to water supplies makes it plain beyond the shadow of a doubt that it would be dangerous to the health of the citizens of San Francisco to divert the water of the McCloud River and take it directly through aqueducts to the bay cities without detention in large reservoirs for a sufficient period to kill off any pathogenic germs."[53] He then emphasized how the McCloud, which did have a large natural flow, differed significantly from his proposed Hetch Hetchy/Tuolumne system, because a "single careless case of 'walking typhoid' on the shores of the [McCloud] river might become a serious element of danger, and either a large storage and detention reservoir or a system of filtration must be provided as a part of this project in order to make it comparable in the sanitary way with the proposed Hetch Hetchy system."[54]

As far as Freeman was concerned, no detailed survey of a possible McCloud River aqueduct was necessary given the requisite need either to erect a large retention reservoir or operate a filtration plant. Better to just build the Sacramento River pump/filtration plant and forgo the expense of tapping directly into the McCloud at a distance more than two hundred miles from the city.

The second group of alternative rivers that flowed west out of the Central Sierra Nevada did not need to cross the Golden Gate or the Carquinez Strait to reach San Francisco; on that score, they differed significantly from the northern sources. However, compared to the Tuolumne they were all relatively small streams; specifically, US Geological Survey gauging of the Cosumnes River in the twenty-first century shows an average annual discharge of 492 cfs, the Calaveras at 225 cfs, the Stanislaus at 225 cfs, and the Mokelumne at 996 cfs. In contrast, the Tuolumne registers an average flow at 2,343 cfs, more than double the flow of the Mokelumne and by far the largest of the Central Sierra sources investigated by San Francisco.[55]

With the Mokelumne, for example, Grunsky clearly could have undertaken a much more vigorous assessment of the river's capacity as a viable alternative. In its original form, Grunsky's report presented to the Army Board on August 1 offered little more than what he had completed a decade earlier, wherein he projected the Mokelumne's supply capacity at a mere 60 mgd and proposed a non-gravity-flow system that required extensive pumping. Freeman recognized this estimate as too low, and an additional statement tacked onto Grunsky's report (as published in Freeman's *Hetch Hetchy Report*) referenced 200 mgd as a plausible supply quantity.[56]

Later on in 1913, the question of the Mokelumne as a possible source would arise (see chapter 6), so it is useful to draw at some length from Freeman's report to provide context for his ambivalence regarding Grunsky's analysis of the stream:

> Mr. Manson happened to have made brief studies and an adverse report on these Mokelumne sources 6 years [ago]. . . . [U]pon Mr. Manson's disability by illness . . . the Mokelumne investigation was turned over to Mr. C.E. Grunsky, who had himself also studied the river as a possible source for San Francisco 11 years ago. . . . Previous investigations had so plainly brought out the disadvantages of the Mokelumne that Mr. Grunsky evidently was impressed with the unwisdom of spending any large sum of money at the present time for further field work in detail, and so bases his statement upon the facts already on record. Moreover, there was not time for any extensive new field work after Mr. Grunsky was called in to take up the work which Mr. Manson had not completed at the time of his illness.[57]

Not wishing to appear totally disengaged from the assessment of the Mokelumne's supply capacity—after all, he was including it as part of his own report on the need to dam Hetch Hetchy—Freeman emphasized how he had pushed Grunsky for a further estimate on a design comparable to what was being proposed for the Hetch Hetchy Aqueduct: "At my request Mr. Grunsky has since made studies for avoiding the expensive pumping at Altamont, by delivering the water from the Mokelumne into a long, low gradient Coast Range tunnel like that of the Freeman project."[58] By late August, however, Freeman had reached the limit on how far he considered it necessary or appropriate for the city to go in further analyzing the Mokelumne. As he stated in his report, "I do not believe it advisable to expend the $15,000 to $30,000, more or less, which explorations and complete surveys for thoroughly working out the best possible project for a municipal water supply from the Mokelumne would cost."[59] By extension, this perspective could apply to all of the alternatives that the city could further investigate, as Freeman did not need any additional cost estimates of such alternatives to justify the superiority of his Hetch Hetchy plan. And he certainly did not see any value in badgering a slow-moving Grunsky to complete such work.

As far back as 1911, Freeman had complained to Manson regarding Grunsky's work ethic, but in the rush to meet Fisher's deadlines, Freeman had little choice but to accept what the former city engineer offered, no matter how it rankled him.[60] Thus, on August 1 when Freeman delivered material to the Army Board, he pointedly, but privately, recorded in his diary, "Grunsky trailing along an hour later with some hasty half baked reports."[61] A few weeks later, as the publication deadline loomed, he again confided to his diary, "Reviewing the weak and inefficient reports of Grunsky and putting in shape for publication."[62] And just before he left San Francisco, he personally confided to the editor of *Engineering News,* "Grunsky has been a great disappointment—extremely suave, pleasing of address, honest but after all chiefly 'hot air' or 'sweetened wind,' who tries to make easily flowing language take the place of careful investigation and who never gets down to brass tacks."[63] But as hasty and half-baked as Grunsky's alternative studies may have been, Freeman did not believe that their imprecision and incompleteness could undermine the overall advantages offered by his Hetch Hetchy system, especially with its sizable hydropower component.

The lack of consistency and rigor in how various alternatives were assessed no doubt constitutes a major weakness of Freeman's *Hetch Hetchy Report.* And this deficiency later provided the city's opponents with reason to believe that Freeman had consciously tried to rig the deck and unfairly argue for the necessity of damming Hetch Hetchy.[64] To Freeman, however, this constituted unjustified criticism, because he held fast to the Hetch Hetchy project's unique ability to both foster a 400 mgd water supply and provide a reliable hydroelectric capacity of more than 150,000 hp. This was something that no other source could match. At least privately, Freeman may have recognized that the analysis of alternative water supplies could have been much more thoroughly and carefully carried out. But in the end, he held firm in the belief that further investigations would not upend the basic calculus underlying his report and that this plan for a Hetch Hetchy supply would remain the best option for meeting the city's long-term needs.

Spring Valley Water Company

One of the key features of the Hetch Hetchy system that Freeman proposed was that it could be readily merged with the Spring Valley Water Company's existing water supply and distribution infrastructure. Many in the Bay Area assumed that the company and the city would eventually come to terms on a fair-value purchase price for Spring Valley's assets. This was hardly a far-fetched expectation as, in November 1909, the two parties successfully negotiated a price of $35 million. However, when the agreement was put to city voters it failed by a few hundred votes to meet the approval of a two-thirds majority. As a result of this shortfall, the transfer of assets was not consummated and the Spring Valley company continued to operate, and expand, its system.

When negotiations on the sale of the company heated up again in the summer of 1912, it was reasonable to assume that Spring Valley's owners would expect/demand more than $35 million for their growing assets. But how much more? And how high a price would be acceptable for a super majority of voters to approve? It was a tricky problem for city leaders, because they did not want to suffer another failed referendum. But to expect that Spring Valley would accept a relatively modest price to facilitate voter endorsement was not a viable negotiating strategy. Not surprisingly, the company wanted to push the city as far as possible in reaching a maximum sales price.

Freeman was involved in the negotiations as an advisor to the city; this placed him in a thorny position vis-à-vis Spring Valley and their consultants, William Mulholland and J. B. Lippincott. In early 1912, the Los Angeles–based team of Mulholland and Lippincott had been engaged by Spring Valley to assess the water supply capacity of the Alameda Creek system and, specifically, the subsurface supply held in the Livermore gravels. Freeman was skeptical of the long-term capacity of the Alameda Creek watershed, and this placed him in conflict with the interests of Spring Valley. In essence, the company believed that a more generous assessment of the Alameda system would equate to a higher valuation of their assets. In his *Hetch Hetchy Report,* Freeman estimated that the future capacity of the overall Alameda watershed was "not more than about 25 or 30 million gallons [per day] additional to what [was] already being drawn" by the company.[65] In contrast, Mulholland and Lippincott believed that the development of surface and subsurface resources within the Livermore Valley alone could provide an estimated "51.5 MGD [of] continuous flow" beyond what the Alameda system was already providing.[66] This represented a significant difference, one that could affect how the value of Spring Valley's overall assets might be assessed going forward.

In late June, Freeman met with Mulholland, Lippincott, and Spring Valley vice president S. P. Eastman in San Francisco, and the company proved willing to allow Mulholland to provide Freeman with technical data related to the

Los Angeles Aqueduct.[67] However, when it came to the capacity and value of the Alameda Creek system, the interaction between the company and Freeman became more confrontational. As tensions rose, Freeman assured Spring Valley's leaders, "I have tried in every way to promote mutual understanding on water supply matters between the City and the Company, and have repeatedly urged . . . that the City and the Company should work toward a friendly understanding and purchase."[68] But through the course of the summer and into the fall, Spring Valley would be wary of the city's (and Freeman's) efforts to place a valuation on the company's assets.

Over the summer, Freeman devoted much energy to advising city officials on the prospective purchase of Spring Valley. On June 22 he began preparing a proposal for the city's use and the next day devoted much time to "offer of purchase." A week later he spent "7 to 11:30 [p.m.] at mayor's office on discussion of city's offer of $39,000,000," and the next morning he worked "on draft of request of S.V.Co. to sell." On July 3 Freeman met with "Mayor Rolph at his request." According to Freeman, "He says some members of [Board of] Supervisors would stand for 39 million but he can get harmony at 38. He asks my advice." In mid-July Freeman and Rolph "discussed purchase" of Spring Valley. Freeman noted, "I set forth earnestly the importance of an offer. Mayor seemed impressed more than before."[69]

By the beginning of August, the city was ready to act, and on August 7, Freeman worked on "new draft of letter to Spring Valley Co." Freeman noted in his diary that during a later meeting with Rolph, "I repeated my views that $38,500,000 was very lowest I could subscribe to."[70] The city's board of supervisors aligned with Freeman's minimum recommendation and approved an offer of $38.5 million for all of Spring Valley's assets. In making the offer the city pointed out that, if accepted, each Spring Valley "shareholder [would receive] an increase of about $10.53 per share, or about 17 per cent, above present market value of his stock."[71]

After a few days, word of the proposed purchase reached the city's newspapers on August 13, and with headlines such as the *Call*'s "Mayor Bids for Spring Valley, Offers $38,500,000 for Entire Plant," city residents became aware of the projected transaction.[72] Spring Valley, however, moved slowly and waited over a month before responding with a counteroffer. On the surface, the company appeared willing to accept the proffered price, but only if it maintained ownership of large tracts of valuable land in the city's Lake Merced district near the Pacific coastline.[73] The company was evidently in no rush to strike a deal, and when Freeman was back in Rhode Island and got word of their counterproposal, he dismissed it: "one of the most absurd propositions that ever came under my notice. It plainly looks as though [Spring Valley] intended to make an answer which was merely polite and plainly impossible for acceptance."[74]

Despite all of the effort Freeman devoted to the city's prospective purchase of Spring Valley's assets, little progress was made on consummating the sale. The company went into the fall with a desire to further publicize the scale of their assets, including the bountiful future capacity of the Alameda Creek system. For Freeman, this posed a problem, because the company's generous estimates as to the capacity of the Alameda watershed were in part based upon the analysis offered by Mulholland and Lippincott. Freeman did not want to alienate the engineer whose work on the Los Angeles Aqueduct provided a foundation for his own Hetch Hetchy plans; however, he could not stand by and—in his view—let the company trumpet unrealistic projections tied to the proposed value of Spring Valley assets.

Evidence of how he tried to walk the tightrope in his relationship with Mulholland appears in a letter he wrote to his erstwhile colleague just before leaving San Francisco in early September. Referring to his recently completed report on Hetch Hetchy, he advised Mulholland, "You will see that I have quoted largely from your work [on the Los Angeles Aqueduct] and you may be able to note that I have frankly and in accordance with my belief criticized some of your findings from what I believe insufficient data for Alameda Creek. Nevertheless, I want to testify to my continued esteem. I told the City Attorney that while it was possible for you to be mistaken it was impossible for you to make a statement which you did not believe to be true."[75]

Perhaps Freeman truly believed that, with this letter, he could mollify Mulholland as to the *Hetch Hetchy Report*'s implicit reproach of his consulting work for Spring Valley. But later in the fall, as the water company prepared for Fisher's "show cause" hearing, it would become clear that Spring Valley's leadership, and Mulholland, would not accept such criticism with grace and forbearance.

The Final Push

The most important deadline set by Secretary Fisher required Freeman to submit to the Army Board, by July 15, a complete report on the city's newly planned Hetch Hetchy Dam and Aqueduct (featuring an ultimate supply capacity of 400 mgd). In essence, much of what appeared in the first 150 pages of Freeman's published *Hetch Hetchy Report* is what was submitted to the Army Board to fulfill Fisher's demand. In August, while preparing the final version of his report, Freeman would update, revise, and supplement what he had submitted to the board in July. But the basic form and structure of the *Hetch Hetchy Report* remained largely intact. Freeman and his staff had only been in San Francisco for five weeks before reaching the deadline, and as Freeman later described, preparing the submittal required tremendous effort: "[On Monday, July 14,] I worked steadily from 5am until 11pm and on the following day when our report had to be in the hands of the Army Board, we kept at it until ten minutes of two [on the morning of July 16]. Col. Biddle very kindly sat in my room all the time after about nine pm and most kindly refrained from looking at his watch so that the report was received officially before midnight of the day specified."[76]

The report presented to the Army Board was quickly passed on to the local press; soon a multitude of newspaper articles appeared, describing key elements of Freeman's dam and aqueduct system, including how it re-envisioned the Garfield Permit. As the *Call*'s headline blared, "Engineer's Report Sets Forth the Need for an Absolutely New Permit."[77] Similarly, the *Chronicle* reported, "John R. Freeman Discards All Previous Studies in His Voluminous Work," and in a subheading the *Evening Post* declared Freeman's intent to "Enlarge Scope of Garfield Permit."[78] The *Post* further explained, "Freeman urgently recommended a complete reversal of the terms of the original grant. He would develop Hetch Hetchy first and begin work at once and then utilize Lake Eleanor and Cherry Creek as need demanded. . . . Freeman recommends a complete change from the original plans in that he urges the elimination of pumping plants."[79]

In addition, the city papers picked up on Freeman's professed desire to not destroy Hetch Hetchy but instead make its beauty more accessible. As the *Post* explained, "Freeman makes a particular feature in his report of the beautification that would be possible in the valley [as] he outlines a system of boulevards, reservoirs, dams and scenic adornment. . . . In this he meets the arguments of 'the nature lovers' who have claimed the valley would be despoiled."[80] Under the subheadline "Scenic Highway Planned," the *Chronicle* quotes Freeman's desire that "scenes from Norway" be replicated in roads making Hetch Hetchy more accessible: "A great boulevard will start at the dam and arch about the lake. Scenes such as are seen in Norway will be duplicated. The road will bore through the rock and there will be overhanging portions just as may be seen in the old world."[81]

With his report on the Hetch Hetchy Aqueduct and Dam submitted to the Army Board, Freeman turned his attention to making sure that Grunsky was, as best as possible, prepared to submit the final batch of alternative source reports to the officers by August 1. He also worked steadily on drafting and editing text for his published *Hetch Hetchy Report* and also proofreading copy that had been prepared by the Rincon Publishing Company. Keeping track of various edits, as well as assembling and organizing more than 130 photos, illustrations, and drawings, was no simple task, and in mid-July, he confided to his diary, "Much of day spent in bad mix up proof—some sheets of first draft gotten mixed in with final copy."[82] But he kept moving, with diary entries reflecting his progress: "Worked until 12 pm putting illustrations into proof" (July 17); "Working all day on review of proofs" (July 19); "Up at 5:15—at work revising Alameda report for printer" (July 21); and "Up at 5—at work correcting proofs of Alameda run-off (August 2).[83]

By the end of the first week of August, his schedule allowed him to take a one-day trip up to the Great Western Power Company's Big Meadows Dam, then under construction across the North Fork of the Feather River. Returning on the evening of August 6, he focused on Hetch Hetchy matters for a few days, before attending the Bohemian Grove "hijinks" organized by San Francisco's Bohemian Club. Dating back to the 1870s, this all-male club brought together the city's economic, social, and political elites; every summer, they and their guests retreated to a rural enclave in Sonoma County for a weekend of artistic revelry and male bonding. Ostensibly, the "hijinks" were to be a time in which business affairs were left behind and replaced by socializing and festive merriment. But, of course, politicking and networking were never far from the surface.[84]

Freeman was not known to be a particularly avid member of any social clubs or fraternities, but he apparently enjoyed his two-day visit to the Grove's encampment. On Friday afternoon, August 9, he journeyed north on "the Club's Special Train" and while at the camp used his time "making new friends and seeing old ones." His engagement with the Bohemian Grove elites was not specifically tied to Hetch Hetchy, but his presence gave him further visibility among San Francisco's ruling hierarchy. On Sunday afternoon, he and Percy Long headed back to San Francisco in an automobile owned by former mayor James Phelan.[85]

The next day, Freeman took the night boat to Stockton, where he accompanied Mayor Rolph, Percy Long, and the Army Board on an outing to Hetch Hetchy.[86] Apparently Rolph had never made the trek to the valley, and for Freeman, it was a good opportunity to spend time with the army officers out in the field. Travelling by automobile from Stockton to Hog Ranch, the party stopped at the site of the future Moccasin Creek power plant with Freeman "point[ing] out features of [the aqueduct route] on the way." The next day, the party switched to horses

Accompanied by Mayor James Rolph (*standing left,* in white suit) and the Advisory Board of Army Engineers (on horseback), Freeman made his third visit to Hetch Hetchy Valley in August 1912. (John Ripley Freeman Papers [MC 51, box 67], MIT Libraries, Cambridge, Massachusetts)

for the ramble into Hetch Hetchy. On August 14 the party visited the lower valley and "inspected city's boundaries and dam site." After lunch they "went to Head of Lake" where Freeman "pointed out paint marks showing flood lines [of the proposed reservoir]." After camping in the valley overnight, the Army Board trooped off to Tuolumne Meadows, while Freeman and Rolph returned to Hog Ranch. At 1:30 p.m., they left by automobile for the Bay Area. Despite getting lost crossing the San Joaquin Valley, they reached Oakland at 1 a.m. on August 16.[87]

For Rolph, the trip to Hetch Hetchy was a great success, and upon his return, the *Examiner* celebrated the expedition with an article headlined "Rolph Back and Enthuses over Water Plans—Hetch Hetchy Makes Mayor Hurrah over Project to Supply San Francisco."[88] A further chance to boost the Hetch Hetchy plan came a week later when Freeman gave a noontime talk at the Commonwealth Club. His presentation attracted the attention of city newspapers and allowed him to promote once again the merits of his gravity-flow system. And he again took a shot at the nature lovers who opposed his plan. Not for the last time, Freeman called out the paucity of visitors to the valley, with the *Chronicle* reporting his comments: "Many who oppose the project say that Hetch Hetchy should be kept sacred to tourists. Under present conditions tourist traffic there is practically nil. . . . It is our plan to build roads all through the valley which, when completed and no longer needed to convey supplies to the construction force, will be turned over to the public.[89] In a similar vein, the *Call* reported, "Rolph's and Freeman's record return automobile trip from Hetch Hetchy demonstrated how when the city constructs its road to haul material for the dams the trip can be made in twelve hours at the most. Freeman sees Hetch Hetchy as the mecca for many pleasant weekend automobile trips."[90]

After his Commonwealth Club talk Freeman was "completely worn out," but there was no time to relax, as he needed to complete his report and get everything to the printer within a week.[91] Secretary Fisher was to pass through San Francisco at the end of the month and wanted to obtain a printed copy of the *Hetch Hetchy Report* before boarding a steamer for Honolulu. The night before addressing the Commonwealth Club, Freeman had confided to his diary, "Report all in printer's hands at 11pm except 2 or 3 pages final polishing off."[92] But this "final polishing off" did not come so quickly. The extent of his final revisions is reflected in the pagination numbers, which after page 160 includes nineteen pages denoted as 160a through 160s. In the closing push to complete the report, Freeman kept adding in material until the last minute, such that there was no time to repaginate the more than two hundred pages of appendixes (starting on page 161) already set in proofs.

Fisher embarked for Hawaii on August 31 with the *Hetch Hetchy Report* in hand, and Freeman had hoped to leave San Francisco the next day. But he was laid low by a "mild touch of ptomaine poisoning" and spent a few more days in the city recovering and writing letters. At last, at 10:20 p.m. on September 3, he set off for Seattle and, the next day, was "on train all day [enjoying] fine view of Shasta Country all green and beautiful." His summer in San Francisco was over.[93]

Manson Out, O'Shaughnessy In

When Marsden Manson collapsed in mid-May and ceded responsibility for the Hetch Hetchy project to Freeman, he did not formally resign as city engineer. Mayor Rolph permitted him to take a leave of absence, allowing that he might someday return to his post. But as the summer wore on, Manson's standing with city officials became ever more tenuous. Freeman's interactions with Rolph kept him abreast of this festering dissatisfaction, and in early July, he noted in his diary that the mayor "says Manson will be fired for incompetence soon."[94] By early August, Rolph was ready to move on. As reported in the August 7 *Evening Post* (under the headline "Manson Quits Office When Under Fire by Mayor"), the decision to resign was, at least officially, made by Manson and then accepted by Rolph.[95] Noting that "Manson will remain in office until September 1," the *Post* reported that the mayor "refused to comment on the resignation of Manson [except to say that it] . . . comes unexpectedly." However, the *Post* observed that "Manson has been on the carpet a great deal since Mayor Rolph took office." Despite pro forma protestations that the resignation was "unexpected," it seems more surprising that Manson lingered as long as he did.

For Freeman, Manson's official departure was hardly distressing. While he offered no public rebuke of the one-time city engineer, Freeman never expressed any great respect for Manson. In an acknowledgment included in his *Hetch Hetchy Report,* he notes Manson's "earnestness" and expresses how he enjoyed "the pleasure of camping in the mountains with one so well versed in knowledge of the forests."[96] But a more revealing assessment came in a personal letter to the editor of *Engineering News* at the end of the summer. In the letter, Freeman expressed an unvarnished take on the former city engineer's bureaucratic failings: "Poor old Manson has had to walk the plank. He is most utterly incompetent for his job in all that pertains to organizing skill and engineering ability, but remained honest and steadfast all thru the crooked administrations. He simply didn't know how."[97]

With Manson's announced resignation in early August, it was left to Mayor Rolph to appoint a new city engineer. Freeman was not closely involved in the search, although Rolph apparently raised the possibility that Freeman himself might take the job. There was little chance, however, that Freeman would give up his consulting practice and his insurance work to be San Francisco's city engineer. He was happy to have someone else fill the position. As his diary for August 22 records, "Mayor asked my advice on City Engineer and I suggested he advertise. He asked if I would take charge of new work as Chief Engineer—I told him he would be afraid of my price."[98]

By the end of the next week, Rolph had found Manson's replacement: Michael Maurice (M. M.) O'Shaughnessy. Born in Limerick, Ireland, in 1864 and an engineering graduate of the Royal University in Dublin in 1884, O'Shaughnessy had come to California in March 1885. He had a wide-ranging career in the Golden State, working for railroads and the Spring Valley Water Works before heading to Hawaii in 1899 to take charge of water supply/irrigation systems. After returning to California in 1907, he became chief engineer for San Diego's Southern California Mountain Water Company; in this position he oversaw construction of the massive rock-fill Morena Dam, completed in 1912. O'Shaughnessy could get big projects built, and he was ready to move up and become San Francisco's city engineer. He accepted Rolph's offer and officially started work for the city on September 1.[99]

Given how closely O'Shaughnessy is now associated with the Hetch Hetchy system (the dam itself was formally designated O'Shaughnessy Dam in 1923), it is notable that he had no meaningful relationship to the project before being appointed city engineer. And prior to August 31, it does not appear that he and Freeman had ever met or corresponded.[100] But over the course of the next year and a half, Freeman and O'Shaughnessy became a team of sorts, developing a strong working relationship. Although they did not always agree, they shared a resolute desire for San Francisco to succeed in building the Hetch Hetchy Dam and Aqueduct along the essential lines laid out in Freeman's report. With O'Shaughnessy's arrival and his success as city engineer, it was easy for San Francisco's leadership (as well as Freeman) to forget about Manson.[101]

$48,989.36

Upon returning to his Providence office, Freeman finally took time to calculate what was due to him for all his (and his team's) labors going back to the beginning of the year. When Freeman arrived in San Francisco in early June, he had intended to present a bill to the city covering the first five months of the year, but the rush of events pushed this to the side. It meant that one very large sum would come due in the fall.[102]

On September 23, Freeman submitted his invoice covering work from January through early September.[103] Including expenses for travel, lodging, and food, and also the salaries and expenses of his staff, the total came to $48,989.36, of which $34,000 was for his personal services. Of this, $9,300 was for work related to the "valuation" of the Spring Valley Water Works, and the other $24,700 was for everything involved in preparing the *Hetch Hetchy*

Report.[104] For his time in San Francisco, Freeman billed the city at a rate of $200 per day. But this was for a seven-hour day, and with a work schedule that commonly stretched from 5 a.m. until 11 p.m., his fee for any given day could exceed $400. For example, during the month of July he billed the city for 68 "seven-hour days," totaling $13,600.[105]

Freeman understood that his skilled labor did not come cheap, and he feared that city officials would "feel greatly disturbed at paying such a bill to any mere engineer." Nonetheless he felt justified in being properly compensated for his services, declaring, "In the Hetch Hetchy Aqueduct layout I did the best job of my life and certainly if one compares it step by step with the Grunsky and Manson projects he will have little difficulty in finding where I have saved the City considerably more than the amount of this bill."[106] Freeman further made the point that his devotion to the city's cause came at great cost to his other professional work: "If I could have quit the city's services six weeks earlier I would now be much better off than with this bill paid in full, because I was compelled to decline one after another, some attractive engagements at Seattle, at Washington D.C., at Schenectady, N.Y. also two important dam designs for Ford, Bacon, & Davis [Engineers] of New York and could not give the attention requested either at Pacific Gas and Electric's Lake Spaulding Dam or at the dam of the Great Western Power Company, and in each of these cases a good retainer would have attached to the per diem."[107]

For their part, the San Francisco officials who had been left hanging by Manson's collapse were thrilled with Freeman's *Hetch Hetchy Report;* they made no complaint as to the amount of the invoice. There was some publicity given to the bill in the local press, and the city auditor raised questions about Freeman's practice of billing more than one seven-hour day for a given calendar day.[108] But once the *Report* became widely available in mid-October, it offered compelling evidence that Freeman had indeed expended an enormous amount of skilled labor over the preceding summer.[109] The bill was not a scam; it could be readily justified in terms of the final work product. At the end of October, he got word that the board of supervisors had approved full payment.[110]

No doubt Freeman was well rewarded for all his strenuous work. But money alone cannot explain his fervor in championing the Hetch Hetchy Dam. His reputation and his ego were on the line. Or as he stated to Mayor Rolph upon submitting his bill, the experience of his "long hours" of service in San Francisco's cause "is not one I would again go through *at any price.* I did it in your case not so much for the pecuniary consideration [but] as a matter of professional pride in meeting the task."[111]

Professional Pride

For Freeman, it had become unthinkable that opponents of *his* dam and *his* aqueduct might somehow carry the day and consign *his* work to the dustbin. Of course, San Francisco would one day build and operate the Hetch Hetchy Aqueduct as a municipally owned system. But it would be *his* plans and *his* vision as set forth in his report that would make it all possible.

Remarkable evidence of just how much Freeman cared about getting credit for the city's project can be found in one of the last letters he wrote before leaving San Francisco. In an instruction to Andrew Wood of the Rincon Publishing Company, he directed with great specificity how the cover of his report should be labeled: "Have the back stamped with big letters—1912, San Francisco Water Supply—Freeman; and on the face of the cover—The Hetch Hetchy Water Supply for San Francisco, 1912, Report by John R. Freeman. The letters on the back [or spine] should run lengthwise and be about half an inch in height, of a bold-faced form that can be seen half way across the room as the book stands in the case."[112] The book designers at Rincon did not disappoint in meeting this request. Today, if the cloth-bound *Hetch Hetchy Report* is placed on a shelf amid a crowded library, a quick glance makes its subject and authorship eminently clear: "1912—SAN FRANCISCO WATER SUPPLY—FREEMAN."

By the start of September, the drafting, editing, and layout of the *Hetch Hetchy Report* had come to an end. It was all in the hands of the printer, awaiting widespread distribution in October. Quickly resetting his task agenda, Freeman began advising city officials of myriad politicians, engineers, educators, and—in twenty-first-century parlance—cultural influencers, who he thought should be given free copies of the weighty tome.[113] As he well knew, the job of winning approval for the city's plans was far from over. In fact, it was just beginning. With the report finding its way into the public consciousness over the course of the fall—and drawing the ire of the Spring Valley Water Company, farmers in Modesto and Turlock, and nature lovers devoted to the beauty of Hetch Hetchy—it soon attracted great scrutiny. In concert with that scrutiny, it would become the foundation and centerpiece of Secretary Fisher's soon-to-be-convened "show cause" hearing in Washington, DC.

CHAPTER 5

The Fisher Hearing and
the Army Board

During the summer of 1912, Freeman focused his energies almost exclusively on Hetch Hetchy affairs. Aside from a day trip to the Big Meadows dam site in early August, his other work remained on hold for three months.[1] But once Freeman's *Hetch Hetchy Report* was in the hands of the printer, he quickly shifted gears.[2] After leaving San Francisco on September 3, he first consulted with the Vancouver Power Company on hydroelectric projects in British Columbia. He then crossed the Canadian Rockies and toured the University of Minnesota in search of design ideas for MIT's proposed "New Tech" campus. Next up was Keokuk Dam, where he reviewed progress on one of the world's largest hydroelectric projects of the pre–World War One era. After three days in Keokuk, he stopped in Chicago, meeting with utility entrepreneur Samuel Insull and consulting with staff of the "Chicago Branch" of the Manufacturers Mutual Fire Insurance Company. On September 16 he boarded the 20th Century Limited bound for New York City, where, the next morning, he convened with the Board of Water Supply on the Catskill Aqueduct and with Great Western Power Company executives on Big Meadows Dam. That evening, after an absence of more than one hundred days, he arrived home to wife and family in Providence.[3]

Barely catching his breath, he spent the next day in his Manufacturers Mutual office "cleaning up [a summer's worth] . . . of accumulated mail." A day later he was in Boston, attending to personal financial affairs, meeting with Stone & Webster executives on Keokuk, and conferring with MIT president Richard

Maclaurin on plans for the school's projected Cambridge campus—what he called his "New Technology studies."[4]

Other than Hetch Hetchy and Big Meadows, MIT's proposed expansion constituted the only big project that competed for Freeman's attention in the fall of 1912. Emboldened by a $2.5 million gift from photography tycoon George Eastman, the school was poised to move from Boston's Back Bay across the Charles River to Cambridge, and when President Maclaurin asked him the previous January to "take charge of preparing designs for the New Technology," Freeman had happily agreed.[5] But over the summer his Hetch Hetchy efforts precluded him from pursuing "New Tech" possibilities. Fall would give him a chance to make up for lost time; upon returning from San Francisco, he invested much personal capital in the "New Tech" project. On September 19 he spent the afternoon with Maclaurin and a few weeks later was back in Boston: "Corporation meeting at Tech. Brought boxful of reports . . . and spoke for 15 minutes to Corp. on progress and scope of studies."[6]

Freeman's technocratic goal was to create a modern, efficient campus suitable for a world-class university. Ever confident in his abilities as an engineer, he saw little need to collaborate with a traditional architect on this formidable task.[7] Such enthusiasm notwithstanding, some MIT leaders were unconvinced that the school should construct a new campus without a classically trained architect at the helm. In a meeting with Maclaurin at the end of October, Freeman got wind of such sentiment and, a few weeks later, revealed in his diary that the president desired "that I transfer my notes to the architect."[8] But despite Maclaurin's warnings—and despite all the work that Hetch Hetchy would require in the coming weeks and months—Freeman held fast to his dream.

Dust Thrown in People's Eyes

Planning for MIT's new campus and pushing for Great Western's abandonment of the Big Meadows multiple-arch dam engrossed Freeman during the early fall of 1912, but he had not forgotten about his work for San Francisco. When the *Hetch Hetchy Report* became widely available in mid-October, the city's plans for a Yosemite water supply entered a new phase, one in which Freeman's plans, analysis, and arguments dominated the discourse. Not content to wait for cultural thought leaders to discover his report on their own, he arranged for San Francisco officials to send copies gratis to a wide range of influential engineers, academics, politicians, and organizations in California and the Pacific Coast environs. He also arranged for copies of the report to be sent to his Providence office so that he could personally mail them to more than one hundred men—no women, they were all men—of influence in New England, New York, and other points east.[9]

A large batch of the printed report reached him by October 21, and he quickly began to distribute the fruits of his summer labor. The recipients were a diverse lot including, for example, the New York–based editors of *Collier's* and *Outlook,* his former mentor Hiram Mills, Carnegie Foundation president Henry Pritchett, Boston lawyer and future Supreme Court justice Louis Brandeis (Freeman knew him through his insurance work), and Harvard University president A. Lawrence Lowell.[10]

Along with personally inscribing copies of the report, Freeman also sent out cover letters to many of the recipients. These epistles are invaluable to historians because they offer insight into how Freeman framed the discussion of Hetch Hetchy when engaging with his colleagues and associates. They also likely reflect the arguments, details, and phrasings he called forth in oral conversations. Notably, these letters espouse how, in his view, hostility to the city's plans was driven by something other than detached engineering analysis or dispassionate public policy. For example, he told a Providence business associate, "This [the Hetch Hetchy initiative] was an extremely interesting problem from many points of view and the studies in psychology are almost as interesting as those in engineering. The private water company [Spring Valley] has the city by the throat and is disposed to make the most out of its grip."[11]

Positioning the Spring Valley Water Company as a malevolent force guiding the opposition was a persistent theme in his letters, and the pernicious image of the city being held by the throat appeared more than once. For example, Freeman advised Pritchett, "I became so profoundly impressed with the justice of the city's cause, and the fact that the water company had it by the throat and was showing no mercy that I worked to my limit. . . . It is popularly believed in San Francisco that the arousing of the 'Nature Lovers' and the irrigationists in opposition to San Francisco utilizing the Hetch Hetchy was prompted on the quiet by the officials of the Spring Valley Water Company."[12] Similarly, he assured Brandeis, "The water company literally has the city by the throat and the city officials believe that it was at the bottom of stirring up the country-wide agitation among the nature lovers."[13]

In his cover letters, Freeman's former clients at Spring Valley were portrayed as Machiavellian fiends, manipulating park advocates and guileless farmers to do their bidding in protecting the water company's lucrative franchise. Thus, he advised Hiram Mills that "the 'Nature Lovers' and others have been used by shrewd and selfish men."[14] Overall, Freeman held little sympathy for tactics employed by so-called nature lovers, but he could adopt a patronizing tone that characterized them as naïve puppets of "strong and selfish interests." In this vein he explained to a Providence business colleague, "There is not the slightest question of the sincerity of the great majority of the nature lovers, who oppose the use of the Hetch Hetchy, but I have been told . . . that a great deal of dust had been thrown in good people's eyes by strong and selfish interests."[15]

"Dust thrown in people's eyes" is a memorable phrase, provocatively suggesting how his opponents were being blinded to the righteousness of the city's cause. In a similar form, it appeared in Freeman's letter to Harvard's President Lowell and likely was a staple of the verbal rhetoric he used to defend his plan: "A great deal of dust has been thrown, partly innocently, partly by design, into the eyes of the public on behalf of the nature lovers and in making a national protest against the utilization of the Hetch Hetchy Valley."[16]

Beyond complaining about Spring Valley and the nature lovers, the economic attributes of Hetch Hetchy as a reservoir site were never far from Freeman's mind; for example, he assured Lowell, "It is my profound belief that Greater San Francisco is already in urgent need of a new water supply . . . [and that] alternative sources would involve an economic waste of fifteen to twenty million dollars."[17] But saving money was not the only factor energizing his plans, as he also depicted the nature lovers' anti-dam crusade as a spur to his own progressive effort to open up the Sierra landscape to tourism. Thus, he counseled Desmond Fitzgerald, "I believe I am right in recommending that the city spend upward of half a million dollars in making this valley more accessible to its citizens."[18]

In the letters accompanying his report, Freeman depicted the city's opposition as falling into three camps: Spring Valley, irrigationists, and the "nature lovers." Each of these presented distinct problems and issues that Freeman needed to counter in defending his Hetch Hetchy plan. Of the three, the preservationists' anti-dam arguments are the best known, at least to environmental historians, but the other two were likewise of concern to Freeman, forcing him to guard the city's interests in what became a three-front campaign. Although confident in his scheme for Hetch Hetchy, he was watchful for surprises because, as he told O'Shaughnessy, there was "no telling what the opposition may spring."[19]

The Opposition: Spring Valley

Freeman's attack on the Spring Valley Water Company aligned easily with progressive tropes that championed publicly owned utilities as correctives to rapacious privately owned monopolies. But Spring Valley's owners did not see themselves as exploitive enemies of the people; in their view, the company's expansive water supply system provided a valuable public service. The company was not unalterably opposed to selling its assets to the city—in fact, in November 1909 it had agreed to such a sale for $35 million, only to be rebuffed when the city electorate failed to approve the deal by a two-thirds majority. Since then, Spring Valley had continued to develop its properties (Freeman's work on the Calaveras Dam was part of this effort), and in any possible future sale, corporate leadership wanted the price to reflect their increased investment. They also wanted any sale price to properly reflect the supply capacity that the company's properties could sustain when fully developed.

On this latter point, Spring Valley worried that city assertions claiming the Bay Area faced a looming water famine unfairly denigrated the capacity of the Spring Valley system. The company was not opposed to the Hetch Hetchy project on the grounds that it would destroy a beautiful mountain valley. But they did object to any contention that a deficiency in *their* system necessitated a rapid deployment of the Hetch Hetchy system. So, to defend their interests, in the fall of 1912 the company published a 506-page report titled *The Future Water Supply of San Francisco* that mimicked Freeman's *Hetch Hetchy Report* in size, font, and format.[20] The goal was not so much to criticize Freeman's Hetch Hetchy design as it was to publicize how the company's resources could fulfill municipal needs for decades to come. Perforce this would call into question the need to immediately fast-track the Hetch Hetchy Aqueduct. And it would also enhance the company's bargaining position when negotiating a sale price for its assets.

As of 1912 Spring Valley was providing about 40 million gallons of water per day (mgd) to users in the Bay Area, but the company and its consultants determined that the ultimate capacity of the system would exceed 200 mgd.[21] In contrast, Freeman's report estimated that the company could develop 78 mgd from its holdings (or about twice the amount the company delivered as of 1912).[22] This significant difference (200+ versus 78 mgd) had ramifications in terms of both the monetary value of the company and the city's (supposed) "need" to rush ahead on the Hetch Hetchy project.

The difference between Spring Valley's and Freeman's projections derived primarily from, first, how the two parties evaluated the existing streamflow

The Spring Valley Water Company did not believe that Freeman's *Hetch Hetchy Report* properly estimated the amount of water that the company's Alameda Creek system could provide for the city. This photograph shows a January 1911 flood along Alameda Creek at Sunol Dam that Freeman used in analyzing the flow available for storage during periods of heavy runoff. (Freeman, *Hetch Hetchy Report*, 184)

records for the Peninsula and Alameda watersheds and, second, how much of the annual precipitation seeping into the Alameda watershed's Livermore Valley could reasonably be drawn out by pumping. Streamflow records went back about twenty years, but there were questions as to their reliability. Of particular importance was the volume brought during times of heavy floods, when the water gauges at Niles Dam and Sunol Dam were required to cope with much higher levels than normal. There, the geographic layout of the stream, the impact of a rocky, irregular streambed, and the effects of backwater and turbulent eddies prompted Freeman to question whether the Alameda Creek gauges overestimated the amount of floodwater available for storage in reservoirs yet to be built (such as Calaveras Dam).[23]

The contest over the ultimate capacity of the Livermore Valley aquifer assumed a personal character after Spring Valley hired William Mulholland and J. B. Lippincott to assess the subsurface supply capacity of the Livermore gravels. Freeman had previously worked with both men in 1906 when he served as a consultant for the Los Angeles Aqueduct, and they had become professional colleagues and friends. However, after reviewing their initial report on the Livermore aquifer in February 1912, Freeman began to have reservations regarding its conclusions. Making no effort to sugarcoat his reaction, Freeman professed in his report that "it is absolutely impossible by means of the utmost practical development of wells and pumps to make these gravels play the part that Messrs. Mulholland and Lippincott have reckoned they can be made to perform."[24]

Although Freeman had relied upon Mulholland's Los Angeles Aqueduct as a template for his proposed Hetch Hetchy Aqueduct, he now exhibited a patronizing, if not condescending, attitude toward the Los Angeles engineer. In mid-October 1912, he cattily advised Long that "Mulholland is at heart honest and simply 'had a few slipped over on him' by our enterprising S.V. friends."[25] In response, Long shared, "I am told that [Mulholland] is somewhat peeved over a remark which was reported to him as having been made by you to the effect that he was not scientific."[26]

Behind Spring Valley's objections to San Francisco's Hetch Hetchy plans lay no small amount of bad blood directed toward Freeman. Only a few months earlier, Freeman had been working for the company as the prospective designer of the Calaveras Dam. But when Freeman began overseeing the city's Hetch Hetchy project in the spring of 1912, his ties to the private utility were cut. And over the summer he had served as a key adviser to the city in formulating a proposal to buy the Spring Valley system. This shifting of allegiance rankled the water utility's corporate hierarchy, and Freeman was aware of their animosity. As the Fisher Hearing neared, Freeman confided to his colleague Allen Hazen, "The Spring Valley officials have fixed up a bulky report of similar appearance to mine. . . . The[y] have done a most artistic job of window dressing . . . [and] have not hesitated at violent personalities. The whole organization appears to be in the state of mind toward me that one imagines from a fond mother if I had ruthlessly scratched the face of her baby."[27]

The Opposition: Turlock and Modesto

Farming interests tied to the Turlock and Modesto Irrigation Districts also raised objections to Freeman's report and to the city's plans for an enlarged Hetch Hetchy project, but this was hardly surprising. Since filing their first water claims more than twenty years earlier, the districts always believed the Tuolumne to be a resource that should be reserved for watering arable land bordering the lower stretches of the river. Not coincidentally, this land closely aligned with the two districts' boundaries. In a November 1 letter to the Army Board, the districts' attorneys reinforced the point, strenuously opposing "any further grant to any of the waters upon the Tuolumne River to San Francisco. . . . [We] believe that there is not sufficient supply of water to irrigate the lands that can be irrigated from the Tuolumne River in the San Joaquin Valley."[28]

The districts further emphasized the magnitude of existing irrigation development in the districts: "The number of property owners has increased from 1,211 in 1901 to 6,317 in 1912 and the assessed valuation [has risen] from $3,745,719 in 1901 to $16,970,675 in 1912."[29] They also underscored the implications of Freeman's system as compared to the city's earlier plans: "In 1901–02 Mr. Grunsky proposed that San Francisco, alone, should secure a supply of 60,000,000 gallons per day. . . . [T]oday it is stated by Mr. Freeman that San Francisco should secure a supply of 400,000,000 gallons per day, capable of being developed to 500,000,000 gallons per day. . . . [T]he proposal to draw ultimately 500,000,000 gallons per day continuously from this same source will mean the use of 27% of the mean recorded annual discharge of the stream and 57% of the minimum recorded annual discharge."[30]

While forceful in defending their rights, the districts did acknowledge that "on the whole the data used [in Freeman's *Hetch Hetchy Report*] is full and reliable [and] we shall not question the data submitted, [however we] shall deny the truth of certain assumptions and conclusions deduced from them."[31] The irrigationist bloc would always consider the city to be an interloper invading its hydraulic domain, but in the runup to the Fisher Hearing, their objections remained relatively muted.

The Opposition: Preservationists

For the community of anti-dam preservationists, publication of the *Hetch Hetchy Report* brought much distress. For more than two years they waited for San Francisco to make a case for needing Hetch Hetchy. During this time, John Muir set off on a months-long expedition to South America and Africa, and although his followers maintained a passion for Hetch Hetchy, the anti-dam momentum of 1908–9 flagged.[32] Although Freeman's summertime presence in San Francisco was hardly a secret, the scale and scope of his completed report came as a shock, with Robert Underwood Johnson lamenting that he was "appalled by the enormity of the Freeman document."[33] No longer could dam opponents confidently assert that San Francisco had offered no case for needing Hetch Hetchy—Freeman's labors had defused that argument and placed the "nature lovers" on the defensive. Now they needed to awaken and respond to the threat posed by Freeman's vision of a 400 mgd system.

Despite discouragement, preservationists quickly marshaled rejoinders that sought to highlight inconsistencies and inaccuracies within Freeman's report. The most important of these was the "Brief of Sierra Club in Opposition to Grant of Hetch Hetchy Valley to San Francisco for a Water Supply," attested to by both John Muir (as Sierra Club president) and William Colby, the San Francisco lawyer who also served as club secretary. While other club members signed the brief, Colby was primarily responsible for its content.[34]

The retort is familiar to environmental historians because Holway Jones quotes it at length in his book *John Muir and the Sierra Club*. In particular, Jones highlights those sections of the Colby letter that critique Freeman's claims that the Hetch Hetchy Dam would not require extensive sanitation regulations affecting camping within the upper Tuolumne watershed; "fanciful" pictures in the report; and inconsistency in how cost estimates were calculated for alternative water supply systems. Underlying Colby's criticisms was his view that Freeman was nothing more than a hired gun who, in pursuit of a lucrative consulting fee, would tell clients precisely—and only—what they wanted to hear: "There is little doubt but that Mr. Freeman with his recognized engineering ability could have made an equally good case for any one of the dozen alternative sources available for San Francisco, by suppressing all of the unfavorable features and enlarging on the favorable ones. This is the result of paid advocacy the world over."[35]

In contrast, the Sierra Club and other anti-dam groups could draw upon only limited resources in their fight, something that in Colby's eyes affirmed their selflessness and virtue. Colby also took pains to portray the preservationist cause as separate from any other parties, presumably including the Spring Valley Water Company, who held a financial interest in opposing the Hetch Hetchy system: "While some of our worthy opponents are strongly inclined to the belief that our campaign is presumably financed in part at least by those who have a source of their own to sell, their inference is without foundation. . . . [Our opponents] utterly fail to appreciate our motives which are solely to save for the people of this nation and for future generations, a great national playground."[36]

In heralding Hetch Hetchy as a "great national playground," Colby's brief did not seek to safeguard the pristine isolation of Hetch Hetchy. Instead, he anticipated a future in which tens of thousands of visitors would journey into the valley every year, in both sun-drenched summer and snowy winter.[37] Far from a wilderness refuge, Hetch Hetchy offered an ideal site for a vast

To counter Freeman's contention that few visitors had ever visited Hetch Hetchy, preservationists proposed the construction of hotels and other amenities to attract tourists to the remote valley. Efforts to block the dam were not focused on protecting "wilderness" but rather on finding ways to bring more people to the upper Tuolumne watershed. Opened in 1927, the Ahwahnee Hotel in Yosemite Valley shown here provides a plausible model for what would have been built in Hetch Hetchy if dam opponents had won the day. Author's collection.

multitude of campers to pitch their tents and, with pack mules, set off for tramps into the park's northern highlands. It was also a place where hotels would thrive. In Colby's words, "Hetch Hetchy is the only other valley in the Park that ranks with Yosemite in point of availability of hotel sites. . . . Here are spacious, beautiful camping grounds for *thousands* besides the smoothly flowing river. . . . As soon as a good road is built to Hetch Hetchy (not by the city, but by the Federal Government) and transportation facilities are provided, hotels will spring up and the tide of tourist travel . . . will turn to Hetch Hetchy in both winter and summer."[38]

Colby was not alone in championing Hetch Hetchy as a site for expansive campgrounds and at least one hotel. The Chicago-based poet Harriet Monroe—a nationally known guardian of the valley—also made the point in a letter to Secretary Fisher: "The meadows and forests of this valley, watered by the great Tuolumne River, are necessary to the purposes for which the park was created, the valley being the only large camping ground and future hotel-site in that [northern] portion of the park."[39]

On the issue of tourism and resort hotels, Freeman's barbed complaints that Hetch Hetchy was but rarely visited had hit a mark. By spotlighting the paucity of visitors that had trekked to the valley, Freeman impelled preservationists to present competing initiatives for increased tourism, including hotels. With his provocative jabbing, Freeman forced his opponents to make their own utilitarian arguments for how the valley would be used—and not simply preserved—in the future.

One other preservationist response to the *Hetch Hetchy Report* deserves special mention, not simply because its author came to hold a special relationship with the Providence engineer but also because of issues it raised involving both the calculation of cost estimates and the ultimate value of the Fisher Hearing. Edmund Whitman was a Boston attorney who had previously testified in opposition to the city at the 1909 House Public Lands Committee hearing and had attended Ballinger's brief "show cause" hearing in May 1910.[40] As president of the "Society for the Preservation of National Parks, Eastern Branch," he provided Fisher with a detailed critique of Freeman's *Hetch Hetchy Report*.[41]

Whitman recognized that other sources of water were available to San Francisco and that Freeman's recommendations rested upon a financial argument favoring Hetch Hetchy. In Whitman's words, "It is now entirely clear, from the report of Mr. Freeman . . . that there are numerous sources of water upon which the city can draw for municipal supply. . . . [T]he question[,] as Mr. Freeman puts it, is purely one of cost."[42] Challenging Freeman's statement that alternatives to Hetch Hetchy would require "some ten or twenty million dollars extra (p. 60 in the Freeman *Report*)," Whitman honed in on the vague ways that the capacities and cost of alternative water sources were compared to Freeman's plan. And he specifically noted the wage discrepancy underlying different cost estimates presented in the report: "[For the Hetch Hetchy Dam and Aqueduct] Mr. Freeman bases his rate of wage at $2.25 a day for eight hours, for unskilled labor (pp. 223, 240, 251) and on page 251 he gives elaborate data upon which this estimate is based. And yet in figuring the Sacramento [River] water cost, Mr. Hazen figures on a minimum of $3.00 for eight hours. It is no wonder that it is easy to figure out the costs of competitive projects much higher than that of the Hetch Hetchy."[43]

Whitman was not an engineer, but he was a perceptive lawyer who could identify and exploit weak points in his adversaries' arguments. Freeman knew that the differences in wage rates tied back to the less-than-organized way in

which Marsden Manson oversaw the analysis of alternative sources and to Freeman having based his own estimates for the Hetch Hetchy system upon data
drawn from Mulholland's Los Angeles Aqueduct. However, acknowledgment
of Manson's shortcomings (not to mention his own lapses) was not something
that Freeman wanted to publicly air.

In his response to the *Hetch Hetchy Report,* Whitman defiantly questioned
the legitimacy of the Fisher Hearing, deftly following the logical consequence
of Freeman's claims as to the problematic character of the 1908 Garfield Permit:
"Mr. Freeman is entirely emphatic that the Garfield Permit is antiquated and
says 'The Garfield Permit is not broad enough for present needs (p. 140),' 'is
outgrown (p. 144),' and the provisions of the permit 'are now found contrary
to the public interest' and to carry them out would be 'an economic blunder
of the worst kind' [p. 149].' . . . [Furthermore,] 'the outlook and state of the
art [of hydraulic engineering has] moved so fast that the Garfield Permit has
become practically worthless for the needs of the city, and a new permit should
be drawn' [p. 150]." For Whitman, this was too much, as he believed that "the
granting of a new permit is beyond the scope of this hearing" and professed,
"We submit, therefore, that technically this hearing can be closed at once on the
admissions of the city, which should be relegated to Congress . . . to ask [for]
new legislation."[44]

In concluding his response, Whitman made the telling point that "this whole
matter is one for Congress. When it comes to flowing [i.e., flooding] the Hetch
Hetchy Valley back for a distance of nine miles . . . there is no congressional
authority warranting the Secretary of the Interior to do any such a thing."[45]
In other words, Freeman's plans to immediately build the Hetch Hetchy Dam
required such a drastic change to the Garfield Permit that the issue—at least
in Whitman's analysis—should be left to Congress for specific authorization.
Whitman's prognostication would not stop Fisher from holding his hearing,
but it would prove prescient in how the secretary ultimately handled his "show
cause" responsibility.

Whitman and Freeman

Whitman was certainly as passionate in his defense of Hetch Hetchy as his
more famous compatriots. But he stood alone in being party to one of the most
incongruous intersections of the Hetch Hetchy saga, in which he maintained
a professional relationship with Freeman on an issue far removed from San

Francisco's water supply plans. From seemingly out of the blue, Whitman came to serve as Freeman's lawyer in a byzantine legal case spawned by turn-of-the-century Massachusetts water politics.

The son of a prominent abolitionist and Civil War officer in the Union cause, Edmund Whitman was born in 1860; he graduated from Harvard College in 1881 and Harvard Law School in 1885. Upon entering the bar, he subsequently joined the Boston law office of Samuel Elder.[46] In his capacity as Elder's associate, he came to represent Freeman in a convoluted civil action related to a valuation of the Gloucester Water Supply Company that was, in part, rendered by Freeman in the late 1890s.

The details of the Gloucester case are not germane to the Hetch Hetchy controversy, but briefly stated, they involve a lawsuit brought by the son of the owner of the city's private water company, an heir who felt shortchanged by the $600,000 valuation placed upon the company by a state-authorized arbitration board that included Freeman as one of its three commissioners.[47] Freeman's work with the board concluded in 1899 and there seemed to be no legal problems attached to it, until suddenly in 1905 Freeman was served with a lawsuit relating to his fees as a board member and the fact that, for some unknown reason, the City of Gloucester had not contributed any money that was paid to the commissioners (coming to about $8,000 each). Thence a convoluted legal drama worthy of Dickens transpired over the ensuing years, eventually instigating Freeman to engage the services of his longtime colleague Samuel Elder—they were schoolmates at Lawrence High School—to defend him. In October 1912 a civil trial appeared to be imminent, and at the end of the month, Freeman conferred with Elder and his younger partner regarding the lawsuit.[48] In a letter to Percy Long, Freeman described the meeting: "Very curiously, I ran into the Mr. Whitman of Boston, who has taken such a violent interest in preserving the Hetch Hetchy, and suddenly discovered that he was identical with the partner of my old school mate Samuel Elder, the foremost jury lawyer of Boston, who is just now preparing my defense in the suit of the Gloucester Water Supply Company."[49]

In this remarkable encounter, Freeman gained insight into the intensity of anti-dam opposition, observing, "[F]rom our little talk I fear [Whitman] is fanatical on this particular question and that he is going to Washington to continue the fight against the city. . . . [H]e tells me he has visited the Hetch Hetchy Valley twice and says my report demonstrates so completely what the nature lovers contend for that the hearing ought not to occupy more than half an hour." Although Freeman was not cowed by Whitman, he did not dismiss him as a feckless lightweight and feared that the city faced a battle that would "be no walk-over." In particular, he warned Long, "[Whitman] is something of a wild eyed fanatic on this particular question and is to be reckoned with. I shall try to reason with him once or twice prior to the Washington Meeting, but there is no stopping a Boston reformer when he has at once got started. . . . It is my growing belief that you are going to need the city's heaviest artillery for the final siege."[50]

As the Fisher Hearing approached, Freeman could feel the pressure building, and on Saturday, November 16, he confided to his diary, "Pretty nearly exhausted nervously & insomnia last night." The following Tuesday he breakfasted with Percy Long in New York City, and the next two days were spent in Providence, where he logged twenty-one hours "steadily working on review & preparation for Hearing." On Friday morning he gathered his files and set out for the nation's capital.[51]

Presidential Politics

Before we turn to the Fisher Hearing, it is important to acknowledge that 1912 was a presidential election year, one in which the incumbent, President Taft, was battling against both Theodore Roosevelt as the Progressive Party candidate and Woodrow Wilson as the Democratic nominee. Ostensibly, the Hetch Hetchy controversy was a subject of national interest, and one might think that it could have been featured as a subject of debate. But there is no evidence that the controversy had any impact on the election, even in California. A year hence, when the US Senate was poised to approve a storage reservoir at Hetch Hetchy, the anti-dam coalition would energize a national campaign to oppose the city's plan. But in the fall of 1912, these plans sparked no great public concern, or at least no such concern made its way into presidential politics.

In the spring of 1912, Taft had held off Roosevelt's challenge to win the Republican nomination. It proved a pyrrhic victory, however, once Roosevelt decided to run as the Bull Moose/Progressive Party candidate and split the Republican vote, opening the way for a Democratic win. On election day (November 5), Wilson took 435 electoral votes, Roosevelt 88, and Taft only 8. In California, Taft was not even on the ballot, with Roosevelt beating Wilson by fewer than 200 votes out of more than 600,000 cast.[52] In a national context, the political landscape had been dramatically transformed by the election; by the time Fisher convened his "show cause" hearing later in the month, Taft was a soon-to-be-ex-president who, along with his interior secretary, was but a few months away from leaving office.

The Fisher Hearing Convenes

After leaving Providence on November 22, Freeman spent part of the next day escorting the San Francisco delegation to the White House for a meeting with the recently defeated President Taft.[53] Newspaper reporters accompanied the delegation, with the *Chronicle* noting how the president "laughed and joked with [his] visitors for some time." More directly, the jovial Taft "listened attentively [but] said he could not express an opinion concerning the justice of the claim of San Francisco. . . . [T]he decision, he declared, rested solely with Secretary Fisher."[54] Following this ceremonial session with an amiable lame duck president, Freeman then spent "all day [Sunday] in conferences and preparing for hearing."[55]

On Monday morning, November 25, Freeman took center stage in the secretary's office before Fisher and the Advisory Board of Army Engineers. He was supported by San Francisco's Mayor Rolph, former mayor Phelan, City Attorney Long, City Engineer O'Shaughnessy, City Clerk John Dunnigan, and engineering consultants including J. H. Dockweiler, Allen Hazen, Desmond Fitzgerald, and Harvard professor George Whipple (notably, former city engineers Marsden Manson and C. E. Grunsky were not present). In opposition, the preservationist bloc was well represented, with anti-dam advocates Edmund Whitman, Horace McFarland, William Bade of the Sierra Club, and John Muir's compatriot Robert Underwood Johnson attending at least part of the hearing; Muir, however, stayed in California, unwilling to travel to Washington. In addition, there were attorneys representing the Spring Valley Water Company and the Turlock and Modesto Irrigation Districts.[56]

Whitman may have believed that the hearing would be brief, but his prediction proved woefully off the mark. In fact, Fisher's opening statement by itself took up close to an hour, setting the stage for long interlocutions during which the secretary would cavil and press witnesses for answers to a lengthy stream of questions. It proved to be a long week, broken only by a recess for Thanksgiving on Thursday. And then they were back at it for two more days, until adjournment came on Saturday night.

At the start, Fisher set forth the "two main questions" that he believed should guide the proceedings. The first focused on whether alternative sources available to the city could obviate the need to dam Hetch Hetchy. The second involved the possible impact that use of Hetch Hetchy for a municipal reservoir would have on the rest of the park. Would a reservoir require the closing of the entire upper Tuolumne watershed to campers and tourists? And if camping was allowed, what kinds of restrictions would be placed upon such use?[57]

Snapshot view of the Fisher Hearing in November 1912 that Freeman glued into
his diary. Secretary Fisher is behind the desk at right. The man in left foreground is
believed to be San Francisco's city clerk, John S. Dunnigan, who worked with Freeman
and helped coordinate the city's Hetch Hetchy lobbying efforts. (John Ripley Freeman
Papers [MC 51, box 3], MIT Libraries, Cambridge, Massachusetts)

In regard to the first question, Fisher was not willing to accept Garfield's
assertion that there was no "need to pass upon the claim that [Hetch Hetchy] is
the only practicable and reasonable source of water supply for the City." Fisher
most decidedly wanted to address this claim and quickly began to probe Freeman as to the available alternatives. So that readers might get a direct sense of
how Freeman responded to Fisher's questioning, their interaction is quoted here
at some length:

> SECRETARY FISHER: Have you personally examined the alternative sources
>   of supply?
> FREEMAN: I have, a number of them, not all of them. I am familiar in a
>   general way, with California conditions, and have done a good deal
>   of engineering work in the Sierras during the past six or eight years,
>   so that I have a general knowledge of the Sierra River[s].
> SECRETARY FISHER: Have you examined personally any of the competing
>   sources or alternative suggestions?

MR. FREEMAN: That part of the work was made on the part of Mr. Grunsky. There was more work than could be done by one engineer, and so the work was sub-divided.

SECRETARY FISHER: Have you personally gone over those alternative sources of supply, shown by the reports of Mr. Grunsky and others, so that you feel qualified to express really an engineering opinion about it, as to the feasibility and cost?

FREEMAN: I have. I have gone over the early reports and the recent reports and have discussed them orally with Mr. Manson, the former city engineer, with Mr. Grunsky, the man who made the original investigations for a Sierra supply [in 1902], and with the assistant engineers who were detailed to go over the ground.

SECRETARY FISHER: Are there any of those sources of supply which in your judgment are physically capable of supplying this metropolitan district, ignoring for the present the cost?

FREEMAN: An engineer cannot properly ignore cost in estimating.

SECRETARY FISHER: He can, in answering a question, can he not? (laughter) . . . I have made it very clear that I intended you to ignore the question of cost, so that you would not mislead anybody hereafter. I want first to know what the physical capacity of those sources of supply [is] and their physical availability.[58]

Freeman knew that the efforts of Grunsky and Manson in analyzing alternative sources were less than stellar and vulnerable to criticism. Notably, after the hearing had adjourned, he admitted to Mayor Rolph, "I must confess that Mr. Grunsky's reports on the alternative sources were not so full of detail as they should have been . . . [and] several of these were ground out overnight."[59] In the midst of Fisher's hearing, however, Freeman could not publicly critique any of these studies without calling his entire *Hetch Hetchy Report* into question.

Freeman had long believed that filtering water from the Sacramento River was the most viable alternative to Hetch Hetchy. More problematically—and as Edmund Whitman had previously pointed out to the secretary in his rejoinder to Freeman's report—the Sacramento River study prepared by sanitation engineer Allen Hazen was founded on an assumption that differed from what Freeman proposed in his estimate for the Hetch Hetchy system. Specifically, Freeman had estimated common labor to cost $2.25 per day while Hazen had assumed a daily wage of $3.00. Fisher was aware of this discrepancy and pressed Freeman for an explanation:

SECRETARY FISHER: What is your idea of the total cost of the Hetch
  Hetchy [system]?

FREEMAN: It is given on page 160[o], of my printed report, that the total
  cost of a supply, varying from a maximum of 240,000,000 gallons
  daily to a minimum of 160,000,000 gallons, daily, in most extreme
  drought, would be $36,000,000. . . .

SECRETARY FISHER: Assuming that the equivalent supply from the Sacramento River were furnished, what would that cost? I want to get them
  on a parity; I want to get these estimates on the same basis, so that
  they are comparable.

FREEMAN: To get that on the same basis one must factor into the cost the
  expense of the supply from the Sacramento River. . . .

SECRETARY FISHER: And how does that estimate [on the Sacramento sup-
  ply], as to cost of units, compare with these on the Hetch Hetchy?

FREEMAN: In general, higher. . . .

SECRETARY FISHER: Why so?

FREEMAN: In the first place, [Hazen] took the prevailing wages in the city
  of San Francisco, which is $3.00 a day for common labor.

SECRETARY FISHER: What did you take?

FREEMAN: I took the going rate of wages in the Los Angeles aqueduct . . .
  where they have been paying common labor $2.25 to $2.50 per day
  during the past year.

SECRETARY FISHER: Suppose you reduced Mr. Hazen's estimate, to your
  basis, what would it be?

FREEMAN: I have not figured that through.

SECRETARY FISHER: You see, Mr. Freeman, it does not help us at all to
  have an estimate based on $3.00 for laborers against another esti-
  mate of $2.25, and have you tell us that one cost more than the other.
  Of course, we know that. . . . At the present time we are not much
  interested as to whether the city ought to pay $3.00 or $2.25. We are
  interested in having the same wage scale applied to the two projects,
  which we want to compare.[60]

With this exchange, Fisher had gotten on record how the estimates for Freeman's Hetch Hetchy design and Hazen's Sacramento River supply were not
directly comparable. Next he demanded that Freeman determine how this difference could be reconciled in a revised analysis—"I want them figured on the
same basis."[61]

Although Freeman avoided a cataclysmic disaster as to the efficacy of his report, he was chastened in being forced to accept Fisher's demand for more data relative to cost comparisons. For their part, the preservationists were pleased with how Freeman had been put on the defensive. Hopeful that Freeman's struggles would continue, that night Bade joyously wired Colby in San Francisco, reporting that the city's consulting engineer had been badly "balled up" by the secretary.[62]

The initial questioning of Freeman occurred before lunch, and when the hearing reconvened at 2:30 p.m., he was given a reprieve. Fisher shifted his attention to *Century Magazine* editor Robert Underwood Johnson, explaining, "[B]ecause of an unexpected event Mr. Johnson is called back to New York this afternoon. . . . [I shall] ask him to make his statement at this time."[63] Fisher gave all hearing participants ample time to express themselves, and Johnson soon intoned a rambling, self-congratulatory introduction that eventually reminded everyone of his role in the origins of Yosemite National Park. Johnson lectured Fisher on his view of the park's history, the supposed invalidity of the 1901 Right-of-Way Act, and eventually the deceit of San Francisco's claim that no other water supply source was available. On this latter issue, Johnson asserted, "[Unless] they have candidly and impartially and fairly investigated all the other sources of supply, we must consider that they have not made their case."[64]

Up to that point, Johnson had expressed what preservationists long believed to be true: Freeman's *Hetch Hetchy Report* notwithstanding, the city had yet to demonstrate that no other viable sources were available. Fisher then pushed Johnson to be more explicit as to what might constitute a reasonable alternative:

> SECRETARY FISHER: Let us see what you mean by that. Do you mean that if there is a source of supply other than the Hetch Hetchy which it is possible to use, irrespective of cost [or] quality of water, that that should settle the controversy?
>
> MR. JOHNSON: I do, because I think San Francisco in its proposed use of the [Valley] is sacrificing more in dollars and cents than the cost of the Hetch Hetchy system. . . .
>
> SECRETARY FISHER: Let us see if you mean just what you have stated. We could, of course, distill the water of the Pacific Ocean for a water supply. That is obviously true. Would you say that this ought to settle this question?
>
> MR. JOHNSON: If, after you have exhausted the other sources of supply from the Sierras, you find that no one of these sources would be

adequate and you do know that by distilling water from the ocean you could supply that, I should say undoubtedly it was the duty of the authorities to do that.

SECRETARY FISHER: Rather than take Hetch Hetchy water?

MR. JOHNSON: Certainly.

SECRETARY FISHER: Even though it might cost—to make it sufficiently absurd, we will say $100,000,000 a year?

MR. JOHNSON: It would be well invested.

SECRETARY FISHER: I wanted to see just how far you would go.[65]

Over the course of five days, the hearing transcript would encompass many thousands of words, many less than memorable. For Freeman, this brief interchange would be the one he most wished to carry to the outside world. But for other preservationists, Johnson's willingness to consign the citizens of San Francisco to a prohibitively costly water supply dependent upon distilled seawater constituted a worrisome burden. As Bade lamented in a letter to Colby, "Johnson (R.U.) did not help us much" at the hearing.[66] Other dam opponents were not as strident as the *Century* editor, but Johnson's status as self-proclaimed "originator" of Yosemite National Park and intimate of John Muir made it impossible for his provocative pronouncements on Hetch Hetchy to pass unnoticed.[67] Freeman would make sure of that.

Fisher's "second question" for the hearing concerned how a possible reservoir—and regulations that might accompany its use—would impact use of park land in the upper Tuolumne watershed. Many preservationists had long insisted that San Francisco's inundation of Hetch Hetchy for a municipal water supply would inevitably require draconian restrictions on the entire 459 square miles of park land tributary to the dam. Freeman derided this notion in his report, but preservationists remained adamant. To bolster its case, the city arranged for a cadre of renowned sanitation engineers to appear at the hearing, including Allen Hazen, Desmond Fitzgerald, and George Whipple. All three were dismissive of claims that deployment of the reservoir would have a dramatic impact on enjoyment of the northern reaches of Yosemite National Park; they also believed that water stored in the reservoir would not require filtration prior to human consumption. Of the three, Whipple was the most succinct and direct in his testimony. He started with a brief review of his relevant experience:

MR. WHIPPLE: I had charge of the laboratory for water analysis in Boston, and in New York my connection was practically the same. I was eight years in Boston; six years in New York. Since then I have been in private practice and last year I was appointed Professor of Sanitation Engineering at Harvard University.

SECRETARY FISHER: Can you qualify as an engineer?

MR. WHIPPLE: I am a graduate engineer . . . [knowledgeable in] chemistry and bacteriology and allied sciences. . . .

SECRETARY FISHER: Tell us what your view is as to the requirements that would be insisted upon in the practical sanitation of this watershed if the [Hetch Hetchy] valley were used by the city?

MR. WHIPPLE: My idea is that the requirement would be just about the same, whether the water supply for San Francisco is taken or not. . . . Contamination in a running stream would be the dangerous feature[:] . . . pollution going to a stream and then a reservoir would reach San Francisco after many weeks or months time during which the dangerous bacteria would die.

SECRETARY FISHER: Well; how long does it take typhoid fever germs going down to a reservoir to be destroyed?

MR. WHIPPLE: I cannot say in round numbers to be exactly. I can give you some idea however. In about ten days about ninety per cent of any typhoid germs discharged in water would die. During the next ten days ninety per cent of what is left would die, and during the third ten [days] ninety per cent of what is left [would die], so that after [a] period of a month or two the number of germs would be reduced to an infinitesimal number. Therefore water stored two months [in a reservoir] may be considered substantially free of danger. Experience has shown and science has demonstrated it in recent years.[68]

A few minutes later, Fisher followed up on the need to regulate camping in the watershed above Hetch Hetchy.

SECRETARY FISHER: Do you think it would be necessary to impose any regulations as far as the city is concerned?

MR. WHIPPLE: So far as the city is concerned, I do not believe any regulations would be necessary except sentimental ones. I would not permit campers bathing in the reservoir itself.

SECRETARY FISHER: Would you put any other restrictions whatever on the use of water on the sanitary conduct of the campers?

MR. WHIPPLE: So far as San Francisco is concerned I would not, but I
    think it necessary to provide restrictions for the disposal of fecal
    matter and for the direct discharge [of] fecal matter and urine into
    the streams themselves. I think that would be necessary in any case,
    whether San Francisco used the water or not.[69]

Preservationists struggled to undermine Whipple's position, drawing upon
Seattle's and Portland's watershed restrictions as a portent of what San Francisco would ultimately demand. But Whipple held firm: "I see no reason for
such a ruling, unless that water is used directly from the stream. Recent pollution is dangerous pollution. If you use water from a running stream, you have
to protect it. If you have storage, you do not have to protect it. A long storage is
practically as efficient as filtration. In San Francisco in my judgment, they will
practically never deem it necessary to filter [Hetch Hetchy water] in any material way, or restrict the use of this [upper Tuolumne] water shed."[70]

Preservationists never stopped raising the specter that San Francisco's use of
Hetch Hetchy for a reservoir would seriously impede, if not forbid, camping or
recreation in the upper Tuolumne watershed. But the basic arguments espoused
by Whipple as to why future sanitary regulations would not require any onerous
sacrifices on the part of visitors to Yosemite National Park have held force for
the past one hundred years.

The Hearing Goes On . . .

Many more interchanges from the hearing transcript could be quoted, but by
now readers should have a sense of how Fisher administered his inquiry. For
five days all sides of the controversy were given a chance to make their views
known and engage with Freeman and San Francisco officials. If Fisher was to
err in his handling of the hearing, it would not be because he cut people off or
limited their expression. For example, he let Edmund Whitman unwind a verbose legal argument positing that the principle of res judicata required Fisher
to accept Interior Secretary Hitchcock's original rejection of the city's application; not surprisingly, Fisher demurred on this request.[71] Whether the testimony
offered at the hearing was always on point or helpful did not trouble Fisher. For
him, more was better than less.

As represented by attorney Edward McCutcheon, the Spring Valley Water
Company accused Freeman of significantly underestimating the future supply

capacity of its system. But William Mulholland and J. B. Lippincott—whose report projected a great subsurface supply in the Livermore Valley gravels—were not present, and no great fight between the Los Angeles engineers and Freeman ensued.[72] Objections were raised, testimony heard, and Fisher and the Army Board took it all under advisement.[73] The issue of how revision of the Garfield Permit might impact the projected sale of Spring Valley's system to the city also sparked about two hours of legal jousting between McCutcheon and City Attorney Long, but nothing was resolved and Fisher eventually moved on.[74]

Attorneys for the Turlock and Modesto Irrigation Districts were also allowed to argue that all of the water in the Tuolumne should be reserved for irrigating land in the San Joaquin Valley. In response, Freeman and the city asserted that the districts' claim to 2,350 cubic feet per second of Tuolumne flow was more than sufficient to meet all the districts' irrigation needs. Fisher and the Army Board took it all in, listening to various claims and counterclaims, but made no determination as to the merits of the districts' position.[75]

During the hearing's second day, Fisher made a point of spending at least some time on each of the various alternative sources. Freeman was generally successful in his defense of the need to dam Hetch Hetchy. However, he failed to convince either Fisher or the Army Board that the city could dismiss the McCloud River as a viable alternative. Lying more than two hundred miles north of San Francisco, the McCloud arises from the southeastern slopes of Mount Shasta, flowing into the Pit River (and thence the Sacramento River above Redding). In his report, Freeman gave little attention to the McCloud, largely because it would require an aqueduct much longer than the proposed Hetch Hetchy Aqueduct. In addition, the McCloud aqueduct would have to pass under the Sacramento River at the Carquinez Strait and connect to San Francisco via a lengthy pipeline lying beneath the deepest part of San Francisco Bay. There also appeared to be no good reservoir sites in the McCloud watershed above the aqueduct intake, and there was no hydroelectric power potential attached to a McCloud Aqueduct. For Freeman, the McCloud River held no allure.[76]

Despite seeming problems with the McCloud as a viable source, prior to the hearing a proposal was presented by R. P. Doak, president of the Mount Shasta Aqueduct Corporation. The pitch was not new—Freeman's *Hetch Hetchy Report* described a version proposed to the city in 1911—but Doak reiterated the company's claim of holding water rights on the McCloud capable of supplying the city with 400 mgd and averred that it was willing to sell/transfer these

During the "show cause" hearings in November 1912, Secretary Fisher and the Army Board berated Freeman for not devoting enough attention to the McCloud River as an alternative to damming Hetch Hetchy. Despite Freeman's objections that the lack of a storage reservoir site, the presence of ongoing logging operations, and the river's great distance from San Francisco obviated its viability as a municipal water supply, Fisher insisted that the city provide more information on a possible McCloud River aqueduct. This circa 1920 photograph shows the McCloud River Lumber Company's logging pond with Mount Shasta looming in the distance. Author's collection.

rights.[77] Freeman and city officials had no interest in Doak's offer, but members of the Army Board were intrigued. In contrast to all the other mountain streams proposed as alternatives, the McCloud was the only one that the Army Board believed might match the scale of Freeman's Hetch Hetchy proposal. At the hearing, Colonel Spencer Cosby of the Army Board questioned Freeman regarding the McCloud:

COL. COSBY: [Do] you think then [Mr. Freeman] that the city has given
     [the McCloud River] the full consideration necessary in order to . . .
     enable us to [compare] that source of supply with the Hetch Hetchy?
FREEMAN: I believe it has. I think it has gone far enough to show that
     there is a preponderating advantage in favor of the Hetch Hetchy. . . .
COL. COSBY: What we are anxious about is to get a comparison of the
     McCloud with the Hetch Hetchy, delivering four hundred million
     gallons [per day]. We have no statement as to what it would be. . . .
     [W]e want to compare the costs of the alternative sources.
SECRETARY FISHER: That is right. That is what we want.[78]

Later in the hearing, the McCloud alternative was further parsed, and Fisher
chastised Freeman's willingness to dismiss its viability without careful study:
"I have a very great deal of confidence in your judgment, Mr. Freeman . . . but
what I fail to see yet is the proof of your having given the McCloud source
and the possibilities of developing it that consideration to which it is entitled.[79]
Freeman had not anticipated that Fisher and the Army Board would become so
enamored of the McCloud River. But the river loomed large in their minds, and
they expected the city to supply more data before they could acknowledge its
limitations.

In publicly defending the city's assessment of alternative water sources,
Freeman put on a brave face. However, no matter what he might say, he knew
that the work undertaken by Grunsky (and Manson) was less than stellar. Of
course, this did not mean that his Hetch Hetchy plans, and the water power that
they could develop, were a scam or an illusion. But Fisher and the Army Board
were hardly unreasonable in pressing him for more, and clearer, data to justify
the city's case.

An Inconclusive Conclusion

Late on Saturday afternoon, Fisher stepped back and began to consider where
things stood after more than thirty-five hours of testimony. Responding to closing comments made by San Francisco's Mayor Rolph, Fisher ruminated as to
his own state of mind: "I have not made up my mind upon either of the main
propositions which I stated at the beginning. . . . There are very important links
in the chain of evidence . . . which to my mind have not yet been furnished."[80]

As he chattered on, Fisher reflected on the suitability of the 1901 Right-of-Way Act as a foundation for a municipal water supply: "The act under which this permit [may be] granted is a most unsatisfactory act . . . [with its] provision that the grants made under it can be revoked at any time by the authority granting them. That provision is unsound and uneconomic in every respect."[81] Harkening back to issues earlier raised by Whitman, Fisher questioned the overall value of the Garfield Permit and, in a remarkable admission, bluntly advised city officials on what they should do to best defend their future interest in Hetch Hetchy: "You ask me what I would advise the city to do. I would advise the city to get an act of Congress. I mean, whether I grant the [revised] permit or whether I do not."[82]

Given this ambivalent bit of advice, Freeman's dismay over the additional work that Fisher and the Army Board expected him and the city to undertake is not surprising. On this point he complained to O'Shaughnessy about the "[i]njustice of putting so great a burden on [San Francisco]" and added, "[W]hile I have the utmost respect for the personality of the Army Board, I feel that they have not studied faithfully the data presented or they would not impose this additional burden."[83] Nonetheless, as the hearing sputtered to an indecisive close, Freeman knew that little good could come if he or the city went public with their frustrations. So he later counseled O'Shaughnessy, "[K]eep cheerful and [we'll] make the very best presentation of costs and comparisons possible."[84]

Freeman and the city might have suffered some embarrassment at the Fisher Hearing, but they had not buckled in the face of arguments from preservationists, Spring Valley, irrigationists, and interests tied to the McCloud River. In addition, Freeman had gotten Robert Underwood Johnson to stake out an extreme position vis-à-vis the city's obligation to accept any alternative to a Hetch Hetchy supply. He recognized the public relations value of Johnson's posturing, and when he published a description of the hearing in *Engineering News,* he sought to ensure that Johnson's declaration would not soon be forgotten: "The 'Nature Lovers' were the [city's] most persistent opponents. They claimed that none of this land [Hetch Hetchy Valley] dedicated to the whole American people should be diverted so long as another source was available. Robert Underwood Johnson of New York City, went so far as to say that rather than give the city the rights asked for it should be compelled to distill water from the Pacific Ocean."[85] While the hearing brought Freeman unwanted headaches in terms of demonstrating "parity," Johnson's testimony came as a welcome gift, one that might offer future dividends for dam proponents.

A New Dress

Once the Fisher Hearing adjourned on Saturday evening, November 30, Freeman conferred with San Francisco officials for a few hours before catching the
night train to New York City. After breakfast the next morning he set out into
Manhattan for "an hour photographing monumental buildings as examples for
New Technology." Only then did he head home.[86]

Freeman had promised Fisher that he would rectify problems with "parity" affecting cost estimates for various alternative water sources and present a
report prior to Christmas. However, in early December he was focused on his
"New Tech" plans for MIT. So he assigned assistant engineer Horace Ropes to
draft the Hetch Hetchy addendum for submittal to Fisher and the Army Board.
In California, O'Shaughnessy and his staff were to handle surveys and estimates for the McCloud River aqueduct.[87]

During the next two months, Freeman split most of his professional time
between Hetch Hetchy and "New Tech," but he also found time to attend the
American Society of Mechanical Engineers annual conference in New York
City, join in Board of Water Supply meetings, review plans for PG&E's Spaulding Dam, visit Niagara Falls in anticipation of a new hydropower plant, and
handle insurance-related lobbying in both Washington, DC, and Columbus,
Ohio. San Francisco's affairs remained a priority, but they did not eclipse other
obligations that crowded onto his schedule. For Freeman, "New Tech" was of
special import, and with a cadre of assistants and draftsmen, he soon prepared
plans for what he called his "Study No. 7."[88] On December 11, he traveled to
Boston and "attended Corporation Meeting & presented plans for New Technology Study No. 7."[89] Three days later he was back, meeting with the school's
executive committee.[90] All this attention to MIT, however, distracted him from
Hetch Hetchy, and as Fisher's pre-Christmas deadline neared, Freeman became
concerned about Ropes's progress.

On December 17 he met with his assistant in New York City, becoming
"scared at [the] slow progress"; he quickly "returned to Providence to start
[himself] on rival estimates."[91] On Sunday, December 22, the rush of work
prompted a revealing diary entry: "Ropes all in and has been no good on the job.
My report would have been worthless had I depended upon him. . . . So practically all the report had to be worked out by me in the past four days."[92] Although
the report on parity was not in great shape, on December 23 Freeman traveled
to Washington and handed it over to Fisher. Their meeting was cordial, but
Freeman soon advised O'Shaughnessy that he was "troubled over Secretary's

vacillating attitude."[93] After a quick review, the Army Board determined that Freeman's submittal was insufficient. Specifically, they complained that "while the costs for the different projects are scattered through [Freeman's December 23 report,] there is no table of consolidation which would readily show the relative costs of the various projects."[94] When Freeman got word that the Army Board was unsatisfied, he protested to O'Shaughnessy, "[I] am so exasperated by this latest communication. . . . I will see what I can do in the further effort to satisfy this insatiate bunch . . . [who] seem to enjoy making the most of their present 'strangle hold.' "[95]

A few days later he made a similar complaint to Percy Long: "I can't believe that either the Army Engineers or the Secretary are really suffering for lack of additional tabulation. . . . I am personally disposed to regard them unreasonable in the details for which they ask."[96] He followed this with a churlish description of the Army Board and (to him) their petty demands: "I believe that as between teaching a kindergarten class in a Sunday School and teaching a class of Army Engineers in municipal water supply, the former job is far preferable[;] . . . still, I suppose it is the rule that we should utter no protest and try to look pleasant."[97]

As much as Freeman was irritated by the Army Board's insistence on more data, he implicitly understood that this was largely an instance of a regulatory body making a supplicant dance to its tune, thus demonstrating that its support could not be taken for granted. So, when Freeman responded to Fisher on January 7 he held back and offered a benign excuse for the ongoing delay: "The matter which has suffered delay because of the illness of Mr. Ropes and myself, is chiefly that of a re-arrangement and presentation in tabular form with discounts at compound interest to present value, and not a matter of original presentation of data. I trust no real harm to the city's case has been done. I regret any inconvenience that may have been caused by the delay of the table."[98]

It took a few more weeks to complete the promised "tabular" comparison, and on January 23 Freeman personally delivered copies to Fisher and the Army Board.[99] Significantly, the board did not require that every one of the possible alternatives be analyzed; they requested data only on the alternatives that they believed could reasonably supply at least 400 mgd upon full development— these included the Hetch Hetchy/Lake Eleanor/Cherry Creek system, the Sacramento River, the McCloud River, a combination of Lake Eleanor/Cherry Creek with the Mokelumne and Stanislaus Rivers, and a combination of the Mokelumne with the American, Cosumnes, and Stanislaus Rivers. In his comparisons, Freeman relied upon similar wage rates for labor and calculated all the estimates in terms of a "present value" pegged to 1920, using an annual interest rate of 4.5 percent.

By the end of January, Freeman had brought together all the "parity" investigations, reports, and tables related to Hetch Hetchy that had been drafted since November and, bundling them under a "Letter of Transmittal," forwarded them to Fisher. Not surprisingly, his belief in the superiority of the Hetch Hetchy–based system had not diminished. Perhaps a bit smugly, he advised the secretary, "With the exception of the new data on the McCloud project [which O'Shaughnessy projected would cost twice as much as Freeman's Hetch Hetchy plan] . . . I find but little of importance brought forward in these papers submitted since the hearing in Washington. They appear to me largely in the nature of replies to rejoinders to rebuttals and largely comprise the re-thrashing of old straw and the putting of facts previously known into a new dress."[100]

Yes, Freeman had jumped through the hoops of "parity" demanded by Fisher, but the fundamental character of the city's plan remained unchanged. No doubt there were alternatives to Hetch Hetchy, but by his calculations they would all cost more and none could promise the same amount of hydroelectric power. Same as it ever was, but this time in a new dress.

Rejected: Study No. 7

During January, and while assembling his final submittal to Fisher, Freeman continued to devote energy to MIT's "New Tech." In mid-December he had gotten seemingly positive feedback from the executive committee, but just before the New Year, President Maclaurin informed him that "the [anonymous] giver of the 2 ½ millions [is] anxious that responsibility be put on some architect with a national reputation."[101] Maclaurin intimated that the project would require the leadership of an experienced architect, and after meeting with the executive committee on January 6, Freeman feared, "I am being crowded out after having done most of the preliminary work."[102]

At this point, a less driven personality would have realized the futility of pushing ahead, but Freeman would not concede. A week and a half later he attended a meeting of MIT alumni in New York City at which he "[t]alked for about 1 hour on Bldgs. for the New Technology to audience of about 400." Later that evening he attended a "Smoker" where he cajoled his fellow alums to support him in his quest to take charge of the new campus.[103] In early February he wired Maclaurin and offered "to furnish all services of design and supervision regularly covered by architect's commission for not exceeding one half the regular architectural commission and possibly only one third—and will include employment of competent consulting architects upon decorative and artistic features."[104] It was a generous proposition, but he heard nothing for more than a week. Then, in a letter Freeman received on February 17, Maclaurin sent word that despite the committee being "[u]nanimous in its appreciation of the fine spirit displayed in your action in this matter . . . [they] were decidedly of the opinion that the institute ought not proceed in the way that you suggested. It did not seem . . . that the circumstances justified so wide a departure from accepted methods." To fend off further entreaties, Maclaurin informed Freeman that New York architect William Bosworth had been hired to take charge of the design. The decision was final.[105]

Loss of the New Tech design commission represented a very public defeat, at least among MIT alumni and the New England engineering community, and Freeman was not used to suffering such an embarrassing professional failure. With the Army Board's decision on Hetch Hetchy hanging in the balance, the Providence engineer was ill prepared to experience another disappointment so soon after the MIT debacle. The day after he received the news from Maclaurin, he headed off to Washington for one last round of lobbying on behalf of his plan for San Francisco.

The Army Board Report

On Wednesday, February 19, Freeman "called on Col. Taylor at War Dept," learning that the board's report had been delivered to Fisher earlier that day. Although denied access to the report, Freeman told Percy Long, "[I]n the course of our conversation [Col. Taylor] expressed himself so freely regarding the absurd position and extreme views of the Nature Lovers that it made me very hopeful."[106] He then visited the secretary and had "a ¾ hour interview in which," as he explained, "I pleaded hardest I knew how for city." During their meeting, it became apparent that Fisher felt overwhelmed by the work on his desk; Freeman noted, "[Fisher] says almost a physical impossibility for him . . . to find time to properly consider this [Army report]."[107] Unsure of what the secretary might do, the next day Freeman "conferred with Congressman [William] Kent . . . about visiting president [Taft] to urge [on] him the importance of finishing the Hetchy case." But Freeman made no attempt to visit the White House before heading back to Providence.[108]

Early that weekend, the Army Board's report became public, and by Sunday, Freeman began to receive "congratulatory telegrams on Hetchy Army Board decision." Percy Long wired, "[The] Army Board indicates complete approval of your views in Hetch Hetchy matter. . . . [Please] accept my sincere congratulations on your victory," and Mayor Rolph expressed his "jubilation."[109] The city had good reason to be pleased because, in all essential matters, the board's report embraced and endorsed Freeman's plan. Most importantly, in their report the army officers acknowledged, "Sooner or later . . . the demand for the use of Hetch Hetchy as a reservoir [will prove] practically irresistible. The board does not think that a delay of a few years in transforming the Hetch Hetchy Valley into a reservoir is of importance, and therefore does not think it necessary to require delaying construction of the [Hetch Hetchy] reservoir until the Lake Eleanor and Cherry [Creek] sources have been fully developed."[110]

In a huge victory for the city, the board thus dispatched one of the key stipulations embedded in the Garfield Permit. As far as the officers were concerned, the damming of Hetch Hetchy need not await the construction of large-scale dams on Eleanor and Cherry Creeks. As Freeman had long argued, and the Army Board now acknowledged, the most efficient development of the upper Tuolumne could occur by first building a large reservoir at Hetch Hetchy.

Of almost equal importance, the board sided with the city in rejecting the preservationist view that, if a dam were built, the northern reaches of Yosemite National Park would be forever blocked to campers and other visitors: "The board believes that the regulations proposed by the city will be found sufficient to protect the waters from pollution, and that these regulations will tend toward the protection of campers and others using the park and will not be onerous upon them." And the board rejected objections raised by irrigation interests in the San Joaquin Valley, believing "that there will be sufficient water if adequately stored and economically used to supply both the reasonable demands of the bay communities and the reasonable needs of the Turlock-Modesto Irrigation District for the remainder of the century."[111]

The army officers did not accept as gospel everything that Freeman advocated or professed. For example, in judging the future capacity of the Spring Valley water system, they refused to endorse either Freeman's estimate of about 78 mgd or the company's estimate of more than 200 mgd, opting instead for a compromise figure of at least 131 mgd. But the board agreed that greater San Francisco would, by the year 2000, need an additional supply of 400 mgd to support a population of about 3.6 million and that local supplies in the Bay Area were insufficient to meet such demands. Thus, all the arguments relative to the capacity of the Livermore Valley gravels or the Alameda watershed were rendered moot, or at least devalued as having any determinative effect on the city's need for a mountain water supply.[112]

What unequivocally tipped the balance in favor of Hetch Hetchy was the huge amount of hydroelectric power that could be generated by Freeman's gravity-flow aqueduct. Specifically, the Army Board ascribed a valuation of $45 million to the hydropower potential of Freeman's scheme.[113] Other sources could be developed such that the long-term water supply needs of San Francisco might be met. But no other alternatives could supply large quantities of water *and* provide as much hydroelectric power for the city. In the words of the Army Board, "The Hetch Hetchy project is about $20,000,000 cheaper than any other feasible project for furnishing an adequate [water] supply. The only exception is the filtered Sacramento [River] project, which in actual cost is about thirty millions greater than Hetch Hetchy project, but by discounting to 1914 becomes only $13,000,000 greater. The Hetch Hetchy project has the additional advantage of permitting the development of a greater amount of water power than any other."[114]

For Freeman, the Army Board report was as good as he could have hoped. A few weeks earlier he had complained privately that the board was an "insatiate bunch," dictating "unreasonable" demands.[115] But he had kept his anger in check, avoiding any outbursts that might disadvantage the city's cause. In the end, his restraint was rewarded. The board members endorsed every essential element of Freeman's ambitious proposal to provide "something better for the city." He could not have asked for more.

Fisher Balks

The Army Board's report in hand, Fisher faced a decision on the Garfield Permit and the issue of "show cause." With Woodrow Wilson and his new administration set to take power on March 4, no further delay or equivocation was possible. After all the *Sturm und Drang* that had transpired since Ballinger first issued his "show cause" order in February 1910, and after Fisher had caviled and parsed for five arduous days in November 1912, the secretary balked.

On February 26 Freeman returned to Washington and was told by Fisher that "issuance of [the] permit [was] physically impossible before March fourth," whereupon Freeman "begged him to shape matters for congressional action early in special session."[116] Two days later, Freeman, Long, and O'Shaughnessy met with Fisher, whereupon the secretary confirmed that he "could not finish the permit."[117] Freeman was incredulous at this news but nonetheless upbeat given the Army Board's endorsement. In a telegram to Mayor Rolph describing the meeting, he signed off, "[E]verything satisfactory except Fisher's intensely logical, quibbling, indecisive brain."[118] The next day Fisher made his (non)decision official, advising Mayor Rolph, "I have reached the conclusion that a permit for this purpose [damming Hetch Hetchy] should not be issued by the Secretary of the Interior under the existing law . . . [because] the only statutory authority under which such a permit could be issued is the Act of February 15, 1901 . . . [which provides that] these permits 'may be revoked by him or his successor, in his discretion.'" For Fisher, this feature of the 1901 law was seriously flawed and "most unsatisfactory as a basis for the important administrative actions that can be taken under it." As a consequence, he refused to take any action on the Garfield Permit and held the city's application in abeyance until "the application can be made by the city to Congress for such action as Congress may deem proper."[119]

Fisher's final decree before stepping off the national stage offered no resolution to the controversy, proclaiming that the city's use of Hetch Hetchy as a reservoir could not reasonably proceed until and unless a new federal law was enacted specifically granting such rights. Technically, the Garfield Permit and Ballinger's "show cause" dictate remained in place, but in truth they had been consigned to the dustbin of history. The fate of Hetch Hetchy would be left to a new Congress and a new president.[120]

Given the anticlimactic conclusion to the Garfield Permit/"show cause" saga, it may seem as though it had all been a big waste of time. Presumably Secretary Ballinger could have simply revoked the permit in February 1910 and then the city, freed of the permit's constraints, could have engaged with Congress and—just as it was to do in the spring of 1913—set out to get a law passed that was completely distinct from the 1901 Right-of-Way Act.

But how likely was it that the city could have prevailed in such an attempt? Not likely at all given that, when in 1908–9 the city sought to obtain congressional authorization to take control over public land to be inundated by the Hetch Hetchy reservoir, it failed miserably to overcome opposition brought by John Muir and his anti-dam cohorts. This opposition largely centered around arguments that the city had failed to demonstrate a compelling need to dam Hetch Hetchy and that there were viable alternatives that the city could develop.

So here we can see why the city's effort to comply with Ballinger's "show cause" order was so important in overcoming the anti-dam arguments that held sway in Congress in 1908–9. Bear in mind it was not City Engineer Manson's effort to "show cause" that proved of any importance but rather the work of Freeman, who, upon taking charge in May 1912, made an expansive case that advanced the city's cause.

What transformed the city's Hetch Hetchy initiative was Freeman's insistence that the Garfield Permit was woefully inadequate and that only an aqueduct with an ultimate capacity of 400 mgd could offer "something better for the city." Freeman had always wanted to move beyond the limitations of the Garfield Permit, but he and the city recognized that that would only be possible if, after Freeman completed his 421-page report, they confronted the "show cause" demand head-on and defended their case before the interior secretary and the Army Board, because only by seeing the "Show Cause" process through could there be a report issued by the Army Board capable of independently affirming the city's justifications for damming Hetch Hetchy. With the Army Board officially supporting Freeman's plan, and the Garfield Permit brushed aside, the city could build upon this success and look to Congress for new legislation authorizing a Yosemite water supply.

CHAPTER 6

The Raker Bill in the House

After a weekend in Providence, Freeman returned to the nation's capital on Monday, March 3, the day before Woodrow Wilson's inauguration. Freeman's political connections were largely tied to the Republican Party, and the ascendance of a Democrat to the presidency (the first since Grover Cleveland in 1893) was, in the abstract, unlikely to attract his attention. However, the issue of Hetch Hetchy and how Fisher's successor might impact the city's plans was enough to draw Freeman to Washington. By midday the announcement came: Franklin K. Lane—at that time, chairman of the Interstate Commerce Commission and former city attorney for San Francisco—would be Wilson's secretary of the interior. This was welcome news for city officials, with Freeman joyfully recording in his diary, "At noon learned that F. K. Lane was to be the new Secretary of Interior and greatly rejoiced."[1] In a letter the next day, he offered context for the city's elation: "We feel jubilant . . . because Franklin K. Lane, the new Secretary of the Interior, was the former [San Francisco] City Attorney, under whom the Hetch Hetchy proceedings were begun. He knows the situation thoroly [sic] and will without doubt complete the good work begun by the Army Board."[2]

Indeed, Lane had been city attorney in 1902 when San Francisco mayor James Phelan first filed a Right-of-Way Act application to dam Hetch Hetchy.[3] With Interior Secretary Ethan Hitchcock unwilling to approve the permit, Lane visited Washington in April 1903 to plead anew the city's case. Hitchcock was unmoved, and Lane advised a friend, "I stayed three or four days in Washington, where I found the Department of the Interior pretty well stacked against me."[4] An active member of California's Democratic Party as early as the 1890s (he campaigned for San Francisco's new city charter, instituted in 1900), Lane had ambitions extending beyond the post of city attorney. However, he fell short in runs for governor of California (1902), San Francisco mayor (1903), and US senator (1904). Despite these defeats, he was an able advocate of progressive causes and, during his 1903 trip to Washington, had caught the eye of Teddy Roosevelt. While in the nation's capital, Lane "saw the President twice and lunched with him."[5] Two years later the connection bore fruit when Roosevelt appointed him to the Interstate Commerce Commission (ICC). From 1906 through early 1913, Lane served on this prominent and influential commission, becoming its chair in January 1913. Soon after, Wilson tapped him for interior secretary.[6]

Although Lane's appointment would have a significant impact on the battle over Hetch Hetchy, San Francisco's plans to build a reservoir in Yosemite National Park appeared to have little bearing on his selection to head the Interior Department. In describing the vetting process for prospective cabinet members, Wilson's political strategist Colonel Edward House later revealed that, soon after Wilson's election victory, Lane demurred from cabinet consideration. But following conferences with House in January, the future secretary began to rethink his position. At first, he was favored by the president for secretary of war. However, after Cleveland mayor Norman Baker turned down an offer to be interior secretary, Lane came to the fore as the top candidate for the post. In recounting the politically complex, if not chaotic, process followed by Wilson in creating his cabinet, House makes no mention that Hetch Hetchy played any role in the appointment.[7]

For the city, Lane's appointment was a serendipitous windfall. Within days of becoming secretary in early March, he conferred with Percy Long in the company of Richard Watrous of the American Civic Association and made his position on Hetch Hetchy clear. Following the meeting, Watrous raised the alarm among the preservationist community, warning that with Lane they would be "up against a staunch advocate of San Francisco's [plans] . . . and that means a prodigious fight to get Congress to decline the permit."[8] Of course, the issue to be considered by Congress was not a revocable "permit" akin to what could be authorized under the 1901 Right-of-Way Act but free-standing legislation authorizing construction of the Hetch Hetchy project. Watrous, however, correctly perceived that Lane would project a different attitude than his predecessor, Walter Fisher. In relation to Hetch Hetchy, Lane would be a political exponent, not a vacillating adjudicator.

San Francisco's position on Hetch Hetchy was strengthened not simply because a friendly administrator had taken charge of the Interior Department.

Of equal note, both houses of Congress and the presidency had come to be controlled by the Democratic Party. As a result of the schism brought by Teddy Roosevelt's Progressive Party challenge to Taft, the Republicans lost control of the Senate for the first time since 1895; for most of the 63rd Congress, the party breakdown in the Senate would be 53 Democrats, 42 Republicans, and 1 Progressive. In addition, the Republican minority in the House of Representatives shrank dramatically with the 1912 elections, to the point that Democrats held the majority with 291 members, more than double the Republicans' 134 (there were also 9 Progressives and 1 Independent). During the first two years of the Wilson administration, the Democratic Party was positioned to control federal lawmaking with limited concern for Republican interests.[9]

The two key elements of the Democratic agenda for 1913 were to establish a federal income tax (and thus lower tariffs) and overhaul the nation's banking system and establish what became the Federal Reserve system. To accomplish these objectives, Wilson called for a legislative session to start in the spring of 1913 and extend to the fall. A summer in Washington was not something that Congress looked forward to, but a desire to push through the revenue and banking bills gave impetus to Wilson's initiative.[10]

Hetch Hetchy had not been a national issue in the 1912 presidential campaign, and there was seemingly little reason for it to receive special congressional consideration. However, San Francisco officials perceived that, with support from Secretary Lane and key congressmen, they might get their Hetch Hetchy legislation onto the docket of the House Committee on the Public Lands and, with luck and Senate support, get a bill favorable to the city enacted by the end of the year. To justify action on the Hetch Hetchy initiative, claims were trumpeted that San Francisco suffered from an impending water crisis and needed immediate legislative relief. The existence of a compelling water emergency was, of course, overstated; it would likely take a minimum of five years before any water from the Sierra could reach the Bay Area, and in that context, a delay of a few months would be of minor consequence. But the Democratic Party controlled the legislative agenda, and if the Democratic leadership said there was an emergency, then there was an emergency. Using the levers of power politics, San Francisco pushed the Hetch Hetchy project forward on Capitol Hill.

When the 63rd Congress convened on April 7, Congressman John Raker of California (whose district included Hetch Hetchy) immediately submitted a bill (H.R. 112) calling for federal approval of the city's planned Yosemite water supply.[11] Serious consideration of what would become known as the Raker Act would not be taken up by the Public Lands Committee until June. But mere

President Woodrow Wilson addressing a joint session of the 63rd US Congress in April 1913. These were the congressmen and senators who, later that year, would vote on the fate of Hetch Hetchy. Author's collection.

weeks after Secretary Fisher had postulated the necessity of direct congressional action, federal legislation on Hetch Hetchy was in the pipeline.

A Good Vacation

Immediately after learning of Lane's appointment on March 3, Freeman returned home. Over the next several weeks he devoted little attention to Hetch Hetchy, focusing instead on other consulting projects. He took trips to Keokuk Dam and to Chicago (where he became engaged in litigation regarding water diversions from Lake Michigan) and also spent time in New York City with the Board of Water Supply.[12] Eclipsing everything else on his docket, Freeman's big new project for the spring was a large hydraulic-fill earthen design for the Great Western Power Company's Big Meadows Dam in northern California.

The previous fall he had convinced Great Western's corporate leadership to abandon work on John Eastwood's multiple-arch dam and replace it with a massive gravity design (this story is told in *Building the Ultimate Dam*).[13] His next step was to schedule a trip to Big Meadows to examine the site in detail and devise an earth dam design of his own making. Heading west on April 2, he made stops in Chicago, Calgary, and British Columbia, before reaching San Francisco midmonth. He then set out for Big Meadows, staying at the damsite for almost a week. After returning to San Francisco on April 25, he left for the East Coast the next day.[14]

During his brief time in San Francisco, he found time to meet and dine with City Attorney Percy Long and City Engineer O'Shaughnessy but otherwise did little to engage in Raker Act plans.[15] He remained on good terms with the city's political and engineering leadership (in mid-May, Long confirmed the city's payment of $12,824.76 to cover Freeman's services from September 1912 through March 1913) but nonetheless made a purposeful decision to avoid any immediate involvement with the Hetch Hetchy project.[16]

Freeman had long planned a summer visit to Germany with a delegation from the American Society of Mechanical Engineers. Using this trip as a catalyst, he decided to turn the excursion into an extended family adventure. As he explained to Long, "I have been working altogether too hard during the past year and beyond all doubt or question owe it to myself and family to take a good vacation, and also to renew my acquaintance with several members of the family, and the safest way to get free from interruption for two or three months is to be on the other side of the Atlantic."[17] During the summer of 1912, Freeman had spent more than one hundred days away from his wife and progeny, mostly in San Francisco working seventeen-hour days. The summer of 1913 would be different.

On June 10 the Freeman family departed from New York City on the SS *Princess Victoria Louise,* disembarking at Hamburg on June 19.[18] Over the course of the early summer, his itinerary took him through Germany in the company of a binational assembly of mechanical engineers.[19] Once these professional responsibilities were fulfilled, in mid-July he and his sons journeyed down the Danube by steamboat to Vienna (wife and daughter stayed in Germany). After a few days in the heart of the Hapsburg Empire, they headed north to Stockholm via Berlin, thence east to St. Petersburg, and then by rail to Moscow, Kiev, and finally Odessa on the Black Sea. Although fearful of cholera outbreaks, he and his sons nonetheless sailed through the Bosporus Straits to Constantinople, into Greece and Italy, then over the Alps through Switzerland and back to Germany. After Freeman reconnected with his wife and daughter, the family decamped to Britain, leaving Southampton for New York City on August 21.[20]

During this remarkable summer tour—wherein he experienced a cosmopolitan Europe soon to collapse in the brutal maelstrom of World War One—Freeman remained far distant from San Francisco's efforts to obtain rights to a reservoir in Yosemite National Park. Only when he returned stateside in late summer would he again become a forceful advocate for the city's cause.

Nature Lovers and the Irrigationists

Upon learning of Secretary Lane's intent to support the city's plans, anti-dam preservationists realized that they faced a difficult future. The only bright spot in the Army Board's report was its acknowledgment that, despite being more expensive, there were viable alternatives to a system centered on Hetch Hetchy. But even the cheapest alternative (filtered water from the Sacramento) was estimated to cost $13 million more than Freeman's Hetch Hetchy scheme—and that comparison failed to take into account a $45 million valuation of the Hetch Hetchy system's hydroelectric power capacity.[21] Seemingly, the best strategy for the preservationists was to lie low and then, at a propitious moment, unleash a public relations campaign founded upon stirring pamphlets that—they hoped—would arouse a multitude of impassioned citizens to write their senators and congressmen and demand that San Francisco find some source of water other than Hetch Hetchy. A comparable effort had proven effective in 1908–9 when Congress first considered bills to support the Garfield Permit. Perhaps it could work again. As William Colby explained to Horace McFarland in mid-March,

> We should issue a final pamphlet somewhat akin to the earlier Muir pamphlet [of 1908–9] containing a statement of the case to date, backed up by quotations and containing a few good illustrations of Hetch Hetchy Valley. The front page would be occupied by an open letter from Mr. Muir appealing to the American People . . . [and suggesting] that those interested write to their [Congressmen]. . . . If Congressmen get appeals from all parts of the country along this line it would seem to me that is going to accomplish a great deal of real good. . . . I hope that this will be our final "Big Gun" in this Hetch Hetchy fight, but it seems to me worthwhile to fire it at an opportune moment.[22]

The preservationists knew they held a weak hand in the face of Secretary Lane and the Democratically controlled House and Senate. They certainly lacked the resources to mount any campaign that might quash the Raker Act in its early stages of legislative gestation. But perhaps as the bill advanced, they could bring out a "Big Gun" centered around mass letter-writing and, at an "opportune moment," sway the result in their favor. Clear-eyed but ever an optimist, Colby held out hope: "We have not as yet lost, especially when the views of the American public are still to be counted on as a factor." Only time would tell if this strategy would prove successful.[23]

The arguments of nature lovers in opposition to the city have long attracted attention from environmental historians. However, while John Muir's impassioned screeds against temple destroyers and park invaders have acquired great notoriety and sympathy over the years, the objections of farmers in the Turlock and Modesto Irrigation Districts (and more generally farmers in the Central Valley) posed the most serious political challenge to San Francisco. The resistance of irrigationists had nothing to do with preserving the scenic splendor of Hetch Hetchy. Their opposition had everything to do with controlling the resources of a river that they believed should be used only for their benefit. To them, San Francisco was not so much a park invader but a watershed invader, a foreign aggressor seeking to steal Tuolumne River water and carry it to a faraway metropolis. While their concerns were rooted in economic considerations, they also possessed a cultural dimension. This aspect of the battle over Hetch Hetchy might be characterized as one in which urban elites locked horns with salt-of-the-earth country folk, people who were determined to defend the sanctity of family farms and America's agricultural heritage.

On the surface, San Francisco, with a population in 1910 of more than 400,000, would appear to have much greater political clout than approximately 6,400 farmer/property owners served by the two irrigation districts. In early twentieth-century America, however, the political potency of irrigation settlements held great force within the nation's cultural imagination. Proponents of an "Irrigation Crusade" championed how small-scale farmers in the arid West could band together, build dams, reservoirs, and canals, and through hard work create a new form of American pastoralism, one that would serve as an antidote to crowded and morally suspect urban industrial centers.[24]

As noted in chapter 1, the Turlock and Modesto Irrigation Districts were among the first enterprises in the state that promised a way for middle-class (not impoverished) farmers to challenge the hegemony of large-scale landowners such as Miller & Lux (a partnership that controlled more than one million acres of land at its operational peak).[25] In the early 1890s, the districts built the La Grange Dam across the lower Tuolumne, claiming water rights to 2,350 cubic feet per second of river flow; this was more than enough to water at least

250,000 acres of land and, on a grand scale, "make the desert bloom." San Francisco and John Freeman may have believed there was sufficient water in the Tuolumne to serve the interests of both the districts and the growing Bay Area. But why should landowners in Turlock and Modesto accept such assurances? Why shouldn't they be fearful of the urban behemoth a few score miles to the west? Because once they relinquished any of the Tuolumne's flow to sustain economic growth in the Bay Area, there was little chance they could ever get it back. For them, the time to fight was now, when the Raker Act was beginning its journey through the labyrinth of Capitol Hill.

San Francisco's need to compromise with the irrigation districts is detailed in a letter from Percy Long to Freeman sent in late August, just as the Providence engineer was returning from his summer in Europe:

> In the early part of May the [San Francisco] Board of Supervisors decided to send [City Clerk] Dunnigan to Washington in order to expedite the passage of the Raker bill. [Upon] arriving in Washington he interested Secretary Lane, who wrote [House Majority Leader] Underwood urging the immediate passage of the bill as an emergency measure. Congressman Ferris, Chairman of the Public Lands Committee . . . had a partial hearing at which the Irrigationists were represented. The committee decided that they would not hear the matter until the differences between the Irrigationists and the City were adjusted and a settlement provided.[26]

At that point the city realized that it needed more firepower than the city clerk to move the legislation forward. Long explained,

> O'Shaughnessy and I went to Washington and spent the month of June meeting with representatives of the Irrigation Districts and, after finding that no measure could be passed in the House without provision . . . to protect the Modesto-Turlock Districts, [we] agreed to certain provisions which were put in the Raker bill. . . . We worked under great difficulties in Washington and had not concessions been made to the Irrigationists . . . there would have been no opportunity for a hearing which would have produced results. The whole of the San Joaquin Valley was aflame with opposition and the members of Congress had been deluged with telegrams and petitions.[27]

So what were these Raker Act concessions that the city acceded to in order to break the committee logjam? On one level, the city appeared to have actually been quite successful in getting the districts to limit their future irrigation systems to serve a maximum of 300,000 acres (rather than a possible 400,000 acres as postulated in the Army Board report) and to getting the districts to agree that any future storage reservoirs they might build would have to be sited below the intake for the city's Hetch Hetchy Aqueduct. But these were relatively minor stipulations compared to what was buried in section 9, paragraph C. The city had long acknowledged that the districts held, under California law, rights to divert a flow of 2,350 cubic feet per second (cfs) at La Grange Dam. But subsequent to negotiations with the districts' advocates, the revised Raker Act required the city (or "grantee") to "recognize the rights of the irrigation districts to the extent of *four thousand second-feet of water* out of the natural daily flow of the Tuolumne River for combined direct use and collection into storage reservoirs as may be provided by said irrigation districts *during the period of sixty days immediately following April fifteenth of each year*" [emphasis added].[28]

Restated, this clause stipulated that for two months every spring in perpetuity, the irrigation districts would be guaranteed a much larger flow from the Tuolumne than their existing claim to 2,350 cfs. They would be assured an additional flow of 1,650 cfs for 60 days, water that would never be available to San Francisco for storage or diversion into the Hetch Hetchy Aqueduct. The question then arises: how much water is needed to sustain a flow of 1,650 cfs for 60 days, and how does this compare with the 400 million gallons per day (mgd) flow that Freeman's plan proposed to deliver to the Bay Area? This question can be readily answered using some computational math: 1,650 cfs flow for 60 days equals a volume of 196,320 acre-feet (or about 64 billion gallons); if this volume is dispersed over the course of 365 days, it equates to about 175 mgd. (See appendix for how this daily flow was calculated.) As a result, the key concession granted by the city to the irrigation districts during their Raker bill negotiations came to almost 45 percent of the ultimate 400 mgd diversion envisaged by Freeman's Hetch Hetchy plan. In essence, it comprised a substantial quantity of water, conceivably enough to call into question the viability of the city ever safely and steadily diverting 400 mgd from the upper Tuolumne. But as O'Shaughnessy later apprised Freeman, the sacrifice was unavoidable: "We have had a very difficult time dealing with [the irrigationists], and did the very best we could to concede the least possible to them . . . but we are up against a practical situation and except we gave them some consideration we should never get the bill through. . . . [I]t is not desirable to resume negotiations of any kind with the irrigationists. I think those [concessions] made are the very best we could secure in face of a violent opposition to the city."[29]

In the late spring of 1913, city officials realized the need to accommodate the political opposition brought by the Turlock and Modesto Irrigation Districts, fearing that nothing could be accomplished in Congress unless they made meaningful concessions to mollify the irrigationists. But they had no such fears in regard to the community of anti-dam preservationists. Or at least the city made no comparable attempt to reach out to the "nature lovers" and seek some way to bridge the disagreement between the two parties. Of course, the irrigationists were not challenging the idea of a storage dam at Hetch Hetchy; they just wanted a guarantee of more water for their canals. In contrast, the existence of any reservoir at Hetch Hetchy, no matter the ultimate size of the dam or how much water downstream farmers might demand, necessarily blocked use of the valley floor for camping, hiking, hotels, or any other use by visitors to Yosemite National Park. On this elemental point there appeared little possibility that the preservationists and the dam builders could ever find common ground for a political armistice. The city and the irrigationists could find ways to apportion the Tuolumne's flow. For the city and the nature lovers, there could be no sharing of the valley; one side would win, the other would lose.

House Public Lands Committee Hearing

On June 25 the House Committee on the Public Lands opened what would become a five-day hearing on the Raker Act (June 25–28 and July 7). In his opening remarks, the committee's chairman, Scott Ferris of Oklahoma, elucidated the tribunal's origins: "It has been represented to this committee that this bill presents an emergency matter in that there is a great shortage of water in San Francisco. . . . [I]ts enactment into law will relieve the city of its blight to progress incident to its insufficiency of water."[30] Previously, the "San Francisco people and the irrigation people" had hashed out an agreement that, among other things, guaranteed the districts a flow at La Grange Dam of 4,000 cfs for sixty days every spring. On June 23, Raker submitted an amended bill (H.R. 6281) satisfying the irrigation districts' allocation requirements; this became the parliamentary focus when the public hearing commenced two days later.[31]

Interior Secretary Lane was the first witness called, and he did nothing to disappoint the San Francisco delegation. Leaving technical arguments largely to others, Lane introduced his remarks with an emotional plea: "San Francisco needs a new and adequate water supply. . . . The situation in San Francisco now is that there are many homes where sufficient water cannot be had for a bath. . . . More than that, you know the situation that developed immediately after the earthquake."[32] Exactly how a Sierra reservoir more than 160 miles from San Francisco's urban core might help fight fires triggered by future earthquakes was left unstated. Lane only knew that the city needed a new, municipally owned water supply.

The secretary testified for about half an hour, addressing myriad questions related to the Hetch Hetchy, including the project's hydropower potential (mostly in terms of how the federal government would be compensated from municipal power proceeds) as well as concerns about meeting the needs of the irrigation districts.[33] The key element of Lane's testimony was registered in the official departmental letter submitted to the committee, in which he unequivocally declared that "the permission desired by the city and county of San Francisco to secure a water supply from the Yosemite National Park for municipal purposes, etc. should be accorded. The communities on San Francisco Bay . . . are urgently in need of an adequate supply of pure, wholesome water for domestic consumption and fire protection."[34] Lane's committee appearance publicly affirmed the executive branch's position on Hetch Hetchy; barring some dramatic turn of events, the Raker bill was guaranteed support from the Wilson administration and its interior secretary.

Lane was followed by other federal officials who offered their views to the committee (these included David Houston, secretary of agriculture; George O. Smith, director of the US Geological Survey; Henry Graves, head of the Forest Service; and Frederick Newell, director of the US Reclamation Service). Not surprisingly, all fell in line with the interior secretary.[35]

The hearing's first day was confined to administration officials with one exception: Gifford Pinchot. For the first time in almost three and a half years, the former head of the US Forest Service was, albeit briefly, again engaged in the battle over Hetch Hetchy. Since leaving the Taft administration in January 1910, Pinchot had played no role in the Hetch Hetchy controversy. But he had returned, or was at least willing to take a few minutes out of his busy schedule and share his perspective with the committee. He came at the personal invitation of Chairman Ferris, who "wired him and urged him to come." Because Pinchot was "anxious to get away" for some unspecified bit of business, Ferris allowed him to jump the queue and speak immediately after Secretary Lane and Secretary Houston.[36]

Pinchot had never visited Hetch Hetchy, but this did not deter him from being "thoroughly and heartily in favor of [the Raker bill]."[37] In offering this opinion he averred no claim of technical expertise related to the city's water supply. Thus, when Chairman Ferris asked him "Have you any occasion to compare the relative merits of the different proposals [for alternative water supply sources]?" he could only tepidly respond, "I must answer the question by saying that I am not an expert in the matter, and I simply accept the opinions of the engineers who are experts and who know better than I do."[38] Although not an "expert," Pinchot understood that "there is no use of water that is higher than the domestic use." Accordingly, he asserted, "if there is, as the engineers tell us, no other supply that is anything like so reasonably available as this one [at Hetch Hetchy], if it is the best, and, within reasonable limits of cost, the only means of supplying San Francisco with water . . . I think there can be no question at all but that . . . weighty considerations demand passage of the [Raker] bill."[39] Thus, so long as the "experts" told Pinchot what needed to be done, he was more than willing to go along and "demand" that the dam be built.

Pinchot's committee appearance lasted about thirty minutes and then he left, never to return to the hearing. He would support the Raker bill through the rest of the year, but during this time he played only a peripheral role in San Francisco's legislative campaign.

On the hearing's second day, the Army Board appeared and stood as objective experts endorsing the city's plans. Any question as to the importance of the board's report issued the previous February in furthering the city's cause was laid to rest in the testimony of the board's chairman, Colonel John Biddle (and to a lesser extent in the contributions of Colonel Cosby and Lt. Colonel Taylor). Although Biddle seemed a little uncertain at the start, he warmed to the task and, after asking if he could look "at certain pages of the report in order to refresh my memory," defended the board's conclusions in support of San Francisco.[40] Biddle may have done little to expand upon the Army Board report's analysis (although he did spend considerable time discussing the McCloud River alternative), but that was of minor consequence. What mattered is that the board—acting as an independent body—affirmed San Francisco's Hetch Hetchy project as the most reasonable and economically feasible alternative for the city to pursue.

It was one thing for Freeman to advance the need to dam Hetch Hetchy, because, in the words of preservationist William Colby, his opinions could be derided as an instance of "paid advocacy." The Army Board had no financial stake in the matter and presumably could be relied upon to offer an unbiased assessment. The fact that the board had offered a sweeping endorsement of the city's plans carried enormous weight among technically naïve congressmen who would be called to vote upon the Raker Act. To oppose the bill would be akin to publicly rejecting the recommendations of a distinguished panel of army officers. For any legislator on the fence about the Raker Act, the board's report offered invaluable political cover to vote "yes."

While Biddle and his fellow officers always maintained their independence in evaluating the city's plans, this did not mean that they held no opinions as to the scenic value of Hetch Hetchy. Going beyond the officialese of their report, the three officers revealed their own sentiments as to whether Hetch Hetchy was a proper source of water for San Francisco:

> CHAIRMAN FERRIS: If you were a member of this committee, having due regard for the rights of the irrigation people, and having due regard for the rights of the nature lovers, who believe you should not interfere with the Yosemite National Park, and having due regard for the needs of San Francisco, which system would you vote for?
>
> COL. BIDDLE: I would vote for the [city's] Hetch Hetchy system.[41]

His fellow officers also expressed their views as to scenic values and the need for the reservoir:

> CHAIRMAN FERRIS: Does the associate engineer desire to make some statement now, or does he coincide with the views already expressed by Col. Biddle?
>
> COL. COSBY: I fully concur with the statement of Col. Biddle. There is only one small point of difference. . . . I believe that with the lake [the valley] will be more beautiful than it is in its natural condition.
>
> COL. TAYLOR: The first year I was up there I was inclined to think that it would be more beautiful as a lake, but in the second year I was inclined to think that it would be more beautiful as a valley. It will be a beautiful place either way.
>
> FERRIS: Do you think that the [issue of] beneficial use that we have been considering ought to enter into it?
>
> COL. TAYLOR: There is not the slightest question in [my] mind but that [Hetch Hetchy] should be used as the source of water supply [for San Francisco] . . . [as] it is by far the best storage reservoir in that section of the country.[42]

When the committee reconvened in the afternoon, Percy Long took the stand. Much of the city attorney's testimony involved a recitation of the city's filings under the 1901 Right-of-Way Act, issuance of the Garfield Permit, and actions taken to further the city's Sierra initiative (including land purchases). Long also addressed questions related to how the Raker Act could serve the needs of other Bay Area communities if desired. But his most revealing remarks concerned the Spring Valley Water Company and why the private company did not at present object to the Raker Act. Given the hostility between Freeman and the company in the run-up to the Fisher Hearing some seven months earlier, the fact that Spring Valley had no representative at the Public Lands hearing and sent no telegrams or other missives opposing the legislation was significant. In his remarks, Long explained that the controversy between the water company and the city "has now been drawn down to an agreement by which . . . we will try and settle . . . the question of the amount that San Francisco shall pay to the Spring Valley Water Co. for the purchase of its system." He continued, "Just before I came to Washington, Mr. Bourn, president of the company . . . stated in my presence in answer to the question whether the Spring Valley Water Company would make any opposition to the acquisition by San Francisco of the Tuolumne water supply in Hetch Hetchy Valley: 'No we are through. We will not place a single obstacle in its way.' . . . [I believe] we have nothing to fear from the Spring Valley Water Co."[43]

As events played out, the Spring Valley assets would not officially be transferred to the city until 1930, but in regard to the company's position on a municipally owned Hetch Hetchy water supply, Long's testimony proved accurate. Throughout the remainder of the year, Spring Valley never registered any public protests relative to the Raker Act or lobbied to that effect on Capitol Hill. Exactly why the company changed its once-obdurate stance is uncertain, but when the Army Board report estimated in February 1913 that the company's peninsular and Alameda system had an ultimate capacity of 132 mgd (some 50 mgd more than Freeman projected in his *Hetch Hetchy Report*), the company's leaders may have gotten what they wanted all along: greater leverage in arguing for a higher sale price.

On Friday, June 27, City Engineer O'Shaughnessy continued to press the city's case, explaining the importance of building a dam at Hetch Hetchy before developing the Lake Eleanor and Cherry Creek watersheds and referencing the negotiations with the irrigation district.[44] Although the stipulation of 4,000 cfs for sixty days was never specifically mentioned, O'Shaughnessy did emphasize how the city had accommodated the interests of the irrigationists: "For the past ten days there have been earnest conferences between the city's representatives, Mr. Long and myself, and [the irrigation districts' representatives]. . . . We have arranged what I believe is an equitable plan for the handling of the water."[45] Later on at the end of the hearing, James Needham, who had previously served as the congressman representing Stanislaus County, made a brief statement "speaking for . . . [those] who represent the Turlock and Modesto Irrigation Districts." He followed up on O'Shaughnessy's testimony, acknowledging that the districts "are not at this time opposing the bill because it specifically sets forth in the bill conditions which we feel protect the prior rights of the Turlock and Modesto Irrigation Districts."[46]

The committee asked no questions of Needham and did not inquire about his "at this time" caveat, letting the record stand that the irrigation districts offered no objections to the revised Raker bill. However, the possibility lingered that perhaps not all farmers in Stanislaus County would accept the negotiated concessions. Such a sentiment was reflected in a brief appearance before the committee by L. I. Dennent, representing the "proposed Waterford Irrigation District." Lying to the west and beyond the limits of the Turlock and Modesto Districts, the Waterford District had yet to be legally organized, and thus any claim it might make to Tuolumne flow was uncertain. But Dennent's testimony signposted that, regardless of what Needham and his associates had agreed to, disgruntled farmers in the San Joaquin Valley might still organize against the city's plans.[47]

When the Fisher Hearing was held the prior November, O'Shaughnessy had been city engineer for less than three months and had had no direct involvement in the creation of Freeman's *Hetch Hetchy Report*. He possessed an agile mind, however, and in the intervening months he gradually began to put his own imprint on the city's proposed project. With Freeman in Europe for the summer, O'Shaughnessy had the opportunity—and responsibility—to advance the city's cause as he thought best. Overall, he had no problems with the design concept envisaged by Freeman, but he perceived that there were some practical issues demanding attention if the city was to be spared later embarrassment.

The most important of these involved the extent that San Francisco would build roads around the reservoir, paved paths intended to bolster park visitation and induce greater appreciation of Hetch Hetchy's beauty. The question was not one of eliminating all the access roads—that would be political dynamite given how the project was being sold as an antidote to the paucity of valley visitors. It involved adjusting the plan so that some of the most unworkable and expensive roads could be dropped from the scheme.

In particular, the road around the southern shore of the reservoir, which would require extensive drilling and excavation along the steep granite face of Kolana Rock, was seen by O'Shaughnessy as totally impractical. Freeman had featured such a thoroughfare in a fanciful illustration that graced the introductory section of his *Hetch Hetchy Report*. And he had projected a photo of the road rimming Lake Oifjords in Norway as an example of what travelers along the Hetch Hetchy's south shore could behold during a trip around the reservoir. But O'Shaughnessy knew that as city engineer he would be responsible for building a real right-of-way, not simply retouching a photograph. If the city was ever going to escape responsibility for this difficult-to-build road, it needed to act quickly and make sure that the Raker Act did not place undue expectations on the city.

> CONGRESSMAN RAKER: In regards to the roads, Mr. Freeman has made a full report upon the question and a great many have seen it. Do you not desire to change the general outline of the Freeman report?
> O'SHAUGHNESSY: I do. . . . I believe that one of his roads is not feasible.
> RAKER: Which one?
> O'SHAUGHNESSY: The one on the south side of the lake.
> RAKER: Why?
> O'SHAUGHNESSY: Because it is a vertical bluff there, and it would be suicide to go over such a road. The precipice there is 400 or 500 feet high, and I have condemned the plan and [the city's engineering] department does not want it. . . . I believe it would cost a half a million dollars to make that piece of road . . . but we are quite prepared to make a road the whole length on the north side of the lake if it is thought desirable.[48]

Freeman's proposal as detailed in the *Hetch Hetchy Report* would always constitute the heart of the city's Hetch Hetchy project, but alterations and adjustments were made over the course of its development and construction. O'Shaughnessy's move to eliminate the south-shore road around Kolana Rock represented one of the first, and most important, of these changes.

When in early June the House Committee on the Public Lands first decided to expedite consideration of the Raker bill, little notice was given of the hearings that would start in a few weeks. Because a transcontinental rail journey from the Pacific Coast required a minimum of four days, preservationists residing in California were hamstrung in arranging their schedules to attend the

Using this photograph of Norway's Lake Oifjords that he included in his *Hetch Hetchy Report* (p. 19), Freeman had proposed building a roadway around Kolana Rock along the southern shore of the reservoir. But during the House Public Lands Committee hearing, City Engineer O'Shaughnessy testified that such a road was too expensive and impractical to construct. Henceforth, this road was dropped from the city's plans for the Hetch Hetchy project.

Washington hearing. As a result, the only anti-dam proponent to appear and defend the sanctity of Hetch Hetchy was Boston-based attorney Edmund Whitman. Remarkably perseverant, Whitman was in attendance for all four days of the June hearings (but not July 7), and his testimony, after beginning on the afternoon of June 27, occupied much of the meeting the next day.[49]

A key component of Whitman's far-ranging testimony revolved around the Army Board's acknowledgment that, while viable alternatives to the Hetch Hetchy project existed, they were deemed less desirable largely based on cost. The issue of cost had always been of importance to Freeman in his advocacy of the Hetch Hetchy Dam, so this was not a new metric that Whitman had suddenly discovered. However, the Boston lawyer wanted to highlight the fact that the city did not absolutely need to inundate Hetch Hetchy Valley to secure a bountiful water supply. It was simply a matter of money to save Hetch Hetchy, and perhaps the amount of money that city taxpayers might need to subsidize development of an alternative source (or sources) was relatively modest given the great value attached to the valley's scenic wonders. Taking the least costly alternative referenced in the Army Board report (filtered water drawn from the Sacramento River), Whitman argued that the extra $13 million needed to finance this project could be offset by, say, an extra $3 million or so in expenses if water from Hetch Hetchy was ultimately filtered in the decades ahead. So, was preserving the Hetch Hetchy Valley not worth $10 million dollars?[50]

In the end, Whitman did an admirable job arguing his case. But he faced a tough audience of committee members who possessed little sympathy for his message. This was vividly illustrated in a brief exchange with Chairman Ferris:

FERRIS: Yes; I believe you will agree that we had before us the Army
    board, men of standing and high character, who were sent out there
    [to California] to make a personal examination of this matter . . . and
    their conclusion was that this was the most available supply, and that
    it should be done now. Were those practically the words spoken here
    by Col. Biddle?

WHITMAN: His opinion seemed to be that . . .

FERRIS: Do you not think that it is assuming a great deal of responsibility
    for a resident of Cambridge, Mass.—even for a learned and dis-
    tinguished lawyer that you are—to set up your judgement against
    that array of talent . . . who have gone into the question the same as
    yourself?

WHITMAN: So far as it is a question of opinion I agree with you; so far as
    it is a question of fact, I have endeavored to present such consider-
    ations of fact as to show that their opinion has not been based upon a
    consideration of the existing facts.[51]

A few moments later Chairman Ferris conceded to Whitman that "you have presented your side very ably." But after Ferris's pointed question, it seemed unlikely that a preservationist's arguments to block the damming of Hetch Hetchy would find much support within the Public Lands Committee.[52]

A "Suppressed" Report?

During the committee hearings, Chairman Ferris received a telegram from Eugene J. Sullivan, president of the Sierra Blue Lakes Water and Power Company, requesting that the company be allowed to present evidence that a viable alternative to the Hetch Hetchy existed and had not been given proper recognition by the Army Board. Sullivan was known to San Francisco officials, as the Sierra Blue Lakes enterprise had offered to sell the city its purported water rights on the Mokelumne, Cosumnes, and American Rivers as far back as 1911.[53] In approaching the committee, Sullivan intimated that an "unfortunate scandal" could erupt unless the committee acted to expose deceptive treachery by city officials. In his June 27 telegram to Ferris, Sullivan alleged, "Absolutely no water shortage here. Such allegations are framed for political purposes. . . . City officials are merely deceiving your committee as they have already deceived Mr. Freeman and Army Board. . . . Army board accepted city's false data in good faith but did not give sufficient time for personal investigation. Respectfully ask time to complete data and present proof to your committee."[54] Sullivan's message had been preceded by a separate telegram from Taggart Aston, who identified himself as a "consulting engineer" for the Sierra Blue Lakes company and who had been engaged to "investigate their Mokelumne River proposed water supply." Aston reported, "I find that [the company] will have available for San Francisco an economically developed supply of pure mountain water of at least 350,000,000 gallons per day. . . . I am preparing and shall have full data in [a] few weeks that will prove the granting of Hetch Hetchy unnecessary and against public interest."[55]

With these telegrams, an infamous episode in the Hetch Hetchy saga began. Prominently featured in Holway Jones's history of Yosemite National Park, the story is best known as one wherein a city report on the capacity of the Mokelumne was "suppressed"—or kept from—the Army Board and thus removed from objective consideration as an alternative to Freeman's Hetch Hetchy system.[56] Later environmental historians have seen this so-called suppression as evidence of the city's deceitfulness in claiming the need to dam Hetch Hetchy.[57] So what do we know about this supposedly suppressed report and the claims that Sullivan and Aston based upon it?

The report in question was prepared by Max Bartell, an assistant engineer working for City Engineer Manson (and later with O'Shaughnessy) and focused on the water available in the Mokelumne River watershed. Dated April 24, 1912, it was completed in the weeks immediately prior to Manson's mental collapse. As noted in the report's introduction, Bartell was "acting on [Manson's] verbal instructions to investigate the water resources of the Mokelumne River as a probable source of water supply for the City."[58] Specifically, he was to tabulate the "possible ultimate development" of a water diversion taken just below the San Francisco Gas and Electric Company (a subsidiary of PG&E) hydroelectric power plant at Electra, lying at an elevation 655 feet above sea level. In response to this directive, Bartell concluded, "The possible ultimate development of the 537 square miles tributary to Electra after making due allowance for all known water rights, is 250 million gallons daily [mgd]."[59] Calculating the diversion at Electra was important, because at that elevation it was possible to design an aqueduct extending southward to the Altamont pumping station featured in the Tuolumne River aqueduct proposed by both Grunsky and Manson.

As part of his investigation, Bartell also considered Mokelumne River flow measured at the Clements gauge located more than twenty-five miles downstream from Electra at an elevation less than 140 feet above sea level. Based upon those data and at that location, he determined that 432 mgd could "be available to San Francisco."[60] This was the bombshell figure that subsequently attracted the attention of anti-dam advocates, who saw it as suppressed evidence contradicting San Francisco's claims that there was no viable alternative to the Hetch Hetchy supply scheme. But take note that the low elevation of the Clements gauge would preclude any diversion at that site to reach San Francisco via a gravity-flow system (and thus would require extensive pumping).

Leaving the low-lying Clements gauge estimate aside, it is important to appreciate that, while the 250 mgd that Bartell calculated as available for diversion at Electra was far greater than the 60 mgd that Grunsky and Manson had initially proposed for the Hetch Hetchy Aqueduct system, it fell well short of the 400 mgd that Freeman proposed for his Hetch Hetchy Aqueduct. And Freeman's 1912 report had raised the possibility that the Mokelumne could supply 200 mgd to San Francisco through a gravity-flow aqueduct, a figure only 20 percent less than Bartell's maximum figure for a diversion at Electra. In addition, any diversion of the Mokelumne at Electra would provide much less hydropower than Freeman's gravity-flow Hetch Hetchy Aqueduct, where storage would be at an elevation exceeding 3,500 feet above sea level.[61]

If nothing else, Sullivan's initiative in bringing the Bartell report into the public arena highlighted how imperfect, if not incomplete, Grunsky's August 1, 1912, report on the Mokelumne had been. Yes, the description of a possible Mokelumne supply as printed in Freeman's *Hetch Hetchy Report* did offer the possibility that 200 mgd could ultimately be available to the city. But the overall capacity of the stream was minimized by Grunsky: "The limit may, for the present[,] be placed at about 60 million gallons per day."[62] For its part, the Army Board had independently considered the capacity of the Mokelumne and, while skeptically acknowledging the possibility of drawing "200 mgd and even more" from the stream, offered 128 mgd as a more credible figure.[63]

After Sullivan telegraphed Congressman Ferris in late June asking for "time to complete data and present proof to your committee," he was granted this request, and the committee reconvened on July 7 to hear his testimony. To say that it went poorly for Sullivan would be an understatement, as he declined to provide the committee with a copy of Bartell's report and also failed to describe exactly what the Sierra Blue Lakes Water and Power Company was prepared to sell to the city that would obviate the need for the Tuolumne/Hetch Hetchy Aqueduct. In addition, the consulting engineer, Aston, who had promised to prepare a report detailing how the company could "have available for San Francisco an economically developed supply of pure mountain water of at least 350,000,000 gallons per day," did not accompany Sullivan and was not available to take questions or elaborate on the company's plans.[64]

In the final part of Sullivan's testimony, committee members began to ask him about his career as a businessman and, in particular, an agreement made with a Mrs. Maud Treadwell in November 1910 relating to a payment of $30,000 she extended for an interest in the Sierra Nevada Water and Power Company (not Sierra Blue Lakes) that seems to have proven of questionable value. The intent of this inquiry was apparently to unmask Sullivan as a calculating grifter who preyed on the unsuspecting and gullible. And given that Sullivan had appeared at the hearing without a copy of Bartell's report and without any specific plan detailing how the Sierra Blue Lakes Water and Power Company could provide the city with 350 mgd of mountain water, he did not come across as a particularly credible witness.[65]

Of course, if Freeman had been in the States and attending the House committee hearing, he would have immediately been engaged in grappling with estimates offered in Bartell's report and could have explained how they related to Grunsky's (and his own) analysis as presented in his *Hetch Hetchy Report*.[66] Precisely how, in July, he would have addressed Bartell's report remains a subject of conjecture, but in September, when he testified before the Senate Public Lands Committee, he acknowledged Bartell's work while downplaying its significance: "I was with Mr. Bartell for two or three months last summer [1912], and I have taken the trouble to look over the figures in connection with Mr. Bartell's report. . . . It was simply one of a series of estimates in which he figures the greatest possible amount of water that could be obtained by developing all possible storage reservoirs on the Mokelumne River, no matter who might own the site, and taking all that water . . . he figured out the total flow of the stream. . . . It was simply a step in the work. . . . Mr. Grunsky had Mr. Bartell's report . . . [and] the Bartell report is merely part of the groundwork in his report to the Army engineers."[67]

Freeman was being a bit disingenuous in soft-pedaling Bartell's report as "simply a step in the work" underlying Grunsky's Mokelumne report—after all, in his diary entry dated August 1, 1912, Freeman had scathingly referenced Grunsky's work as "some hasty half baked reports."[68] But Freeman would not have believed that Bartell's conclusion that 250 mgd from a diversion at Electra (at an elevation of 655 feet) constituted a serious alternative to his proposed 400 mgd Tuolumne/Hetch Hetchy project, one that also included an estimated 157,000 horsepower hydroelectric power component valued at $45 million. And a prospective flow of 432 mgd at Clements (less than 140 feet above sea level) would also suffer in comparison to Freeman's Hetch Hetchy scheme.

Yes, Sullivan's introduction of a "suppressed" report into the anti-dam narrative did not help the city or Freeman make the case for needing Hetch Hetchy Dam. The revelation, however, did not prove catastrophic, and despite later efforts by Robert Underwood Johnson to publicize Bartell's findings, the city was able to absorb and mitigate the report's impact.

Robert Underwood Johnson: "An Open Letter"

Aston's promised report never made it into the public record, but a copy of Bartell's report did become available to Ed Parsons of the Sierra Club in mid-July and was soon distributed to leaders of the anti-dam coalition, including Johnson.[69] Acting on his own initiative, Johnson decided the time was ripe to energize opposition to the Raker Act and, with the "suppressed" Bartell report as a call to arms, instigate a mass letter-writing campaign to congressional leaders.[70] Colby had discussed such a strategy earlier in the spring, and with Sullivan's claims offering an opening, the time seemed opportune—at least to Johnson—to shine a public light on San Francisco's presumed deceit. In early August, Johnson printed "An Open Letter to the American People" focused on the "Hetch Hetchy Scheme [and] Why It Should Not Be Rushed Through the Extra Session" that was intended as a direct plea to the American people and also as a release "for publication and comment in the press." In the coming weeks and months, his open letter attracted the attention of many newspaper editors and spurred anti-dam articles and editorials across the nation.[71]

Never one to shy away from a bit of bombast in his defense of Hetch Hetchy, in the opening paragraph of his letter Johnson boldly exclaimed, "The language of hyperbole is the only appropriate medium to describe the features of your Yosemite National Park. Better that there had never been a Niagara than that the northern half of the Park should thus be diverted from the use of the public.

The Hetch Hetchy is a veritable temple of the living god, and again the money changers are in the temple."

Here, Johnson returns once again to the sweeping claim that use of Hetch Hetchy for a municipal water supply will require that the entire upper Tuolumne watershed be closed to public use; later in his letter he claims that the city is "endeavoring to rush through a drastic measure that would turn over to the city 500 square miles—half of your National Park." Of course, the sanitation restrictions stipulated in the Raker Act placed no oppressive constraints on use of the upper watershed, and this had been foretold and ably defended by sanitation engineer George Whipple at the Fisher Hearing.[72] Nonetheless, Johnson felt justified in fearing the worst. After all, he had already embraced "the language of hyperbole" when discussing Yosemite National Park. So why not cast a wide net and attack the city's plans in the broadest terms?

Johnson did not describe the Bartell report in any detail, and he avoided any mention of Taggart Aston, but in his open letter he claimed that "the city actually withheld a report showing that the Mokelumne River region will afford abundant resources at a smaller expense." Seeking to highlight the difference between Grunsky's August 1, 1912 report and the Bartell report's most extreme estimate, Johnson ignored the fact that Freeman's *Hetch Hetchy Report* did include an estimate for a 200 mgd gravity-flow aqueduct providing Mokelumne water to San Francisco.[73] Instead, he emphasized that Grunsky had placed "the resources of the Mokelumne at 60,000,000 instead of 432,000,000 gallons daily! This withholding constitutes an important suppression of the truth, and was a wrong to the [Army] Board, to the city's expert, to the members of both Houses of Congress, and to every other American citizen."

In his open letter Johnson acted as a political advocate, and in this capacity, he had little incentive to consider how the 432 mgd estimate was derived from measurements made at an elevation of about 140 feet above sea level—and how this might be categorically different than 400 mgd drawn from reservoirs more than 3,500 feet above sea level in the upper Tuolumne watershed. He wanted to depict the city's position in the worst light possible. It was for others to make their own counterarguments.

In championing the Mokelumne as a bountiful alternative to the Tuolumne, however, Johnson manifestly tied himself to the claims of Sullivan and the Sierra Blue Lakes enterprise. As he proffered in his open letter, "If the [Raker Act] legislation is not railroaded through Congress, an even fuller report of the Mokelumne resources than that of the Engineer Bartell will be presented, along with an offer of rights and sites, by the Sierra Blue Lakes and Water Power Co."

But what if a "fuller report" was never forthcoming? And how to separate the anti-dam movement from the reputation of Sullivan as a promoter/businessman/ grifter seeking to profit from the Hetch Hetchy controversy? Sullivan's charge of a "suppressed" report gave preservationists a new weapon to brandish against the city, but it came with troublesome baggage. By de facto allying themselves with a private enterprise that was seeking a considerable payday from the city, Johnson imperiled the moral high ground preservationists had claimed when blasting Freeman as a paid advocate. Were anti-dam advocates simply a "cat's paw" advancing the interests of businessmen possessing supposed rights to the Mokelumne? Once Johnson's letter was distributed, preservationists could not easily separate themselves from Sullivan's Sierra Blue Lakes scheme and from his sullied reputation.

Near the end of his letter to "fellow owners of the Yosemite National Park," Johnson drew upon an economic argument to close the case: "The American people are asked to subsidize the city's water supply to the extent of the money value of Hetch Hetchy and of five hundred square miles of phenomenal scenery. Put up at auction, what would this wonderland bring? 'What am I bid,' the auctioneer might say, '. . . Do I hear $20,000,000 to start the bidding? Remember that these natural features are priceless.' Will the reader of these lines also remember that fact?" And then he made his plea: "Citizens, will you not help prevent this outrage by writing in protest, however briefly[,] to your Senators and Representative, and to Hon. Reed Smoot, U.S. Senate, and Hon. F. W. Mondell, M.C.[,] Washington D.C., and to the press, and by asking others to do the same? 'They have rights who dare maintain them.'"

Overtly and unrepentantly political, Johnson's letter marked the beginning of the preservationists' final defense of Hetch Hetchy.

The House Approves

A few days after Johnson released his "Open Letter to the American People," Congressman Raker submitted Report No. 41 on the "Hetch Hetchy Grant to San Francisco" that offered a justification for House Resolution (H.R.) 7207 (the final version of the Raker Act). The forty-three-page printed document drew extensively from the Public Lands Committee hearing in making a case for why the committee unanimously recommended its enactment.[74] Cognizant that Sullivan's claims had attracted public attention, Raker included Colonel Biddle's emphatic denial that the supposed "suppression" of any reports had influenced the Army Board. Raker also made the point that "Mr. Sullivan informed the [Public] Land Committee that Taggart Aston, his engineer, is employed on a 10% contingent fee—Mr. Aston's compensation depending upon the sale of the Blue Lakes project."[75] As the bill moved forward in Congress during the late summer and fall, Raker's report served as the primary House document endorsing the legislation.

With Democrats holding a commanding House majority there seemed little chance that the Raker Act would fail to win passage. Nonetheless, the House membership would have an opportunity to debate the bill and offer amendments. On Friday, August 30, the House convened as a Committee of the Whole and began debating H.R. 7207. These deliberations continued for three days with Scott Ferris and John Raker taking charge as the bill's lead advocates. The opposition was led by three Republican Congressmen: Frank Mondell of Wyoming, James Mann of Illinois, and Halvar Steenerson of Minnesota.[76]

Ferris's and Raker's defense adhered closely to arguments presented in House Report No. 41 and would have been familiar to anyone following San Francisco's effort to tap into the Tuolumne. However, Sullivan's charges of "suppression" introduced a wild card into the deliberations and, in Ferris's view, needed to be addressed quickly. In Ferris's opening remarks, the Public Lands Committee chair averred that there was "no truth" in the charges and that Sullivan "tried to fight this [Hetch Hetchy] measure because he had a rival system that he wanted to sell the city of San Francisco. The committee has exploded that proposition.... [T]he hearings disclose that Sullivan was an adventurer in the real estate business, who would like to ... unload a piece of property of doubtful value on the city which she does not want."[77]

When Congressman Mann interjected, "[I] understand that this statement [relating to a suppressed report] is made by Robert Underwood Johnson, a gentleman of the highest character," Ferris acknowledged, "The gentleman is right, Robert Underwood Johnson, in my judgement, is a patriotic and good man," but nonetheless Sullivan and the engineers associated with him "who are on a contingent fee in event of sale, [have] stirred up interest among patriotic men who know nothing of the real conditions."[78] With this, a strategy on addressing Johnson's attack began to coalesce: Johnson (and by extension his fellow preservationists) were accorded respect as "patriotic men," but despite their honorable objectives, they were being manipulated by an "adventurer in the real estate business" who was considerably less honorable.

During later comments on the House floor, Congressman Thomson of Illinois more directly criticized Johnson, quoting his Fisher Hearing testimony in which he professed his willingness to distill seawater if that was the only alternative to using Hetch Hetchy. After reading an excerpt from the hearing, Thomson reinforced the point to his colleagues: "Mr. Johnson is sufficiently unreasonable about this proposition to go to the extent of saying that rather than take the Hetch Hetchy water supply, San Francisco ought to be compelled to distill the salt water of the ocean."[79] "Sufficiently unreasonable" offered a tempered yet censorious description that the long-time *Century Magazine* editor would find difficult to shake as the Raker Act moved forward in the fall of 1913.

Congressman Steenerson of Minnesota was the most overt proponent of preservationist arguments to protect Hetch Hetchy. In his remarks on the House floor, he couched his views in a decidedly anti-urban framework, complaining that Congressman Knowland of California "[y]esterday pointed out the wonderful increase in population in the cities on San Francisco Bay, and he took great pride in it." Steenerson stated, "Instead of taking pride in that, I deplore it. All the great statesmen of our time are deploring the fact of the influx to the city from the country." Steenerson represented a rural district in northwestern Minnesota, and his constituents presumably had little interest in urban culture—or at least he would not suffer politically from chastising San Francisco politicians. The Hetch Hetchy controversy gave him an opportunity to affirm the benefits of country life and decry the baneful effects of city living: "I would encourage the people to go to the national parks, where they can admire nature in its pristine beauty and become imbued with a love of nature. . . . It is said this [part of Yosemite] park is hard of access; that only a few hundred people reach it every year. . . . I would rather have a few see it in its natural glory than in its desecrated form."[80] Here, Steenerson adopted a rather extreme perspective, one that did not easily accord with the more practical economic arguments of preservationists, many of whom backed the construction of hotels in Hetch Hetchy.

A reactionary anti-urbanist, Steenerson expressed a poetic and idealistic vision in defending Hetch Hetchy, one in which sentiment and sentimentality were to be celebrated, not simply endured: "Perhaps some lone, footsore, weary wanderer may find his way into this valley some day and by means of inspiration of these wonderful surroundings will produce something more valuable than money. Suppose he could write a poem like Burns' poem to a mountain daisy? Would you trade it for [the] $45,000,000 that the taxpayers of San Francisco have voted for this [municipal water supply] enterprise? Why, you could not estimate the value of such a contribution to human thought in its refining effects in dollars and cents."[81]

While the Minnesota congressman was presumably heartfelt in his speechifying, it is unlikely that many (any?) of his colleagues would ever contemplate trading $45 million for a prospective poem inspired by Hetch Hetchy. Testimony of this timbre played to an audience of true believers convinced that Hetch Hetchy was a holy sanctuary of nature. But across the broader political spectrum, Steenerson's pleadings likely did little to expand the anti-dam constituency. To someone like Freeman they only reinforced the narrative that the "nature lovers" were out of touch with and insensitive to the real needs of people in San Francisco.

In addition to environmental or "sentimental" issues, a significantly different type of argument was raised against the proposed legislation, involving the relationship of the federal government to the sovereignty of individual states. In the twenty-first century, we largely take for granted that the federal government is often a party to water projects, but in the early twentieth century, water rights and water control were widely considered to be the province of state government. In this context, a controversial element of the Raker Act concerned the stipulation that guaranteed the Modesto and Turlock Irrigation Districts a flow of 4,000 cfs for sixty days every spring. What was controversial was not the fact that the irrigation districts might benefit from such a stipulation but that a federal law might guide how water would be allocated within an individual state.

Congressman Frank Mondell of Wyoming made the most impassioned plea against the bill on the grounds of state's rights: "This Congress has no power to divide the waters of the sovereign state of California. . . . I am something of a 'State's righter' . . . [and] I do not believe this mighty Government can stand unless . . . we shall recognize and guard jealously the reserved rights of all the people of the Commonwealths of the Union."[82] San Francisco officials were not averse to Mondell's position—they would have been happy to forgo the stipulation of 4,000 cfs for sixty days. But stripping out this guarantee would have risked tremendous backlash from agricultural interests and likely unraveled the truce negotiated between the city and Modesto/Turlock farmers.[83] When Mondell offered an amendment that would eliminate any clauses designed to "divide the waters" of California, Ferris and Raker strongly objected. Mondell's amendment went down 6 yeas to 45 nays.[84]

Over the course of a day and a half (August 30 and September 2), more than twenty amendments were considered, and almost all were rejected. A few minor word adjustments were approved, but the final version of H.R. 7207 closely aligned with what the Public Lands Committee had sent forward in early August.[85]

Late in the afternoon on September 2, H.R. 7207 was approved by the Committee of the Whole and sent to the full House. Shortly after noon the next day, the final vote came, and the Raker bill received overwhelming support, with 183 yeas, 43 nays, and 9 votes recorded as "present." While 194 House members did not formally cast a vote, this did not necessarily reflect indifference to the Hetch Hetchy issue. As the *Congressional Record* made clear, 134 members had "paired" their vote with another member who held an opposite view on the bill. This mechanism of "pairing" votes—which was a common congressional practice in the early twentieth century—allowed members flexibility in being away from Congress by making sure that the opposition did not benefit from their absence. Taking into account these 67 paired votes, a more accurate House tally on H.R. 7207 would be 250 yeas, 110 nays, 9 present, and 60 not voting.[86]

Regardless of vote pairing, San Francisco had won convincingly. Now the bill moved to the Senate. For this final stage of the battle, John R. Freeman would be a presence and force on Capitol Hill.

CHAPTER 7

"A Great Victory"

On August 21, 1913, Freeman and his family steamed out of Southampton, England, aboard the SS *Imperator,* docking in New York six days later. After a two-month respite from his business affairs in America, he was back at work the next day, attending a Board of Water Supply meeting and later conferring with Great Western Power Company officials. Upon returning to Providence, he was soon in "[o]ffice all day on correspondence." His summer vacation in Europe was over.[1]

San Francisco's Hetch Hetchy initiative immediately engaged his attention, and in anticipation of his return, City Attorney Percy Long forwarded him two letters detailing the evolution of the Raker bill. Most importantly, Long apprised him of concessions made to the Turlock and Modesto Irrigation Districts to win their support.[2] The key allowance agreed to by the city—the "4,000 cfs for 60 days every spring" stipulation—significantly increased the amount of water guaranteed to the districts (see chapter 6 and the appendix). Long knew Freeman would be concerned about this clause—it conceivably threatened the city's ability to ever draw 400 million gallons per day (mgd) from the Hetch Hetchy system—and he wanted to explain why it was added. He especially wanted to make sure that Freeman did not react publicly in any way that might be construed as criticism: "We had to depart from your original plan of taking all the water [above 2,350 cfs]. . . . [O]nly one who was on the ground and familiar with the sentiment in Congress can appreciate how necessary it was for us to yield. . . . [Y]ou will probably be appealed to by some people here to give your views on this matter. I earnestly trust that you will not give out anything that may seem to be a criticism." Long further emphasized the difficult demands posed by the irrigationists, stressing, "I would not again go through my experience in Washington for anything."[3]

Freeman took heed of Long's plea and, after "studying Hetch Hetchy reports," came to accept the city's conciliation, wiring the city attorney that "of course [I] will not antagonize your necessary compromise."[4] A few days later, he explained his position to O'Shaughnessy: "I have been in business many years. . . . I recognize that pretty much nearly every advance is secured through compromise, and therefore am ready to stifle my objections so long as we are on the eve of a great victory." Although unhappy with the need to negotiate with the irrigationists, Freeman was a pragmatist; he did not want the perfect to imperil the good, acknowledging that "the city is better off a hundred-fold to take the bill as it passed the Lower House in Congress than to get nothing at this session."[5] Perhaps if he had been present at the negotiations, he could have mitigated the concessions granted by the city. But he had not been there, and going forward Freeman would take no action to upend or amend the agreed-upon terms.

Upon arriving in Providence, Freeman quickly reached out to City Clerk John Dunnigan, who was overseeing San Francisco's lobbying on Capitol Hill. After advising, "I have a few friends among the Congressman and Senators," Freeman inquired, "[C]an I be of any service at this stage?"[6] Dunnigan welcomed the offer: "I was mighty glad to get your letter today. I hope you will find it convenient to get to Washington this week. We expect the Hetch Hetchy Bill to pass the House [on] Tuesday. . . . I think it would be worthwhile if you would steal a couple of days and come down here. We are close to final action and need every bit of ammunition available." Dunnigan further beseeched Freeman, stating, "[Y]ou can help me a great deal in the Senate," and counseled that "[Senator] Smoot [of Utah] argues that our recognition of the Turlock-Modesto [water] rights is in effect a distribution of water. Several Senators don't like this provision in the bill."[7]

Freeman soon headed off to the nation's capital. On Tuesday, September 3, he joined up with Dunnigan and Congressman William Kent of California and "attended [House] session while Hetch Hetchy bill was passed 183 to 43."[8] The stage was now set for his work as a Raker Act lobbyist. The next day he returned to Capitol Hill with Dunnigan, visiting Senate offices and recording that "on Hetch Hetchy matters interviewed with Senators Colt & Lippitt of RI (Rhode Island), Sen. Henry Myers (Montana) of Senate Land Committee, Sen. Charles Thomas of Colorado, Sen. Works of Calif., Sen. Perkins of Calif." On Thursday he and Dunnigan were back, calling on Senators Elihu Root (New York), Reed Smoot (Utah), and Atlee Pomerene (Ohio).[9] After a weekend in Providence, he

returned to Washington for the Senate Public Lands Committee hearing scheduled for September 10. But because of "the strenuous tariff campaign," the
meeting was delayed and pushed back two weeks.[10] For the rescheduled hearing, Freeman arrived two days early to continue "interviewing Senators about
Hetchy bill," including Senators Theodore Burton (Ohio), Charles Thomas
(Colorado), and Francis Newlands (Nevada).[11]

Freeman did not describe the tenor and tone of these Capitol Hill interviews
in his diary, but we can get a sense of the issues he addressed as a lobbyist by referencing contemporaneous letters written to Mayor Rolph and O'Shaughnessy.
In the letters, the focus was on the issue of states' water rights and on the preservationist campaign being pressed by Robert Underwood Johnson. Stressing
that he had pledged to the various senators his "profound belief that the bill
was meritorious and that a great emergency really exist[ed]," Freeman had also
come to appreciate the seeming intransigence of the states' water rights bloc,
realizing, "The ideas voiced by Mondell[,] and his amendment in the House,
are sure to come up strong in the Senate; and Smoot told Dunnigan and myself
most positively that the bill could not pass at this special session with all those
clauses permitting Federal management of State right[s] affairs."[12]

The influence of Johnson-led preservationists was also a concern for Freeman, who observed to O'Shaughnessy, "[T]he task before the Senate is not as
simple as I had supposed. . . . [T]he Nature Lovers are as vicious and menda-
cious as ever in issuing their frantic appeals thru the press and presumably by
letters to the Senators. . . . I gather that Senator Root had attached some weight
to the frantic screams of Robert Underwood Johnson and John Muir." On this
latter point, Johnson and his "Open Letter" were clearly having an effect, as
Freeman explained to O'Shaughnessy: "Friends who know of my connection
with the Hetch Hetchy work have been mailing me newspaper clippings from
various parts of the country showing how generally editors took the wild and
vicious misrepresentations of the Underwood circular as being facts."[13]

For Freeman, Johnson's ability to arouse newspapers in defense of Hetch
Hetchy soon struck close to home, with the *Providence Journal* publishing an
editorial on September 5 defiantly headlined "The Hetch Hetchy Grab."[14] Freeman took umbrage at this provocation on his home turf and quickly drafted a
response, terming the editorial a "great injustice" and assuring the *Journal*'s
readership that "a most serious shortage in San Francisco's water supply exists
beyond all reasonable doubt." Emphasizing his personal knowledge of the valley (and its mosquitos), Freeman gave no ground as to whether any beauty or
park access would be lost if the Raker Act passed: "I have myself camped out in

the Hetch Hetchy Valley in three different summers. . . . [A]s I can testify from personal experience, in early summer and mid-summer, the mosquitos, which breed by millions in its swamps, make a night's rest in the valley or a stroll in the tall grass most uncomfortable. . . . The works proposed [by San Francisco] will make this valley even more beautiful than it is today, and will make it accessible to a hundred fold more lovers of nature than can see it under present conditions." Turning to the idea that half of Yosemite National Park might be barred to visitors because of the water supply reservoir, Freeman also assured readers, "There is not the most remote intention or possibility that the public will be excluded from the watershed or from the surroundings of the future lake."[15]

For Freeman, the objections of states' righters like Senator Smoot might pose serious political problems, but he found Johnson and Muir to be personally offensive. On that point, he did not hesitate to express his feelings in print: "I presume your article, like that which I have noticed recently in other journals, was in response to a printed circular letter . . . written by the former editor of *Century Magazine*. . . . I find [the circular] chiefly interesting as a study in psychology, for notwithstanding the distinguished names it bears, I have never seen more of a wild, frantic untruthfulness crowded into a document of equal length."[16]

Taken together with observations made in his letters to Rolph and O'Shaughnessy, Freeman's letter to the *Providence Journal* reveals the kinds of arguments and counsel he likely tendered when interviewing senators. And in these interactions, he did not refrain from denigrating Robert Underwood Johnson's effort to spread "wild, frantic untruthfulness." For Freeman, defending the city's plans was business. But it was also becoming quite personal.

Senate Public Lands Committee Hearing

On September 24, the Senate Committee on Public Lands, under the chairmanship of Henry Myers of Montana, convened to consider H.R. 7207. The quorum was barely met, as only eight members of the fifteen-member committee showed up. Seven of the eight were Democrats, and the only Republican to appear was the progressive ideologue George Norris of Nebraska. Norris was a forceful advocate of public power (he would later become an influential proponent of President Franklin Roosevelt's Tennessee Valley Authority— TVA's Norris Dam was named in his honor), and he supported San Francisco's Hetch Hetchy project largely because it constituted an effort to build a publicly owned water and power system. Support for H.R. 7207 did not strictly align with party membership, but it was a bill predominantly endorsed by Democrats and opposed by many Republicans.[17]

Interior Secretary Franklin Lane had enthusiastically promoted the Raker bill at the House hearings, and in early September, the *Los Angeles Times* reported that "Secretary Lane was delighted when told that the House had passed the Hetch Hetchy bill."[18] Thus, it might be presumed that Lane would reiterate his support at the Senate hearing. But on September 10 he was stricken by an "attack of angina pectoris" (a heart attack) while attending a parade in Oakland. As the *Washington Post* reported the next day, he "was ordered to bed and will be kept absolutely quiet for several days."[19] He convalesced on the Pacific Coast for almost a month before getting back to Washington, DC, on October 10.[20] As a result, he was three thousand miles away and bedridden during the Senate hearing. Lane's backing of the Raker Act never wavered, but through the fall he kept a low public profile as interior secretary and stayed quiet as the Senate deliberated on Hetch Hetchy.

The most notable committee members absent from the hearing were Republicans Reed Smoot of Utah and John Works of California. Both senators were in their home state in late September and had no plans to return east simply to participate in a one-day hearing dominated by pro-dam Democrats and the maverick Norris. Smoot made no secret of his opposition to the legislation on the grounds that it would bring the federal government into the realm of water rights, but he was little concerned about the flooding of Hetch Hetchy Valley. What he cared about was possible federal intrusion into a legal arena where he believed state's rights should dominate.[21] For Senator Works the issue was not simply about an expansion of federal power but also about the rights of farmers and irrigators to control local rivers and not suffer from diversions directed toward expanding metropoles.

Works was a Republican from southern California (he had been a judge in San Diego) and had no ties to San Francisco and no particular reason to back the city's effort to obtain a mountain water supply.[22] But he did perceive that supporting rural agricultural interests might prove politically advantageous.[23] Works's fellow California senator George Perkins and California members of the House of Representatives supported H.R. 7207 but Works reveled in being an independent iconoclast on this issue. Championing the irrigationist cause, he would steadfastly oppose the bill through the final Senate vote.

During the course of the one-day committee hearing, ten witnesses were called to testify. Of these, four opposed the Raker bill on the grounds of its impact on the beauty of Hetch Hetchy Valley and on the sanctity of Yosemite National Park; they included Edmund Whitman, representing the Society for the Preservation of National Parks, Robert Underwood Johnson as "former editor of *Century Magazine*," Richard Watrous, secretary of the American Civic Association, and former congressman Herbert Parsons, a New Yorker who had opposed San Francisco's Hetch Hetchy plans in 1910.

Watrous's testimony was brief, as he placed into the record a letter from Horace McFarland (who could not attend because of illness) averring that if "San Francisco should prevail in her present contention, it is obvious that there can be no safety whatever for any part of any national park upon which municipal engineers might cast an eye."[24] Adopting a similar preservationist perspective, Parsons objected to the inundation of Hetch Hetchy on the grounds that San Francisco simply did not wish to pay for water from an alternative source of supply. In a sentimental vein, Parsons also lamented that impoundment of the reservoir would not just degrade the park's visual splendor but also diminish the "nature sounds" that filled the valley.[25]

Edmund Whitman certainly held the nature of Hetch Hetchy in high esteem, but he kept a tight focus on criticizing the city's economic justifications for damming the valley. He quickly zeroed in on the Sacramento River's ability to serve as a municipal supply at a small cost over the Hetch Hetchy project. This led to an engagement led by Senator Norris that drew Freeman into the exchange:

> SENATOR NORRIS: Do you contend, Mr. Whitman, that the City of San Francisco ought to be required to take its water out of the Sacramento River?
>
> MR. WHITMAN: It is reported by the Board of Engineers as a perfectly proper source. Mr. Freeman would say so.
>
> SENATOR NORRIS: Would it not be necessary to filter it?
>
> MR. WHITMAN: Yes; most water supplies of this country are filtered, sir.

Picking up on whether or not San Francisco should be required to accept filtered water from the Sacramento, Senator Thomas of Colorado pressed the issue, making a comparison to an East Coast city familiar to Freeman:

> SENATOR THOMAS: Would not that argument apply equally to New York? There is plenty of water in the Hudson River. Why should not New York, instead of disturbing the supply of the Catskills and marring the usual beauties of that section, be required to get its water from the

Hudson River, from which millions of feet are running, as Mr. Lincoln would say, unvexed to the sea?

MR. WHITMAN: The city of New York[,] sir, was left to get its own supply in a business way; that is all.

SENATOR THOMAS: San Francisco is trying to do the same thing. . . .

MR. FREEMAN: I will say . . . as one of the engineers of the New York scheme, that we seriously considered filtering the Hudson River at Poughkeepsie, but absolutely turned it down because manifestly, I will say, it was not so attractive as pure water. The same is true of the water supply of the community of which Mr. Whitman is a resident [Cambridge, Massachusetts].[26]

A few minutes later Whitman revisited the issue of "mountain water," tying it to the hydropower potential of Freeman's gravity-flow aqueduct: "[T]he idea of mountain water has carried the people away. What is San Francisco asking for? She is not only asking to destroy the large use [by campers] of this corner of the park, but she is getting an electric power which the [Army Board] engineers estimate is worth $45,000,000 and which Mr. Freeman says in his report is capable of developing 200,000 horsepower. . . . I have no objection to San Francisco trying any experiment in municipal ownership [of an electric power system] . . . but not at my expense as a citizen of this country."[27]

In concluding his remarks, Whitman summarized a position that he and other preservationists shared in opposing San Francisco's plan: "I am prepared to admit, that at some time in the future the necessity of the people of California will be such that not only will this Hetch Hetchy Valley be used [for water supply] but the Yosemite Valley may be used. The pleasure of the people must give way to the necessities of life. I simply say[,] sir, that that time has not come, and that the city of San Francisco has plenty of water elsewhere which she can get cheaper, if she is only willing to do so."[28]

"That time has not come": by employing that tactic, Whitman sought to postpone a decision on building the dam to some distant date, a time when the Raker Act might well be more reasonable. But perhaps that day would never come, and that was certainly the hope of many preservationists. Delay, in Whitman's calculus, was their friend.

Robert Underwood Johnson had not appeared at the House hearing, but in early August his "Open Letter to the American People" argued that "[t]he Hetch Hetchy Scheme . . . should not be rushed through the extra session [of Congress]." This letter had been widely circulated and, as evidenced by the

Providence Journal piece that had raised Freeman's hackles, fostered numerous articles and editorials criticizing San Francisco's plans. Appearing at the Senate hearing, Johnson had a chance to continue his fight. But he was also placed in a position where he could be questioned about earlier claims and opinions.

As a witness, Johnson was permitted to ramble on, and he soon lectured the senators on his role in helping create Yosemite National Park: "Mr. Muir and I are probably the only men living who were engaged in the project to create this national park. As I have said to you, Mr. Muir fell in with my proposition."[29] But committee members were more interested in questioning him about claims made in his open letter, particularly in regard to the Bartell report and the Mokelumne River. They were also interested in probing his relationship with the water promoter Eugene Sullivan. Senator Norris took the lead:

> SENATOR NORRIS: There is one of your letters[,] Mr. Johnson, and I want to call your attention to it, and that is one of the things which I think shows your enthusiasm.
>
> MR. JOHNSON: Why should we not be enthusiastic?
>
> SENATOR NORRIS: I think sometimes a man gets enthusiastic until he becomes unreasonable on the subject. . . . This is from your letter. . . . You say: "[I]f this legislation is not railroaded through Congress, an even fuller report on the Mokelumne resources than that of Engineer Bartell will be presented along with an offer of rights and sites by the Sierra Blue Lakes Water [and] Power Company. . . ."
>
> MR. JOHNSON: I so claimed; yes.
>
> SENATOR NORRIS: Did you write this letter before you knew that all these things were false? Did you not know that Mr. Sullivan, who represents this company, came before the House committee, and his testimony is in the record, and that it was clearly shown from his own cross-examination that there was nothing to this[?] . . .
>
> MR. JOHNSON: I have not read his testimony.
>
> SENATOR NORRIS: That testimony showed, and I think it was conceded by everybody, that this was in plain words nothing but a humbug. . . .
>
> MR. JOHNSON: I have no brief for Mr. Sullivan in the matter. . . .
>
> SENATOR NORRIS: It is certain that this statement by Mr. Sullivan amounts to nothing.[30]

The other key point that Norris wanted to get into the record involved Johnson's expressed belief that the city should be kept from damming Hetch Hetchy even if it cost $100 million a year to desalinate seawater. After reading aloud Johnson's testimony from the Fisher Hearing, Norris asked if he still agreed with what he had said ten months prior:

> MR. JOHNSON: That latter part about that $100,000,000 between myself and the Secretary was mere pleasantry; but the time has perhaps come when the distillation of sea water for domestic purposes will be possible.
>
> SENATOR NORRIS: Mr. Johnson, the question is whether it is possible now. . . . Do you think we ought to compel San Francisco to go to that length before we give her the Hetch Hetchy Valley?
>
> MR. JOHNSON: Not at all. Now Gentlemen, I do not wish to be driven into a corner on pleasantry of that sort. It is a hypothetical question. I was leaving at that moment for a funeral of a friend who had just died that morning. . . . In all seriousness, if you ask me whether San Francisco should pay that amount for its water I should say no. Take the Hetch Hetchy Valley. . . .
>
> SENATOR NORRIS: That is what I wanted to know, and I am glad to hear that.[31]

Perhaps Johnson was too glib in dismissing his answer to Secretary Fisher as a hastily uttered "mere pleasantry." And perhaps Norris was too keen on pushing Johnson to retreat from an ill-advised answer to an extreme hypothetical. But Norris had made his point that someone like Johnson might well be too "enthusiastic" in arguing for the preservation of the Hetch Hetchy Valley no matter the cost. And by extension, perhaps all preservationists seeking to protect Hetch Hetchy suffered from their own cases of overenthusiasm and had become blind to the reasonable needs of San Francisco.

Preservationist objections to the Hetch Hetchy project were hardly new, and nothing remarkable came from the testimony of Watrous, Parsons, Whitman, or Johnson. They parried with committee members but broke no new ground. This stood in contrast with the testimony of W. C. Lehane, a Modesto farmer who took aim at the water allocation compromise reached between city officials and irrigationists in June. That had been a hard-fought negotiation, with the city making considerable concessions (most importantly, the guarantee of 4,000 cfs flow for sixty days every spring). And the city had depended upon statements made by irrigation district representatives as evidence that it would not divert Tuolumne flow to the detriment of agriculture. But all seemed threatened when Lehane unleashed a determined attack and, asserting that farmers had an unqualified "moral right" to the Tuolumne, demanded that consideration of the Raker bill be put on hold pending further investigations.

The crux of Lehane's testimony centered around three telegrams sent to Myers, the committee chair. The first was from Levi Winklebeck, a water user from Modesto who claimed to represent a large coalition of disgruntled farmers who had "signed a petition to the effect that the lands tributary to the Tuolumne River [were] able and willing to store the Hetch Hetchy waters, and asking and urging that the senate postpone action on the Raker bill and appoint a commission to investigate and report on [these] claims."[32] Lehane followed with similar telegrams from a farmer "representing the Turlock Irrigation District" and from the Chamber of Commerce of Crows Landing, a community lying on the west side of the San Joaquin River; both asked the Senate to forestall action on H.R. 7207.[33]

Most provocatively, Lehane also posited a "moral" argument in opposition to San Francisco's plans: "We [farmers in the valley] have the moral right to the water in the Hetch Hetchy, because we are on the Tuolumne watershed and it is a universally conceded matter that the water should go to the people on the watershed of the streams where they live."[34] This was not explicitly an "anti-urban" diatribe because, perhaps, a large metropolis might someday spring up in the Turlock/Modesto region. But San Francisco and other Bay Area communities were not in the Tuolumne watershed and, by Lehane's logic, could possess no "moral" claim to the stream.

Committee members did not attack the integrity of Lehane the way that House members had gone after Eugene Sullivan. Still, they questioned him about why they should discount previous statements from irrigation district leaders who supported the bill. Lehane offered his interpretation of what had taken place earlier in the summer: "A committee was rushed to Washington [in June]. . . . Word came home that San Francisco had an air-tight cinch on the Hetch Hetchy down here. . . . [Our representatives] wired back that we would decide when they got home . . . [but they] did not get home until the first week of August. . . . We went out into the country with those fellows and debated this with the farmers. . . . [T]hey never got a farmer who was so favorable to the Raker bill that he wanted to stand up and be counted."[35]

Lehane's "moral rights" posturing and his request to postpone deliberation on the Raker bill did not marshal any support among committee members. Instead, the attending senators believed that the House committee had worked in good faith with irrigationists and there was no need to delay the proceedings based on a few farmers' belated protest. To bolster this argument, the committee called upon Congressman Denver Church, whose congressional district encompassed the lower Tuolumne.

Upon being introduced by Congressman Raker, Church told the committee, "[O]riginally I was very much opposed to this plan. I had heard of this Hetch Hetchy matter for years and years." In the spring of 1913, the newly elected congressman from Fresno realized that legislation was in motion; here is how he explained his change of mind: "When I came to Washington this spring the battle was on in reference to the Hetch Hetchy and very fortunately certain representatives of the [irrigation] districts came here. . . . I felt greatly relieved, because I knew that what they decided in relation to the matter would be for the best interest of the districts . . . [and] they entered into an arrangement—an agreement. Those agreements are embodied in this bill. . . . For that reason I withdrew any opposition that I had."[36]

Church further described the reaction "back home" to the final version of the Raker bill, using telegrams to verify his claims. The first, dated August 13, reported upon a joint meeting of the Modesto and Turlock Irrigation Districts and advised that "the Raker bill, as recommended by the House committee[,] was approved."[37] This telegram also referenced support from the Stanislaus County Board of Trade, who resolved "that no further opposition would be made to the Raker bill." But this board's support came with a key caveat: "The bill should be adopted without any material amendment and . . . the strongest opposition should be made to any change in the bill." Congressman Church then read two shorter telegrams emanating from a "mass meeting of taxpayers and irrigators" and "a mass meeting of the Turlock Irrigation District" that also urged support for the Raker Act.[38] After reading these messages into the record, he bowed out of the proceedings: "Gentlemen, that is all I have to say."[39]

The committee declined to question Church about how his testimony clashed with Lehane's pleadings expressed just a few minutes earlier. Farmers in the valley would continue to call for further investigations into H.R. 7207, and local resistance to the bill would grow through the fall. As Church affirmed, however, the city had previously negotiated in good faith with the districts' leadership, and together they had reached agreement over water allocations of Tuolumne streamflow. As far as the Senate committee was concerned, that was on the record and it was persuasive.

The hearing's final phase was given over to witnesses who supported the city's cause. First up was Congressman John Raker, whose district encompassed the damsite (but not Turlock and Modesto). As the author of the bill that carried his name, Raker was familiar with the legislation and could testify with authority as to its purpose and terms: "[The bill] utilizes the park in the proper way. It gives to the city and county of San Francisco the pure water it ought to have, it absolutely guarantees to these [irrigation] districts the permanent supply of water that they themselves by their committee and their representatives state they ought to have." Raker also acknowledged the "states' rights" argument against the bill and addressed the possible fears of the irrigation district farmers on this point: "Your [water] rights, whatever they may be, will not be interfered with by this bill or any provision of it, because it is left entirely to the laws of the State of California to settle and adjust every right and to determine every need."[40]

After Raker stepped down, Congressman William Kent came forth. A prominent progressive, Kent was independently wealthy (his father had made a fortune in Chicago real estate and meatpacking) and, prior to being elected to Congress in 1910, had become famous in conservation circles for buying a tract of coastal redwoods north of San Francisco and donating it to the federal government for preservation as a national monument. In making this gift, he requested that it be named the Muir Woods National Monument in honor of the renowned naturalist.[41] But while he had deep respect for John Muir's life work, Kent differed with him on the issue of Hetch Hetchy, becoming an outspoken supporter of the city's plans. This support went so far that, at the time the Senate committee hearing was being held, both Freeman and Dunnigan were staying as guests at Kent's spacious home at 1925 F Street NW, a few blocks from the White House. Kent was a prominent figure in the nation's cultural life, and his advocacy for the damming of Hetch Hetchy did much to counter positions held by Johnson and other preservationists. Kent spoke to the committee for only a few minutes, but he was forceful and direct: "[I] resent the criticism that we who stand for this bill are opposed to conservation. . . . [W]hen an opportunity comes to give to a great community upward of 200,000 horsepower . . . [and] when it comes to the question of benefiting upward of a million people, then I believe that conservation demands that I do my duty and try to help not hinder such a worthy project."[42]

Kent concluded with a brief broadside, targeting Eugene Sullivan and the way that Johnson, in his open letter, had championed Sullivan's scheme as a practical alternative. Kent branded it all as falsity and propaganda: "This man Sullivan, whom we proved in the House to be a thief[, is] a man who ought to be in the penitentiary. We proved his claims to be absolutely valueless. . . . Every clipping I get from the public press—and I get lots of them—has this same foundation of falsity."[43]

Following Congressman Kent, Alexander Vogelsang of the San Francisco Board of Supervisors refrained from attacking Sullivan or Johnson. Instead, he wanted to calm the waters stirred by Lehane, stressing the city's positive ties with the two irrigation districts: "We are very friendly to the people of the San Joaquin Valley. . . . We have attempted always to treat them as friends, we have guaranteed them always, and always voluntarily, as to the [water] rights they originally claimed." Yes, the city was asserting a claim to the floodwaters of the Tuolumne. But such claims were subordinate to the senior rights of the irriga- tion districts, and that was why the city needed a large storage dam: "We have filings subsequent to theirs which can only be fulfilled by the conservation of the storm waters of that section. We are asking [only for] the right to conserve the storm waters by building this dam at the mouth of Hetch Hetchy Valley." Vogelsang did not brashly demand that San Francisco's desires be met as a mat- ter of municipal (or moral) right. Instead, he came to the Senate with a simple plea: "I have come to ask you to extend a hand to us."[44]

John R. Freeman: The Last Witness

About 5 o'clock, City Clerk Dunnigan introduced Freeman to the committee as "the chief engineer of the Hetch Hetchy project, who you probably know as one of the [most] noted hydraulic engineers of the world." He was available "to answer any question any senator may ask him regarding the engineering or other features of this proposition."[45] With this, the testimony of the meeting's last witness began.

Immediately, Senator Thomas of Colorado raised a question made "in con- nection with the Spring Valley works" and a possible "available daily supply of 230,000,000 gallons of water" for the city. Freeman deftly responded, assuring the senator that "that matter has been investigated most carefully" and that any effort to develop such a supply was "so utterly impractical for the supply of the city that it ought never to be considered." Senator Thompson of Kansas then broadened the question and asked, "[I]s there any other place where they can get such a supply as at Hetch Hetchy?" to which Freeman replied, "No."[46]

But rather than let this terse response to Thompson's question suffice, Free- man then set the spotlight on Robert Underwood Johnson and complained, "I am sorry that so good a man as Mr. Johnson should feel that he should try to discredit the motives of others."[47] With this, Freeman implicitly criticized the impact made by Johnson's recent open letter, in which the former magazine editor harped on the city's lack of interest in the Mokelumne River. To Freeman, Johnson was the proximate cause of all the negative publicity being spread across the nation; it was Johnson and his cohort Whitman who were responsible for disseminating ill-informed propaganda about the bountiful alternatives that the city—and its leading consulting engineer—were ignoring. Freeman took such attacks personally and, especially in regard to the Bartell report and the claims made by Eugene Sullivan, wanted to call his bête noire to account:

MR. JOHNSON: Mr. Freeman, please specify, will you kindly?

MR. FREEMAN: I will. . . . Have you not, at least, made the statement that a report by an assistant in the city engineer's office was suppressed that should have been given to the Army Board?

MR. JOHNSON: I did make that statement.

MR. FREEMAN: Are you still so sure about that?

MR. JOHNSON: I think so.

MR. FREEMAN: Was it your friend Mr. Sullivan?

MR. JOHNSON: Not at all. I beg of you that you will not speak of Mr. Sullivan as my friend.

MR. FREEMAN: He seems to be the chief author of your circulars.

MR. JOHNSON: That is offensive. I have said nothing offensive to you. . . . I do not like to have you asperse my motives. . . . I beg of you that you will not speak of Mr. Sullivan as my friend. I have not said or done anything here that could in the slightest way be offensive to you gentlemen. I wish to be entirely square.

MR. FREEMAN: Well, I wish to be fair also.

SENATOR NORRIS: Mr. Johnson[,] is it or is it not true—I want to get what the facts are—is it or is it not true that in this letter I [earlier] quoted from your source of information was from Mr. Sullivan?

MR. JOHNSON: No. It was from Mr. Aston. Mr. Taggart Aston.

SENATOR NORRIS: Mr. Sullivan represented the same company here.

MR. JOHNSON: Mr. Taggart Aston. He is the author for my statement. I simply stated what is claimed by everybody. I am not taking my information from Mr. Sullivan at all. . . .

MR. VOGELSANG: I simply wish to make the statement that Mr. Aston, referred to by Mr. Johnson, is an alleged engineer whom we have never met, and who has never appeared before the House committee. The testimony of Mr. Sullivan shows that he and this man Aston were associated together. . . .

SENATOR THOMPSON: They had some private interest which they wished to
    sell to the city.[48]

With this exchange, Freeman—supported by Senators Norris and Thompson—brought the credibility of Johnson into question and reinforced how the
most provocative claims in the open letter were connected to efforts by Sullivan and Aston to market the supposed assets of a private water company. As
discussed in chapter 6, Freeman then offered assurance to the committee that,
while much was being made of the Bartell report by city opponents, it was simply a "step in the work" of estimating how a possible Mokelumne-based water
supply might compare to the city's Hetch Hetchy plan:

MR. FREEMAN: Mr. Bartell was one of the assistant engineers in the office
    of the city engineer. I was constantly in communication with Mr. Bar-
    tell. The report that he made on the Mokelumne source was at my
    suggestion and at the suggestion of the then city engineer, Mr. Man-
    son. Mr. Bartell's report was made to me. He was asked to take all
    the estimates and submit a report relative to the Mokelumne source.
    The report was not in any sense suppressed. I wish to deny that most
    emphatically. . . . Mr. Bartell was available to the Army board. There
    never was the slightest attempt to suppress anything of that kind.
THE CHAIRMAN [MYERS]: Then the report was available for the use of the
    Army board. Is that true?
MR. FREEMAN: It was. . . . I wish to say to the committee that the matter
    [of the Bartell report on the Mokelumne River] has been carefully,
    candidly, and fully presented, to the best of my knowledge, judge-
    ment and belief.[49]

It might be argued that here Freeman overstepped the truth in claiming that
everything related to the Bartell report had been "carefully, candidly, and fully
presented." But the supposition that the report had been purposely suppressed
and kept from the Army Board was something that Freeman could reasonably
deny. And that is what he did.
    As the hearing neared adjournment, Freeman closed by underscoring how it
had been his desire to not just design a major water supply system but also to
enhance the Sierra environment and make it more beautiful through a creative
act of engineering. Noting his "pride in the work," he trumpeted, "I am something of a nature lover myself. I have worked hand in hand with Frederick Law

Olmsted and some of the other noted landscape artists of the country [and] for many years I was consulting engineer to the Metropolitan Park Board of Boston, and I developed the plan for the Charles River basin in Boston and for the drainage of the Fresh [Pond] marshes in Cambridge." With that preamble, he wanted the world to know, "For many years I have been liberal in my time upon these public works because I fully believed in them. I came to this [Hetch Hetchy] work in the same spirit, and I think if anyone will go through my report they can find everywhere the handiwork of a man who tries not to tear down but to leave things more beautiful. [It is] one of the matters in which I take great pride."[50]

In his testimony Freeman offered no technical treatment of how to measure runoff in the Tuolumne basin or assess the groundwater capacity of Alameda Creek, and he did not discuss how his proposed dam design would accommodate geological conditions at the Hetch Hetchy site or how the hydroelectric power capacity of his system could be maximized. Technical issues were put to the side. In his parting words to the committee, Freeman stressed a desire "not to tear down but to leave things more beautiful." He would not allow his opponents to lay sole claim to the issue of beauty and to what constituted the greater public good. Yes, he was an engineer, but his arguments for the dam transcended the mere utilitarian and spoke to broader ideals of social betterment and civic progress.

Once Freeman stepped down, Chairman Myers asked "for a motion to take up this bill [H.R. 7207]." Senator Key Pittman of Nevada quickly complied, and after a unanimous voice vote, Myers "ordered that the committee favorably report the bill." Their work completed, at 5:25 p.m. on September 24 the committee adjourned.[51]

Resistance and Unanimous Consent

With the Committee on Public Lands's endorsement, the Raker bill headed to the full Senate and a presumed final vote. But how quickly it would reach the Senate floor was unclear. The city wanted to move fast, but opponents resisted doing anything to help speed the bill along. They hoped that senators weary from a long summer would push consideration of H.R. 7207 off until the next session of Congress, or perhaps even longer.

After the committee hearing, Freeman stayed overnight in Washington and the next day resumed lobbying on Capitol Hill, where he "interviewed Senator Gronna of No. Dakota for an hour on Hetchy Bill."[52] He then returned to

Providence. In a brief letter, he updated O'Shaughnessy on his recent service to the city: "I came back yesterday from another week of walking the halls of Congress and interviewing Senators and contradicting the infamous lies that have been circulated by the pious, self-satisfied highbrows Muir, Underwood Johnson, and Whitman. It was easy to be a 'lobbyist' on an errand whose righteousness I believed with all my heart."[53]

While praising the city clerk's skillful work, Freeman nonetheless hesitated to claim imminent victory: "Dunnigan has his check list of the senators marked and verified by cross-references so completely that it now seems impossible that the passage of the desired bill can be delayed beyond the first of the coming week, but senatorial courtesy is sometimes a strange and unreasoning power, and Smoot is coming back full of determination to cut out all those features of the bill that tell how the water will be distributed, and I shall not rest easy until the President's signature is affixed."[54]

Dunnigan was hopeful that enactment would come quickly, but at the end of the month he warned Freeman, "[Senators] Borah and LaFollette want our bill delayed." Pressure from the city's opponents was building, with Dunnigan conceding that the "nature lovers have started [a] big crusade in Kansas and other states. This has scared some senators." Nonetheless, he remained upbeat: "We are going to stick to the job, as we feel we have enough votes if we can get the bill up."[55] In regard to Freeman, Dunnigan understood that the Providence engineer had other demands that required his attention—including a three-day trip to Niagara Falls on a hydropower project starting on September 30.

After many days of strenuous politicking, the long-debated tariff/income tax bill passed the Senate on October 2 and freed up the legislative calendar. Two days later, Senator Key Pittman took the floor, moving that "the Senate proceed to consideration of House bill 7207." But Senator Joseph Bristow of Kansas quickly interceded, calling out how "there is a great deal of objection to this measure throughout the country and it is not fair to a tired and worn Senate . . . to precipitate this matter on this Saturday afternoon." Bristow added, "So I hope the senator will not insist upon his motion being put." Pittman politely ignored Bristow's entreaty, and within a few minutes the Senate agreed by a vote of 37 yeas to 15 nays to proceed with considering the bill.[56] However, Bristow had set a marker, letting the bill's advocates know that it would not be an easy afternoon. And once the bill was formally read to the Senate, sitting as a Committee of the Whole, Senator Robert La Follette of Wisconsin immediately demanded a quorum call that, although eventually met, brought deliberations to a crawl before they had even started.[57]

Over the course of the afternoon, substantive questions and concerns were raised about the bill, with Senator Asle Gronna asking about alternative sources available to the city, Senator Miles Poindexter of Washington inquiring specifically about the Eel River, and Senator Harry Lane of Oregon asking about the irrigationist issues Lehane had raised at the recent committee hearing.[58] The anti-dam "scenic" perspective was also brought forth, with Senator Wesley Jones of Washington entering into the record a telegram from Ruth Karr McKee, president of the Washington State Federation of Women's Clubs: "The 5,000 club women citizens of Washington protest . . . against the despoilation of the Hetch Hetchy Valley, believing that it is in the interest of the country as a whole that such scenic places be preserved."[59]

More broadly, opposing senators professed that the bill was being needlessly rushed ("railroaded") and complained that the absent Smoot of Utah and Works of California deserved to participate in the debate. Works was particularly nonplussed that the bill was moving to the Senate floor while he was in California; on October 2 he had wired Senator Myers to register his grievance: "I have satisfied myself that the Hetch Hetchy bill should not pass without further investigation. Ninety-nine percent of water users in the irrigation district are strongly opposed to it and claimed they were betrayed by those who consented to the compromise measure. . . . The bill should not be rushed through at this session under such circumstances."[60]

The only senator with a comparable status to Works was George Perkins, the senior senator from California. As the afternoon wound to a close, Perkins took a different tack than his colleague, urging passage of H.R. 7207 as a "most meritorious and worthy measure." Avoiding any reference to the complaints of irrigators raised by Lehane, he confidently asserted that there was "no possibility of a doubt [that] the agricultural water users who live below the Hetch Hetchy Valley [will continue to hold] the riparian rights to which they are by law entitled." Ignoring preservationist protestations, Perkins also professed how "it is the consensus of opinion that this lake will add to the beauties of this wonderful canyon in a great measure." As far as Perkins was concerned, all objections to the bill were groundless. He was ready for a vote.[61]

Once Perkins relinquished the floor, the presiding officer addressed the Senate: "If there are no further amendments . . . the Bill will be reported to the Senate." Senator Poindexter immediately demanded a quorum call, and with only twenty-nine senators answering, adjournment quickly followed at 5:02 p.m. Further action on H.R. 7207 was put off until the Senate convened three days hence.[62]

Shortly after noon on October 7, the Senate returned to further consider the Raker bill. However, the proceedings quickly took a new trajectory. After Senator Pittman asked for debate to resume, Senator Poindexter pleaded, "Mr. President[,] I should like to have the matter suspended for a moment until I can confer with the Senator from Nevada." A short while later, Pittman brought forth a different plan:

> MR. PITTMAN: Mr. President, I withdraw my request that the Senate proceed with the consideration of House Bill 7207. . . . I should like to have the Secretary read the agreement. . . .
>
> THE SECRETARY READ AS FOLLOWS: The Senator from Nevada [Mr. Pittman] asks unanimous consent that on Monday, December 1, 1913, immediately upon the conclusion of routine morning business, the Senate will proceed to consideration of the bill (H.R. 7207). . . . and that before adjournment on the calendar day of Saturday December 6, 1913 the Senate will vote upon any amendment that may be pending to the bill, any amendments that may be offered, and upon the bill, through the regular parliamentary stages, to its final disposition.
>
> THE PRESIDING OFFICER: Is there objection to the entering of the order as read by the Secretary? The Chair hears none, the unanimous consent decree is entered into and ordered.[63]

With this seemingly simple parliamentary maneuver, the Hetch Hetchy bill entered a new legislative chapter, one guaranteeing that, up or down, the Senate would vote upon the Raker bill by midnight on Saturday, December 6. While further speeches could be offered that day and in the coming weeks, no vote on H.R. 7207 could be taken until after Thanksgiving. In compromising with Senator Pittman, Poindexter provided preservationists with valuable breathing space in their fight to save the valley. Similarly, pro-states' rights and pro-irrigator forces would also have time to strengthen resistance to the bill. But getting this additional time came at a cost. On December 1 the Senate would take up the bill, proceed with debate, and administer a final vote within six calendar days. Because the Senate had agreed to this timetable by unanimous consent, it could only alter or amend the decree by unanimous consent—meaning that the will or action of a single senator could guarantee that a vote would take place before midnight on December 6. This was tremendously important because it eliminated the possibility that a vote on H.R. 7207 could be filibustered ad infinitum. Perhaps the city's supporters would lose and Freeman's plan would be rejected.

Or perhaps not. Regardless, the die was cast. Yea or nay, the Senate would vote on the bill in early December.

Strike Hard and Very Fast

After his return from Europe, Freeman devoted significant time to lobbying for the Hetch Hetchy Dam, making three separate trips to Washington, DC, in September. But other clients also held a claim on his time, and he was not present on Capitol Hill during the Senate's sessions on October 4 or October 7. Although Dunnigan kept him updated, for the immediate future his energies were dispersed over an array of projects stretching from the Smoky Mountains of Tennessee and North Carolina to Chicago, Keokuk Dam, Niagara Falls, and the Bassano Dam in the Alberta prairie of western Canada. Although on the back burner for most of October, the Hetch Hetchy project would regain Freeman's attention by the end of the month.[64]

Following approval of the unanimous consent decree, City Clerk Dunnigan was the only person from the San Francisco delegation to remain in the capital through the latter part of November. He knew of Freeman's travel itinerary and met with him on October 8 as he passed through Washington on his way to Knoxville. Dunnigan informed Freeman, "Poindexter is arguing that [the] city should use McCloud or Eel [Rivers] and leave Tuolumne for irrigators." Turning to the nature lovers, Dunnigan expressed concern about their possible influence: "[Senator] Bristow [of Kansas] has blown up. He got a telegram saying five thousand women protested, and on Saturday he told me that if we could not get water from any other place 'than the Yellowstone,' 'we could move the city to another location.' Of course, I know he meant Yosemite, but this slip shows his state of mind."[65]

With this latter comment, Dunnigan acknowledged how the anti-dam crusade, with Johnson at the helm, had been inspiring citizens to flood their senators with letters and telegrams. John Muir had been corresponding with Johnson since June about how best to oppose the Raker bill; after the House approved it in early September, he continued to support Johnson's efforts.[66] With Muir pressing Johnson, "[W]e have to strike hard and very fast," the two friends rallied their compatriots in both the California and Eastern Branches of the Society for the Preservation of National Parks. The goal was to blanket the nation with thousands of pamphlets and circulars warning people of San Francisco's plans and imploring them to write President Wilson and their senators in mass protest.[67]

In 1908–9 a groundswell of popular support had been stirred by widely distributed pamphlets publicizing the poetic prose of Muir as well as other anti-dam arguments. And it worked. San Francisco's attempt to take ownership of the Hetch Hetchy reservoir site pursuant to the Garfield Permit had been soundly beaten back. The example was clear. Only a few years before, citizens across the nation had rallied in defense of Hetch Hetchy and dealt a blow to the park invaders.[68] Why couldn't they do it again in 1913 and end the dam controversy once and for all?

Sometime after the House had passed H.R. 7207, the California Branch of the Society for the Preservation of National Parks (SPNP) began distributing "Circular Number Seven" to offer guidance on "how to help preserve the Hetch Hetchy Valley and Yosemite National Park." The circular listed all the senators in the 63rd Congress and also provided a sample letter/telegram that could be edited "in your own language and in accordance with your own ideas" as well as a "resolution" that could be used by groups of "spirited citizens" to help them register their objections to the dam. Details about the dispute over the city's plans were largely missing, although the circular proclaimed that "eminent engineers report that this proposed invasion of a national wonderland is wholly unnecessary and that San Francisco can get an abundance of pure water elsewhere." As reflected in this circular, what was to be protected was not some locale of untrammeled wilderness, but rather the "wonderful scenery" and the "great public playgrounds" that constitute America's national parks.[69]

The Eastern Branch of the SPNP took a different tack, publishing a pamphlet titled "The Truth about the Hetch Hetchy" that did not include practical information such as a listing of senators or sample letters or resolutions.[70] Instead, it raised issues related to the city's claim of needing Hetch Hetchy for a reservoir, to the impact on irrigation in the San Joaquin Valley, and to concern that use of the upper Tuolumne watershed ("nearly one half of the Yosemite National Park") would be severely curtailed. In terms of the valley proper, the issue of how the mountain meadow could be used to increase park visitors was further emphasized, because "the Hetch Hetchy Valley is the only large level place in the northwestern portion of the park where hotels and permanent camps can be located." As with the California Branch pamphlet, the concern was not about wilderness but about leveraging the valley's scenery to increase tourism.

Almost certainly Edmund Whitman wrote the Eastern Branch pamphlet; with frequent reference to the "great expert, Mr. Freeman" and to "Freeman's Report," it emphasized the existence of alternatives to the Hetch Hetchy system—such as filtering the Sacramento River—which, while perhaps costing a bit more, were nonetheless viable.[71] Drawing from the language used in John-son's August 1 "Open Letter," the pamphlet boldly proclaimed that "the city deliberately suppresses the Bartell official report ... which shows that from the Mokelumne River alone 432 million gallons of water a day may be had. . . . It is a Sierra water drawn from mountain lakes." While the Bartell report does make reference to a possible flow of 432 mgd, that number was taken from measurements made about 140 feet above sea level. To say that this would constitute "a Sierra water drawn from mountain lakes" is disingenuous and represents the type of statement that Freeman could readily renounce as, at best, a "half-truth."

Along with the pamphlets discussed above, Robert Underwood Johnson drafted a letter for the nationally distributed *Collier's* magazine published on October 25. The themes and arguments covered by Johnson were familiar, but perhaps a new audience could be found for them and thus incite more telegrams and letters of protest. Thus, Johnson decreed to *Collier's* readership that if the dam were built it would have a devastating impact on campers seeking to use the upper Tuolumne: "The plain fact is that if the city takes the Tuolumne and Hetch Hetchy it must have the whole watershed—the whole 500 square miles—to protect itself. The necessary sanitary regulations will exclude the public from the free use of the park." And he again reminded people that, as the Army Board had affirmed, "there were other [water supply] sources which, if combined with the present supply[,] would solve the [city's] problem, which was one *simply of cost.*"[72]

Following up on his earlier "Open Letter," Johnson also claimed that "the Mokelumne watershed could probably furnish 432 million gallons per day." But perhaps because of criticism heaped on him at the Senate hearing, he made no mention of Eugene Sullivan, Taggart Aston, or the Sierra Blue Lakes Water and Power Company. He also avoided referring to the Bartell report as being "suppressed," instead using the more subdued phrase "quietly pigeonholed."[73]

In both the Eastern Branch's pamphlet and Johnson's letter in *Collier's,* the hydroelectric power that could be generated by Freeman's gravity-flow aque-duct was raised as an issue. Freeman considered his scheme's power potential to be a great attribute, with an estimated value of some $45 million, but Whitman and Johnson saw it differently: "We do not quarrel with San Francisco on any plan for municipal ownership of public utilities using electric current, but that should be at its own expense. And not at the expense of the nation." In his let-ter, Johnson went so far as to claim that electric power, not water supply per se, provided the real impetus underlying formulation of the legislation: "Take out of the bill the right to sell electric power and the city will withdraw the measure at once."[74]

Complaining about how the city would benefit from electric power generated by a municipally owned system represented something quite different than objecting to the dam on the grounds that it would destroy park scenery or deprive campers of park access. Perhaps it broadened the argument against the dam, but it also highlighted the enormous amount of electricity that Freeman's plan promised for the people of California. And why was all that hydropower such a bad thing?

As reflected in the thousands of letters and telegrams sent to senators in the fall of 1913, preservation efforts clearly roused many American citizens to register their disapproval of San Francisco's plans. There was undoubtedly an anti-dam movement engaging people across all forty-eight states. But was this protest having an impact on Capitol Hill? In the weeks after the Senate agreed to push off consideration of the Hetch Hetchy bill to December, City Clerk Dunnigan remained in Washington to monitor reaction to the swarm of anti-dam publicity. A week into November, Dunnigan apprised Freeman of the situation. As far as he was concerned, the news was good: "I cannot see that we have lost any ground during the interim forced upon us by certain Senators. . . . I do not think the nature bugs have made a dent in the Senate and, excepting Hollis [of New Hampshire,] there is not a Democrat that is listening to them." Regarding Senator Works of California, Dunnigan was also upbeat: "[He] will insist upon his amendments, eliminating all federal control and the irrigation conditions. [But] I do not think he has a chance to pass his amendments."[75]

In the collective mind of the anti-dam community, the struggle against San Francisco was portrayed as a lopsided affair, one in which the city's resources were much greater than what the preservationists could marshal. In many ways this was true. The city did have a substantial financial base that could pay Freeman's consulting fees and underwrite publication of his 421-page, heavily illustrated *Hetch Hetchy Report.* But when Horace McFarland claimed, in a November 6 letter to the *New York Times,* that the city was maintaining "an expensive and persistent lobby" in Washington, he woefully overestimated the size and scale of the city force that, through the fall of 1913, was promoting the Raker bill[76] In fact, in the weeks after the unanimous consent decree was agreed to in early October, the "Hetch Hetchy Lobby" (as McFarland's letter was headlined in the *Times*) consisted of City Clerk Dunnigan as the lone lobbyist. During much of October and November, when the anti-dam forces were hoping to turn the tide on Hetch Hetchy, there was no large and expensive lobby standing in their way. There was just Dunnigan, deploying arguments garnered and honed during his earlier work with Freeman.

Freeman and Olmsted

Once Freeman was back from his October trip, he was not reticent to fume about the "violent propaganda that Underbrush Johnson and Whitman have been carrying on."[77] But he hoped that some of their authority could be deflated in the public eye if his colleague Frederick Law Olmsted Jr. (whose father had co-designed New York City's Central Park and also served as an early park commissioner of the Yosemite reserve) could be prevailed upon to endorse—or at least not criticize—the city's plans. In his testimony to the Senate committee, Freeman had proudly referenced having "worked hand in hand" with the Massachusetts-based landscape architect, and he used Olmsted as a touchstone for his own standing as "a nature lover" himself.

In October he sent Olmsted a copy of his *Hetch Hetchy Report* and then in November followed up with a letter explaining, "I have been told [that] Senator Hollis of New Hampshire had quoted you as . . . opposed to granting San Francisco this valley." Hoping this was not true, he advised his architect friend, "So violent and misguided a campaign of opposition is being waged by some good men who ought to have inquired into the facts more thoroly [*sic*], that I fear you may be misled by some of their statements. I have now in my pocket a copy of a circular sent to editors throughout the United States by Robert Underwood Johnson, asking them to write editorials against the bill now before Congress, and I have never seen a political document more full of misrepresentations, even to the point of falsehood, in the form of garbled extracts and half truths."[78]

Freeman further emphasized to Olmsted how he envisaged the Hetch Hetchy project as more than a mere water supply system: "I think you know and are ready to trust me as a something of a nature lover myself, and it was this instinct and habit which impelled me to laboriously urge . . . that the city should spend about a half a million dollars extra [on roads] . . . to open up this region to the citizens of California. . . . [I]t is wiser to open up a beautiful valley like Hetch Hetchy to the multitude than to keep [it] as a preserve for the few who love solitude."[79] Here, Freeman told a bit of a half truth himself, as O'Shaughnessy had already gotten the plans changed to excise construction of the road along the southern rim of the reservoir. Nonetheless, the city was still committed to making the valley more accessible to tourists.

A few days later Olmsted responded: "[I]t is true that I told Senator Hollis that it seemed to me unwise to grant the Hetch Hetchy Valley to San Francisco as advocated by you." Olmsted then explained how he had reached this state of mind: "I was not led to this conclusion by the obviously prejudiced and emotional statements issued by some of the opponents of the project, but mainly by a study of your own excellent report. . . . [I] find myself wholly unable to concur in your judgement as to the effect of the project upon the value of the Park for its original purposes. I believe the [dam] project, if executed, would result in a very serious loss to the people of the United States."[80]

Olmsted further advised Freeman that he had written an article for the *Boston Evening Transcript* in which he would be "expressing [his] views" on the Hetch Hetchy project. In closing his letter he tried to strike a conciliatory tone that might mollify his engineering colleague: "I am sorry to differ with you, but I am sure no bitterness will creep into our pleasant relations because of it. . . . I am sure that what we both want is to get at the truth of the matter and advocate a wise policy."[81]

In quick response, Freeman professed hope that Olmsted would be "sure of the facts" because much of what had been written in opposition to the dam had been "either absolute untruths or misleading half-truths." But in defending the city's plan, Freeman himself made an argument that, while it sounded authoritative, he clearly knew to be an exaggeration, if not an outright "untruth": "It has often been repeated that the city had made no thoro [*sic*] investigation of the other possible sources. So I will say to you with the utmost sincerity, that the competing sources [in northern California] have been investigated . . . with fully as great thoroughness as was used by me or by various others when investigating alternative sources before deciding on the Catskill supply for New York."[82]

Allen Hazen's report on the Sacramento River may have been done with "great thoroughness." But for Freeman to avow that many of the reports submitted by C. E. Grunsky were comparable to his work for New York City a decade earlier strained credulity. With the push to get the Hetch Hetchy bill over the finish line, however, Freeman was unwilling to acknowledge any possible cracks in the city's armor. Now was not the time for nuance; it was a time to make bold assertions, even to the point of "untruths."

While Freeman assured Olmsted that he could "respect difference of opinions without end if the man on the other side show[ed] evidence of care in collecting and verifying his facts," his promise on this count came with a veiled threat: "You can be sure that so long as you are careful with your facts, that any difference of honest opinion will not dim my friendship or appreciation for you and that I expect to be ready a week, a month, or a year hence to write just as appreciative a letter recommending you for a large civic problem as I did last week to one of the most influential men in Canada, relative to the civic improvement of Ottawa, or when I urged your qualifications to certain inquiring friends of the Exposition Board at San Francisco, a year or two ago."[83] The message was clear: Feel free to say what you will, my good friend, I just hope that it won't impact any future recommendations I may, or may not, make on your behalf. Olmsted's article in the *Evening Transcript,* published on November 19, is a carefully reasoned essay that weighs the costs and advantages of San Francisco's Hetch Hetchy plan. In the city's defense, Olmsted observes, "Where water supply is the primary purpose to be served it is very often possible to secure incidentally important means of public recreation of certain kinds at a very small cost and with no impairment of the water supply function. . . . [T]he claim that necessary sanitary regulations would interfere seriously with camping in the park [is] a matter that seems to have been somewhat exaggerated by opponents of the reservoir project."[84] Here, Olmsted offers no support for the oft-expressed preservationist argument that creation of a reservoir at Hetch Hetchy would necessitate the closing of five hundred square miles of Yosemite National Park to campers and visitors. On this point Freeman would readily concur. Yes, there should be sanitary regulations, but they need not have a draconian effect on public use of the park.

The effect of the reservoir on the splendor of Hetch Hetchy, however, was not something easily discounted in Olmsted's view. The reservoir would not be a mountain lake, and the landscape architect knew it would incur significant cost: "The proposed body of water would not be a normal mountain lake, but a reservoir regularly subject to depletion, leaving a margin of unsightly banks and mudflats exposed for a considerable part of the shore. . . . [T]he unnatural and disagreeable appearance of a partially depleted reservoir would be apparent by far the larger part of the time." This was not the type of assessment that Freeman hoped for, but Olmsted well understood how the proposed reservoir was very different from a natural lake.[85]

Taken as a whole, Olmsted's article offered a balanced weighing of the pros and cons related to the damming of Hetch Hetchy—it was certainly not a polemic like Johnson's "Open Letter." In his concluding paragraphs, however, Olmsted left no doubt as to the proper course for the nation to take. Believing that "[t]he United States deliberately undertook to preserve the scenery of the Yosemite National Park intact for the enjoyment of all future generations," he feared that using "Hetch Hetchy as a San Francisco reservoir site would be to abandon that purpose by indirection, and would establish a precedent for abandoning the purpose of any and every park in case it conflicts with any considerable utilitarian interests."[86]

Olmsted had made a principled defense of Hetch Hetchy, one that avoided fearmongering over the claim that half of Yosemite National Park would be blocked to campers if San Francisco had its way. However, by taking a more measured approach, the article may have been destined to have less impact or visibility than Johnson's more clamorous "Open Letter." In addition, Olmsted's essay appeared very late in the game, with little chance that it might be reprinted and reach an audience beyond the *Evening Transcript*'s New England readership. By the time it was published, less than two weeks remained before the Senate would convene and begin debating H.R. 7207.

On Freeman's part, there is no evidence that he ever responded to Olmsted's article, either in personal communication or in print. Instead, he apparently chose to shrug it off and not proffer any publicity that might increase its visibility. For Freeman, the less said about Olmsted's anti-dam musings the better.

The Senate Convenes

On Wednesday, November 19, Freeman boarded the Merchants Limited train to Washington, DC, joining Dunnigan and Vogelsang at the Powhatan Hotel. During the next two days he conferred with his San Francisco colleagues, talked with Senator Pittman, and spent time "reading clippings on Hetchy." On Friday night he returned to Providence, spending much of the next week drafting a report on the Bassano Dam in Alberta. But on Thanksgiving Day he was in his office "selecting papers and packing [his] trunk for Washington." That night he headed south, and the next day he met up with Senator Pittman and the San Francisco delegation, which included Mayor Rolph, City Attorney Long, and City Engineer O'Shaughnessy. He would remain in Washington for another ten days, all the way through the final Senate vote.[87]

During the time the Senate considered the Raker bill, Freeman was on Capitol Hill from morning until late at night, ever ready to guide, encourage, and instruct senators as to the righteousness of the city's plans. Unfortunately for historians, he made only limited diary entries spelling out these activities. To construe how he defended the city's need to dam Hetch Hetchy, we are left to draw from commentary and arguments offered in earlier letters to Dunnigan, O'Shaughnessy, Long, and Olmsted, from his rejoinder in the *Providence*

The US Senate chamber circa 1905. This is where the final debate over the Raker bill was held from December 1 through December 6, 1913. During that time, Freeman maintained a vigil in the visitors gallery above the chamber. (LC-DIG-stereo-1s05771, LC-DIG-stereo-2s05771 [*stereoscopic views*], Library of Congress, Washington, DC)

Journal, and from his testimony before the Senate Committee on Public Lands. On December 1, his diary records, "Case opened in U.S. Senate. I had conference with Sen Colt of Rhode Island on merits of the bill. Evening reviewing notes and conference with Vogelsang and [O']Shaughnessy." On December 2 he continued: "In attendance U.S. Senate all day, conference with Sen Poindexter in morning, Listened to Sen Works argument." This was followed up with several summary comments: "In Senate gallery watching course of debate" (December 3), "All day at Senate until 11pm" (December 4), and "All day at Senate until 11pm" (December 5).[88] We do get slightly more insight into his time spent in the Capitol from a letter he wrote upon his return to Providence: "I sat in the Senate galleries almost constantly from Monday morning until Saturday midnight, from ten am until eleven pm, with no noon recess and with only a recess between six pm and eight pm, part of which was commonly occupied in digging up fresh items of truth for friendly senators."[89]

In contrast to Freeman and the San Francisco delegation, Robert Underwood Johnson, Edmund Whitman, and other prominent figures in the anti-dam coalition do not appear to have made their presence known on Capitol Hill during the first week of December. Of course, John Muir was far away in California—in late October, he had informed Johnson that it was "impossible [for him] to go to New York or Washington [in] November or December" to aid in the battle to save Hetch Hetchy.[90] Muir no doubt could have stirred up some anti-dam publicity had he made the trip east to fight for what he called "one of Nature's rarest and most precious mountain mansions."[91] But he stayed on the Pacific Coast, leaving it to Johnson to organize resistance to the park invaders.

In New York City, Johnson sought to rally the troops by presiding over an anti-dam protest held at the American Natural History Museum the weekend before Thanksgiving.[92] In addition, he placed a letter to the editor in the December 1 *New York Times,* calling attention to the "unnecessary vandalism" that was to be wrought by "the flooding of the great Hetch Hetchy Valley." In this letter, he also asked for "checks in any amount" as a counter to "advocates of the scheme [who] have unlimited financial resources."[93] In the run-up to the Senate debate, Johnson attempted to engage the citizenry in his defense of Hetch Hetchy, but he failed to gain much traction in countering San Francisco's plans.

From December 1 through December 6, most of the Senate's time and energy was devoted to Hetch Hetchy, but attention was also given to routine "morning business" and to the impending "currency bill" that created the Federal Reserve Banking system. The senators who took center stage for opposing and supporting the Raker bill were largely those who earlier had spoken for and against the bill in October. Prominent in the anti-dam contingent were Senators Works, Smoot, Poindexter, Walsh, Lane, Borah, Weeks, Gronna, and Bristow. The pro-dam coalition was led by Pittman, Myers, Perkins, Thomas, Thompson, Newlands, and Norris.

A review of the more than two hundred pages of small-font text in the *Congressional Record* that document the Senate debate over H.R. 7207 makes it clear that, while many words were spoken, the nature of the arguments and rebuttals would have been familiar to anyone who had been paying attention to the Raker legislation since the spring of 1913. No new bombshells (akin to Sullivan's earlier claims of a "suppressed" report) were brought forth. The issue championed by the nature lovers to protect both Hetch Hetchy and the sanctity of Yosemite National Park was not ignored, but at least as much attention was directed to whether it was appropriate (and/or legal) for federal legislation to stipulate how water in the Tuolumne should be allocated. Concern that San Joaquin Valley farmers were being short-changed by San Francisco also precipitated much of the anti-dam rhetoric.[94]

Along with extensive speechifying, the *Congressional Record* documents a multitude of letters, telegrams, resolutions, and memorials sent to senators by citizens and organizations from across the nation.[95] There were scores of these missives—both pro-dam and anti-dam—that were either read aloud on the Senate floor or included in the public record. A few examples will suffice to illustrate the character of such communications: "Mr. Thompson presented petitions of the United Trades and Labor Council of Pittsburg, Kans., of the Kansas Society of California, and of the California Club of San Francisco, praying for the passage of the so-called Hetch Hetchy bill." Additionally, Vice President Thomas Marshall "presented a memorial signed by the sundry teachers of Belmont, Mass., and a memorial from the Connecticut Woman's Suffrage Association, remonstrating against the so-called Hetch Hetchy bill. . . . He also presented a telegram in the nature of a petition, from the executive board of the San Francisco district of the California Federation of Women's Clubs, praying for the passage of the so-called Hetch Hetchy bill."[96]

As a last pair of examples, Senator Henry Ashurst of Arizona, upon noting that he had received "4,000 communications regarding the Hetch Hetchy bill," asked for a reading of a brief telegram sent by the "Hon. H.D. Ross, one of Arizona's most valued public servants, a judge of the Supreme Court of Arizona" that consisted of a single sentence: "Personally I am strongly favorable to Hetch Hetchy bill and would like to see you vote for it." Ashurst followed with a second telegram, this one from "(Mrs. B.A.) Ella Q. Fowler" of Phoenix, whom he described as "a "very worthy and cultured lady." Mrs. Fowler (whose husband had served for many years as president of the Salt River Valley Water Users' Association) pleaded to save the valley: "I trust that it seems as sacrilegious to you as it does to me to grant the city of San Francisco for reservoir purposes the beautiful Hetch Hetchy Valley; if so, you will by argument and vote oppose the bill on December 6."[97]

Ashurst was a Democrat who eventually voted in favor of the Raker Act, but he wanted to show courtesy to both a prominent judge—with whom he agreed—and a woman of high social standing—with whom he disagreed. Like many of his compatriots, he was savvy enough not to want to alienate anyone needlessly or give the appearance that he was rudely dismissive of those who did not share his perspective. In essence, Ashurst was, like all his Senate colleagues, a politician.

Given that six working days were allotted to Senate debate over the Raker bill the question is not whether the two opposing sides were given an opportunity to make their respective cases on the Senate floor. Instead, a more pertinent question might be: exactly how many senators actually bothered to listen to the (often verbose) arguments that their colleagues were expounding. This was particularly apparent in regard to Senator Works's speech, which stretched out for almost four hours starting in the afternoon of December 2. The California senator went into great detail on matters involving irrigation district claims to the Tuolumne, legal dangers attached to water allocation stipulations, and the many other water sources that San Francisco could tap into. However, senators were not so transfixed by Works's analytic insight that they wanted to listen to him for hours on end. Over the course of his speech, eight quorum calls were requested because of sparse attendance; although Senators did eventually make their way back into the chamber, only rarely did the calls induce more than sixty senators to appear. Soon after, the assemblage would again begin to dwindle.[98]

Although unable to participate directly in the Senate's deliberations, one congressman who was aware of Works's arguments publicly changed his mind on Hetch Hetchy. This was Denver Church, whose House district included the Modesto and Turlock Irrigation Districts and who, at the Senate committee hearing in September, had dutifully defended the bill. By December, the heat brought by angry farmers like W. C. Lehane was too much, and Church determined that protecting his political future required a dramatic about-face. On December 4, Senator Works took the Senate floor and read a letter from Church into the record that stated, "[Now] I am unqualifiedly opposed to [the bill's] passage. I am opposed to the waters of the Hetch Hetchy Valley being taken away from the farmers on the plains below and given to the city and county of San Francisco. . . . I am sincerely hoping that you will be successful in your opposition to this measure."[99] Works undoubtedly welcomed Church's letter and made a point of reading it on the Senate floor. But at that point in the debate, how many senators were willing to change their minds?

In the early days of the Hetch Hetchy controversy, it was a commonly expressed truism that the Spring Valley Water Company was boosting the anti-dam movement to advance its own selfish interests. This was certainly a centerpiece of Freeman's protestations in the fall of 1912 as he prepared for the Fisher Hearing (see chapter 5). But after the Army Board report appeared in

February 1913, the Spring Valley leadership backed away from opposing the city's Hetch Hetchy initiative, and through the summer and the fall, little was said by or about the company. However, at one point when he had the floor, Senator Thompson of Kansas opined that "the opponents of the measure consist mainly of the Spring Valley Water Company and those affiliated with them in commercial interests, some irrigationists whose rights are fully protected, and the few people who call themselves 'nature lovers.'"[100]

Thompson's complaints about irrigationists and nature lovers were par for the course for those supporting the bill. But calling out Spring Valley as an opponent harkened back to earlier battles and was not reflective of how the controversy was unfolding in 1913. In response, Senator Clarence Clark of Wyoming objected to Thompson's Spring Valley reference, reassuring his colleagues, "I am sure . . . that the Spring Valley Water Company has no interest whatsoever in the passage or defeat of the bill. . . . [The company] is entirely indifferent."[101] Clark's assertion was reinforced by a telegram from Spring Valley President W. B. Bourn that Senator Perkins entered into the record on the final day of debate: "[Regarding statements] made to the Senate by Senator Thompson, to the effect that this company has a lobby in Washington opposing the bill. We do not feel that that statement . . . should go uncontradicted. The company has not made, promoted, or encouraged opposition of any nature to the bill in Washington or elsewhere. . . . When [in prior years] this company opposed the [Hetch Hetchy] grant it did so openly and above board. I request that you have this telegram read to the Senate."[102]

Some senators may have been skeptical of Clark's and Bourn's protestations, but there is no evidence that, in the time after the Raker bill was first introduced in the House in the spring of 1913 through the following December, the company or its emissaries attempted to impede or block the bill's passage.

The belief that private electric power companies, seeking to protect their presumed monopolies over hydroelectric power in California, were covertly opposing the bill persisted in the public imagination. It did not, however, constitute a major argument that was widely expressed during the final Senate debate. Nonetheless, it is notable that Senator Norris portrayed the development of hydroelectric power by the city as a key and vital attribute of H.R. 7207:

> I said that I was in favor of this bill to a great extent for the reason that
> it developed this power [below Hetch Hetchy]. This power will come
> into competition with the various water-power companies of Califor-
> nia. . . . This bill is not giving to a private corporation any power. It is

giving to the people of the locality of San Francisco the right to use a cheap power when it is developed. To my mind this is the very highest type of conservation. Here for ages this stream has been running down from the mountains. . . . without doing man any good, and this proposition is to harness that power and put it to public use and not give it to a private corporation. . . . If the power of the Hetch Hetchy is developed it will come into direct competition with what . . . is a monopolistic control of the hydroelectric power of California.[103]

When responding to a question posed by Senator Smoot, Norris made clear his conspiratorial vision of what would later be termed the "Power Trust": "Power corporations and other kinds of monopolistic corporations never come out in the open when they fight a proposition. They go around behind and, perhaps[,] get some nature lovers who are particularly honest to fight their battles. . . . Defeat this bill and you will receive the plaudits, the acclaim and the praise of every hydroelectric corporation in the state of California."[104]

What is remarkable in Norris's framing of the power issue is how much he foregrounds his antipathy for private power companies as a reason to support the Raker bill. As a corollary, he might have been expected to perceive Freeman—who was involved in the affairs of both the Great Western Power Company and Pacific Gas and Electric—as someone whose motives would be suspect. After all, a key component of Freeman's consulting business involved advising investor-owned utilities (and he was often one of the investors!). But when it came to Hetch Hetchy, there is no evidence that the two men ever clashed over the city's plans or how they were promoted. If nothing else, the ability of Norris—the great champion of public power and enemy of the Power Trust—to work with Freeman—one of America's greatest consultants for private power enterprise—reflects the extraordinary dynamic of how, in his advocacy of the Hetch Hetchy project, Freeman successfully straddled the divide separating the contentious realms of public and private power.

Throughout the course of the week, many senators unfamiliar with California and the nature of water in the American West were called upon to make sense of all the rhetoric that filled the Senate chamber. In the end, some came to support San Francisco's plan, and some chose a different path. To get a perspective on how senators came to rationalize their position on the bill, we can consider how the two senators from Rhode Island—who both conferred with Freeman over the proposed legislation—came to explain their vote. The character of the remarks they made in the hours preceding the final vote also offers insight into how other senators grappled with, and rationalized, their own assessment of the controversial legislation.

Senator LeBaron Colt was a well-regarded New Englander who had long served as a judge on the United States First Circuit Court of Appeals. A Republican, he became Rhode Island's junior senator in November 1912 and was serving in his first term when Raker's Hetch Hetchy bill came before the Senate. On the final night of debate, he took less than five minutes to explain his decision regarding H.R. 7207. To Freeman's chagrin, Colt's stance largely accorded with arguments expressed by Senator Works earlier in the week: "I am opposed to the bill because it seems to me that San Francisco has other available sources for a water supply and therefore there is no public necessity for the passage of this act. . . . [O]n principle, the national parks of this country should remain devoted to the uses for which they were intended, in the absence of some grave public necessity."[105]

There was no bravado in Colt's assertion that "no public necessity for this act" existed, and he did not try to impugn the motives of those whom Muir and Johnson would characterize as "park invaders." Nonetheless, Freeman clearly had no success in convincing Colt that a "grave public necessity" underlay San Francisco's push to dam Hetch Hetchy and provide city residents with a bountiful future water supply.

In contrast, Rhode Island's senior senator, Henry Lippitt (also a Republican), was more than willing to embrace the legislation and all that it entailed. A Rhode Island native whose family heritage was closely tied to textile manufacturing, Lippitt professed a great love of nature. But he also held a deep appreciation of water power as a driver of social and economic growth. As he explained to the Senate, "When this question was brought up in this body, [I was] very strongly moved by two directly opposing influences—my appreciation of the benefits of natural scenery . . . and the importance and value of water." It did not take long for him to express, however, which of these influences proved most persuasive; clearly, he had been swayed by Freeman's assurances: "I believe the theory of the nature lovers that a great scene of beauty is going to be destroyed is not correct. . . . [T]he beauty of the Hetch Hetchy Valley is still going to be there."[106]

For Lippitt, the most remarkable aspect of Freeman's plan was the tremendous amount of water power it would provide, for he saw in water power a larger importance in terms of his state's—and by extension his country's—growth and livelihood: "The prosperity of Rhode Island was dependent upon its water power. . . . It was because there was a water power at Pawtucket that [Samuel] Slater, coming from England, started the first cotton factory in the

United States there." With this frame of reference, Lippitt, like Senator Norris, was enthralled by the power potential of the upper Tuolumne: "The value of that waterpower appeals to me very strongly." But whereas for Norris the great appeal of the Hetch Hetchy hydropower scheme had been tied to its municipal ownership, this was not a concern for Lippitt, as he was "not so much worried about monopolies as some people are." Lippitt was more focused on how a great resource was, day after day after day, ebbing into the sea without any social benefit: "[I want] to see that this stream, which has run for countless generations uselessly down to the oceans[,] is made a servant of mankind."[107]

With Lippitt, Freeman had found a great supporter of his vision for San Francisco. In wrapping up his speech, Lippitt noted his confidence in Freeman's authority regarding the need to dam Hetch Hetchy: "I have made no claim . . . that [the Tuolumne] was the only source of supply [for San Francisco]. . . . I do however understand . . . that her engineers, who are among the most competent of any in the country, have decided, after careful study of the case[,] that this is the best source of supply for San Francisco, and, so far as I go, I want her to have it."[108] Recognizing that there might be alternative sources that San Francisco could tap into, Lippitt nonetheless relied on Freeman's expertise to guide his decision. He trusted in Freeman's engineering judgment as to the "best source of supply for San Francisco" and voted accordingly. Many other senators who supported the city likely shared this view of the Providence engineer.

The Vote

On Saturday evening, senators continued to pitch their views, with Senator Pittman planning to speak in support of H.R. 7207 for much of the evening. However, as Freeman later described, "With the high tension and rapid changes in the closing hours, Senator Pittman concluded that the best strategy was to yield about half of his time to certain opponents of the bill and thus exhibit his fairness."[109]

The talking and pontificating went on and on; no matter how much had been said, there always seemed to be more to say. More than forty hours of debate and commentary on Hetch Hetchy had filled the Senate during the prior week, yet as midnight came closer, for some it was not enough. Senator James Martine of New Jersey, despite professing that he had "no desire to interfere with the unanimous consent order," felt aggrieved that his ability to speak was being constrained by time limitations: "It was my purpose and disposition to express some thoughts on this measure . . . because I believe it to be an iniquity, a monstrosity, and a wrong. . . . I thought I may have had an opportunity to express my views to the Senate . . . but it seems impossible. . . . I feel I would be justified in pursuing any method possible to defeat it, even to talk it out. . . . I shall vote with all the earnestness of my nature and life against this thing as being undemocratic, ungenerous, [and] un-American."[110] Senator Martine's last-minute invective is perhaps most notable in highlighting the significance of Senator Pittman's consent decree, which required that a final vote take place before midnight on December 6. Had this not been unanimously agreed to, the sparring over H.R. 7207 could have continued ad infinitum.

About 11 p.m., Senator John Williams of Mississippi took the floor and advised that the October 7 decree necessitated that the Senate begin to vote, first on proposed amendments and then culminating with a vote on the final bill. Referencing Senate precedent involving prior unanimous consent decrees, Williams demanded that any further debate be curtailed.[111]

Votes were called on five separate amendments, starting with a change proposed by Senator Clark of Wyoming that would have eliminated any of the water allocations opposed by the "states' rights" bloc. It was defeated 23 ayes to 35 nays. Clark then offered a second amendment, designed to eliminate any perceived conflict with California water law. This too failed: 25 ayes to 41 nays. Two more amendments related to water allocations and states' rights were also turned aside. In a final proposed amendment, Senator Weeks of Massachusetts asked that the bill be changed so that the distance in the clause "[n]o human excrement, garbage, or other refuse shall be placed in the waters of any reservoir or stream or within three hundred feet thereof" be adjusted to "fifty feet thereof." His amendment was rejected. With this, the Committee of the Whole concluded its work and H.R. 7207 was "reported to the Senate without amendment."[112]

At last, the time had come for the final vote. Senator Poindexter requested a roll call that appears to have taken about fifteen minutes. The result was not a rout, but it was decisive: 43 in favor, 25 opposed, and 27 not voting. Of those not voting, the *Congressional Record* indicates that 16 senators had "paired" their vote, so that a more accurate tally would be 51 in favor, 33 opposed, and 11 not voting. Although short of an overwhelming landslide, it was far from a nail-biter, in which the shift of a few votes could have transformed the outcome.[113]

As the voting on the Raker Act played out, Freeman remained steadfast in the Senate gallery, watching the drama unfold below. That night he proudly recorded in his diary, "Senate passed Hetchy Bill at 11:57 PM. A great victory for the city and for my plans of development."[114] Three and a half years had passed since he first signed on as a consultant for San Francisco in anticipation of Ballinger's "show cause" hearing. His opponents would disagree, but for Freeman there was no question that, after years of work, he had succeeded in achieving "something better for the city."

The day after the Senate approved H.R. 7207, Freeman lunched with Senator Newlands and spent the afternoon in "sundry conferences with Mayor & City Atty & Vogelsang in which I suggest[ed] settlement of Spring Valley case out of court." That night he attended a celebratory dinner at Senator Pittman's home. On Monday he decamped to Philadelphia for "a conference on Mutual Insurance affairs"; the next day, he was in New York attending to Board of Water Supply business before heading home to Providence. Over the next several days he focused on a range of projects, including insurance affairs, Big Meadows Dam, Bassano Dam in Alberta, Alcoa's proposed Cheoah Dam east of Knoxville, and—for a new client—a municipal dam and aqueduct to serve Hartford, Connecticut. Although he spent some time considering the impact of Raker Act stipulations on "Hetchy flow," overall his work for San Francisco was complete. Freeman was moving on.[115]

Some city officials, including City Engineer O'Shaughnessy, remained in Washington and awaited the signing of the bill by the president.[116] Given Secretary Lane's support, and the endorsement of the bill by many Democratic leaders, there seemed little reason that Wilson would veto the act. But until his signature was affixed anything was possible.[117] Anti-dam forces held out hope that Wilson might take their side, with Johnson suggesting that "before he signs the bill [Wilson should] obtain the advice of the National Fine Arts Commission."[118] Such high-minded protest, however, fell on deaf ears. On December 19, Wilson signed the legislation, noting, "The bill was opposed by so many public-spirited men, thoughtful of the interests of the people, and of fine conscience in every matter of public concern. . . . I take the liberty of thinking that their fears and objections were not well founded. I believe the bill to be, on the whole, well founded."[119]

Senator Works sought to negate Wilson's ratification by quickly introducing a bill to repeal the Raker Act.[120] It went nowhere. San Francisco was victorious. Today, more than 110 years after its enactment, the legislation championed by Freeman and Dunnigan on Capitol Hill in the fall of 1913 remains in force, governing the city's modern-day Tuolumne water supply system.

Did Any Senators Change Their Mind?

As President Wilson recognized, a sizable political contingent across the nation opposed the damming of Hetch Hetchy, and it is reasonable to ask whether the turmoil that enveloped the Senate and the country after the House passed H.R. 7207 had much impact on how senators ultimately voted. One way to approach this question is to look at the vote taken in early October to bring the Raker bill to the Senate floor and compare that with the final vote taken in December. On October 4, 37 Senators voted yea to bring the bill to the floor; 15 nays opposed. Of the 52 senators who voted on October 4, only 4 changed their votes in the final ballot two months later (Democrats Sheppard of Texas and Kern of Indiana switched to nay, and Republicans Brandegee of Connecticut and La Follette of Wisconsin switched to yea). That was it. By this accounting, only 4 votes changed, and they offset each other. In essence, the bottom line did not move.[121]

Viewed through the prism of party politics, the Raker Act may not have strictly been a Democratic bill, but taking into account the documented 16 paired votes, on December 6 only 6 out of 53 Democratic senators cast votes against it; conversely, 11 Republican senators voted for passage.[122] Of these yea-voting Republicans (including paired votes), 5 came from the eastern states of Connecticut, Rhode Island, and Pennsylvania, where Robert Underwood Johnson and Boston-based Edmund Whitman presumably would have had some influence. As evident in his myriad diary entries, Freeman spent much of his time on Capitol Hill lobbying Republicans, and he did not have much luck convincing anti-dam senators like Gronna of North Dakota, Root of New York, or Colt of his home state of Rhode Island to support the city. The goal, however, was never to win every vote. That would have been impossible. The city's goal was to keep as many Democrats as possible in the San Francisco camp and to split off as many Republicans as they could from the preservationist/irrigators/states' rights bloc. In this, Freeman and Dunnigan succeeded. Muir and Johnson and Whitman and McFarland and their devoted followers did unleash an intense public relations blitz against the dam in the fall of 1913. And major newspapers like the *New York Times* bolstered their cause. But when the moment of decision came minutes before midnight on December 6, their efforts fell well short. Freeman won.

CHAPTER 8

After the Raker Act

Although the anti-dam community under the leadership of John Muir and Robert Underwood Johnson failed to protect Hetch Hetchy Valley from inundation, passage of the Raker Act did not establish a political precedent whereby land lying within national park boundaries would forthwith be easily subject to exploitation by municipalities, irrigation districts, electric power companies, logging outfits, or other parties. To the contrary, the battle over Hetch Hetchy gave impetus to a long-standing desire to create an expansive National Park Service that would take the existing national parks and national monuments (including Yosemite, Sequoia, Grand Canyon, Yellowstone, Glacier, Crater Lake, and Muir Woods) and meld their administration into a single agency devoted to their protection. The desire for a comprehensive agency to oversee national park properties had existed prior to enactment of the Raker Act, but with passage of the bill a new urgency for establishing an independent park service system within the Department of the Interior took hold.[1]

Importantly, political advocates pushing for creation of a national park system were not confined or limited to those who had opposed the damming of Hetch Hetchy. California Congressman John Raker, who had sponsored the bill authorizing the inundation of Hetch Hetchy, embraced the National Park Service cause and, in December 1915, introduced a bill authorizing the new federal agency. Congressman William Kent, who had forcefully argued for San Francisco's need for a Hetch Hetchy reservoir and had welcomed John Freeman as a guest in his Washington, DC, home in 1913, signed on as a cosponsor of Raker's park service legislation. And Secretary of the Interior Franklin Lane—a prominent proponent of the Hetch Hetchy Dam—also supported creation of the new agency. By the summer of 1916 Congress was ready to move, and on August 25, 1916, President Wilson signed into law the "Organic Act" creating the National Park Service. With this, at least some of the clouds left by the damming of Hetch Hetchy began to lift for park supporters. As historian Robert Righter notes, "If the Hetch Hetchy fight accomplished anything, it magnified the need for a federal Agency committed to the parks. The [new] agency would administer parks, but more significant, would define their meaning and mission and, when necessary, defend them. Such an agency would help to heal the weak and wounded system and give credence to Muir's hope that the fight for Hetch Hetchy had not been in vain. The 1916 National Parks Act partially fulfilled Muir's hope." Although creation of the National Park Service in 1916 would do nothing to impede San Francisco's ability to create a Hetch Hetchy Dam and Aqueduct, it offered hope among many conservationists that nothing comparable in scale or scope would ever befall a national park again.[2]

San Francisco and O'Shaughnessy

Upon passage of the Raker Act in December 1913, San Francisco set out to develop final plans for building the Hetch Hetchy Dam and Aqueduct. Freeman maintained an interest in the project's construction and operation, but responsibility for building the system fell to City Engineer M. M. O'Shaughnessy. Freeman could have been retained in a formal consulting capacity, offering ongoing advice and helping guide construction of the project during the next decade and beyond. O'Shaughnessy, however, was an accomplished engineer in his own right and considered himself more than qualified to handle the job without extensive assistance from the Providence-based engineer.

But make no mistake, O'Shaughnessy always held Freeman in high regard; in his autobiographical history of the Hetch Hetchy project published in 1934, he both praised his East Coast colleague as "an engineer of great versatility and broad experience" and assigned "much credit [to Freeman] for originating the brilliant conception of raising the project from a miserable 60 million gallon a day application to one of 400 million gallons daily. . . . [F]or this idea alone his services were very well worth double all that he received in compensation."[3] Further acknowledging the magnitude of Freeman's contribution, O'Shaughnessy praised the "splendid vision [Freeman] displayed in the general location of the aqueduct from the mountains to the San Francisco Bay region" and avowed, "[A]ny changes I have recommended in the modification of his plans [for Hetch Hetchy] were due to a more intimate study of the technique [on] the ground after thorough study, and were not intended to be critical of the plan prepared by Mr. Freeman."[4]

The modifications O'Shaughnessy alluded to primarily involved the elimination of the road/trail around the south rim of the reservoir (which the city engineer first proposed during the House committee hearing on the Raker bill in the summer of 1913). They also included the realignment of the road extending to the damsite, construction of the Priest storage reservoir used to regulate flow feeding into the Moccasin power plant, and (eventually) elimination of the tunnel Freeman proposed to carry water directly from the Cherry Creek and Lake Eleanor reservoirs into the Hetch Hetchy impoundment. In addition, O'Shaughnessy and the city never did build an automobile road along the north shore of the reservoir, although a hiking trail from the dam past the Tueeulala and Wapama waterfalls, up to Rancheria Falls, and into Tiltill Valley was constructed.[5]

O'Shaughnessy and Freeman remained on friendly terms throughout construction of the Hetch Hetchy system, and when the American Society of Civil Engineers met in San Francisco in October 1922 (while Freeman was ASCE president), the city engineer proudly accompanied Freeman on a visit to inspect the dam as it neared completion. By that time, O'Shaughnessy had become the public face and unquestioned leader of the project, and later the next year, Mayor Rolph and other city leaders officially designated the massive concrete structure O'Shaughnessy Dam in his honor. Freeman's determined efforts in 1912–13 had made the whole project possible, but O'Shaughnessy would be remembered and commemorated as the system's builder.[6]

In the summer of 1914, the city let the first Hetch Hetchy contract for a road from Hog Ranch to the damsite. This road was designed to be suitable as a right-of-way for the railroad that O'Shaughnessy planned for carrying personnel, equipment, and supplies needed to build the dam and aqueduct. But any hope that construction would proceed at a rapid pace was upended by the outbreak of World War One in August 1914. Although the United States did not formally enter the conflict until April 1917, the hostilities shook the American economy and impelled an inflationary rise in prices unanticipated by the city. Equally important, the war's economic turmoil raised interest rates and made it difficult for the city to sell any of the $45 million worth of bonds authorized to fund project construction; the proposal approved by voters in 1910 stipulated a maximum interest rate of 4.5 percent for the Hetch Hetchy bonds, which, although reasonable at the time, later lay well below market rates. This interest

As president of the American Society of Civil Engineers, Freeman (center) attended the society's annual conference in San Francisco in the fall of 1922. On October 6 he took the train to the Hetch Hetchy damsite in the company of City Engineer M. M. O'Shaughnessy (left) and C. E. Grunsky (right), the city engineer who, in 1902, had developed the initial plan for the Hetch Hetchy Aqueduct. More than ten years had passed since Freeman's last visit to Hetch Hetchy in August 1912. (M. M. O'Shaughnessy Photograph Collection, ca. 1885–1986, BANC PIC 1992.058—PIC, Bancroft Library, University of California, Berkeley)

rate cap complicated and slowed the city's ability to secure project funding through the teens and into the 1920s.[7]

Prior to passage of the Raker Act, the city anticipated (or hoped) that water from the Hetch Hetchy system might reach Bay Area consumers about five years after the beginning of construction. But with the war, inflation, and the rise in interest rates, this proved to be wildly optimistic. In the end, water from the upper Tuolumne would not reach the Crystal Springs Reservoir on the San Francisco Peninsula until 1934. This delay became a point of contention in

San Francisco city politics, but O'Shaughnessy was more focused on building a substantial and durable water supply and hydropower system than in getting water to the city as quickly and as cheaply as possible. By 1934—when Hetch Hetchy water first reached the Bay Area—more than $85 million in city funds had been expended to complete the first phase of the project.[8]

Through the mid-1920s, building the Hetch Hetchy system involved five primary components: First, construction of the 68-mile-long Hetch Hetchy Railroad extending from the Sierra Railway right-of-way east of Modesto (at Hetch Hetchy Junction) up to the damsite. Work began in December 1915 and reached completion fourteen months later. Second, excavation of the 19-mile-long Mountain Division Tunnel connecting Early Intake with the Priest reservoir forebay of the Moccasin power plant. Work on this tunnel began in the summer of 1917 and reached completion in the spring of 1925. Third, construction of a 70-foot-high concrete multiple-arch dam to raise the storage capacity of Lake Eleanor and provide water for a hydroelectric powerhouse used to power construction equipment at the Hetch Hetchy Dam and along the aqueduct right-of-way. Work on this dam began in August 1917 and reached completion ten months later. Fourth, construction of Hetch Hetchy/O'Shaughnessy Dam itself began in August 1919. The massive concrete structure reached an initial height of 344 feet above deepest foundations (at an elevation of 3,726.5 feet above sea level) in July 1923 and provided a storage capacity of 206,000 acre-feet (67 billion gallons); in 1938 the structure was raised 88.5 feet to a crest elevation of 3,812 feet, providing a storage capacity of over 360,000 acre-feet (117 billion gallons). And fifth, construction of the Priest Dam/Moccasin power plant, which began in fall 1921; the plant began generating and transmitting high-voltage AC power in August 1925.[9]

Upon completion of the first phase of construction by 1925, the reservoir at Hetch Hetchy began to provide water to the Moccasin power plant about 30 miles downstream. However, more work was needed before water could reach San Francisco. Completing the lower section of the aqueduct involved work on three key segments: the 15.8-mile-long Foothill Division Tunnel below the Moccasin power plant; the pressurized steel pipelines extending 47.5 miles across the San Joaquin Valley to the Tesla portal; and the 28.5-mile-long Coast Range Tunnel running from the western slope of the San Joaquin Valley to the Irvington Portal near the shore of San Francisco Bay.[10] Water finally flowed through the aqueduct into the Pulgas Water Temple and Crystal Springs Reservoir south of San Francisco on October 24, 1934. But O'Shaughnessy was not there to witness its arrival. Just days before, on October 12, he had succumbed to a heart attack at the age of seventy-two.[11]

In 1923 the initial dam construction phase was completed and, as shown in this ca. 1930 photo, the reservoir at Hetch Hetchy was allowed to fill to a crest elevation of 3,726.5 feet above sea level. The dam would not be completed to a crest elevation of 3,812 feet until 1938; it remains at that height today. In 1923 the dam was officially named in honor of City Engineer M. M. O'Shaughnessy. Author's collection.

Although O'Shaughnessy did not live to see water from Hetch Hetchy reach consumers in the Bay Area, he did witness the City of San Francisco take legal ownership of the Spring Valley Water Company and its holdings. The prospective municipal purchase of the private water company dated back to the 1870s and had been much discussed in the period 1912–13 (with Freeman closely involved in the negotiations). But reaching a price that would be approved by a two-thirds majority of voters proved elusive. In 1915, 1921, and 1927, purchase agreements between the city and Spring Valley came before the electorate only to fall short. Finally in 1928 the stalemate was broken. Voters approved a deal for almost $40 million that, in 1930, allowed the city to take ownership of the company's water supply system.[12] With that, the city was able to combine the Hetch Hetchy system with Spring Valley's assets (including the city's extensive distribution system, the Crystal Springs Dam, the Peninsula supply system the water resources of Alameda Creek, and Calaveras Dam) and establish the enduring foundation of San Francisco's modern water supply infrastructure.[13]

As part of the city's integrated supply system, the Hetch Hetchy reservoir played a vital role in serving urban consumers far removed from the Sierra Nevada. But within the confines of Yosemite National Park the reservoir also had a major impact. The valley floor disappeared under the impoundment, and as Frederick Law Olmsted Jr. had foreseen (see chapter 7), the intermittent release of water from the reservoir to meet downstream demands created a strip of barren land along the shoreline that became ever more visible as water levels dropped. And during times when the reservoir was full and assumed the appearance of a natural lake, only rarely did the water surface present a reflection of the soaring granite cliffs astride the reservoir. Perhaps the reservoir at Hetch Hetchy did possess beauty of its own account, but it was different than a natural Sierra Lake. In terms of restrictions on park use, operation of the reservoir did preclude bathing and also limited nearby camping (and no hotel was ever built overlooking the reservoir). However, fears expressed by preservationists that construction of the dam and reservoir would have a major impact on access to vast stretches of the upper Tuolumne watershed proved unfounded.

Freeman and San Francisco officials had trumpeted the possibility that the Hetch Hetchy reservoir would provide dramatic reflections of the surrounding granite cliffs. But the vast majority of time when the reservoir is filled there is no reflection, as evident in this circa 1940 postcard view. Author's collection.

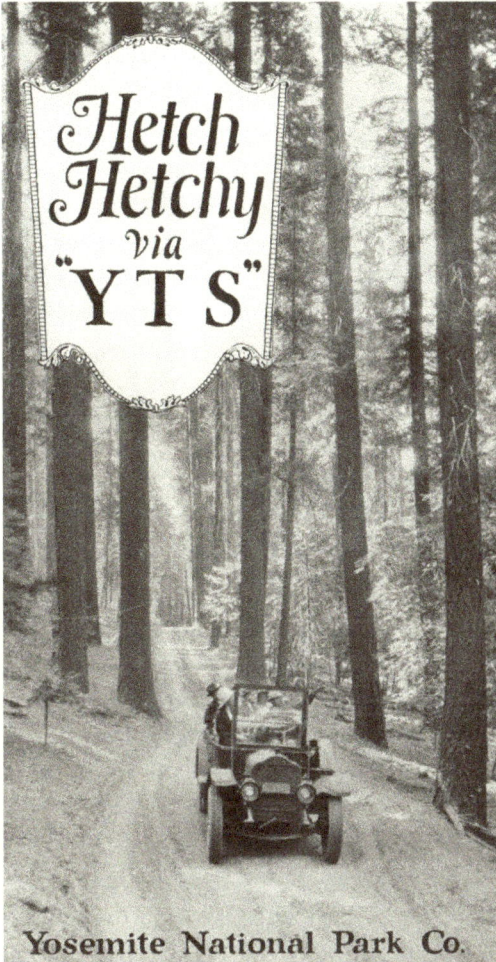

Hetch
Hetchy
via
"YTS"

Yosemite National Park Co.

As Freeman understood, tourism at Yosemite National Park in the twentieth century would revolve around automobiles. In the early 1920s, tourist access to the Hetch Hetchy damsite was provided by the privately owned Yosemite Transportation System, a jitney service based out of Yosemite Valley. Author's collection.

This is especially true in how camping and visitation in Tuolumne Meadows has remained a part of park life for the past century.

Freeman had championed the idea that construction of the water supply reservoir would initiate a great increase in tourists seeking out the northern reaches of the national park. But compared to Yosemite Valley to the south, no tremendous surge of automobile traffic ever made its way to Hetch Hetchy. There is, however, easy road access for anyone who wants to experience the valley, the dam, and the reservoir firsthand, and in the 1920s, tourists began arriving by jitneys operated by the Yosemite Transportation System. And while the valley floor at Hetch Hetchy is inundated and inaccessible, the path that extends across

A postcard view circa 1940 featuring rustic cabins at Tuolumne Meadows Lodge about twenty miles upstream from the Hetch Hetchy reservoir. Sanitation regulations restrict camping in close proximity to the reservoir, but they do not block cabins and human habitation in the upper Tuolumne watershed. Author's collection.

the dam and along the north shore of the reservoir is well used by hikers and backpackers seeking passage to Wapama Falls, Tiltill Valley, Rancheria Falls, and points beyond.

Tuolumne River

City officials always understood that use of the Tuolumne watershed would have to be shared with the Modesto and Turlock Irrigation Districts. After all, these two irrigation districts, formally organized in the 1880s, had completed the La Grange diversion dam across the lower Tuolumne in 1893. By the early 1900s, the districts together maintained rights to a perennial flow of 2,350 cubic feet per second (cfs) at La Grange Dam for use in irrigating more than 200,000 acres of farmland. In planning for Hetch Hetchy's 400 million gallons a day (mgd) water supply system, Freeman had accounted for the irrigation districts' senior water rights. But he had not anticipated stipulations included within the Raker Act that would require the city to guarantee a flow at La Grange of 4,000 cfs for sixty days every year from mid-April through mid-June. For the city, this extra 1,650 cfs for two months every spring constituted no minor burden—in total, it equaled an annual volume of over 195,000 acre-feet—or the equivalent of a constant flow of about 175 mgd spread out over the entire year (see appendix).

The 4,000 cfs concession was negotiated by O'Shaughnessy and City Attorney Long in the spring of 1913, and Freeman expressed concern when he first learned of it. Nonetheless, he accepted the political necessity of the compromise, informing the city engineer that he was "ready to stifle [his] objections" because he saw the situation as "the eve of a great victory." In his view, "[T]he city is better off a hundred-fold to take the bill as it passed the Lower House in Congress than to get nothing at this session."[14] Publicly, Freeman voiced no displeasure with the additional water ceded to the irrigation districts. But privately he remained concerned about its possible impact, and in early February 1914 he confessed to O'Shaughnessy that a preliminary review of "river flow in the low years [appears to] show that this extra grant to the irrigationists [of 4,000 cfs for 60 days] ought never to have been made, and will be a terrible burden on the ultimate development" of the city's Hetch Hetchy system.[15] In his last formal engagement with the city's Hetch Hetchy project (and for which he was paid $1,588.20), Freeman undertook a more complete analysis of Tuolumne River streamflow records to better determine the impact of the districts' additional allowance on San Francisco's proposed system.[16]

In this final study, which he forwarded to O'Shaughnessy in late March 1914, Freeman acknowledged the potential impact of the springtime 4,000 cfs diversion, especially because "this increase of 1650 second feet in the priority [comes] during sixty days of the best part of the flood season."[17] He convinced himself, however, that the overall flow of the Tuolumne was so large that, if the city created an additional storage capacity of 75 billion gallons (about 230,000 acre-feet) in the watershed above the aqueduct's Early Intake, the city could meet the demands of the Raker Act and (ultimately) draw 400 mgd for use in the Bay Area. This extra capacity would require, in Freeman's hopeful analysis, raising the proposed heights of the city's Hetch Hetchy, Lake Eleanor, and Cherry Creek dams, or alternatively depending on additional storage dams across Tiltill Creek (above Hetch Hetchy) or at Poopenaut Valley (lying a short distance downstream from Hetch Hetchy), but he believed it could be done. Freeman avoided discussion of the politics that might come into play if the city proposed increasing the size of its storage facilities in the upper Tuolumne (or adding new ones) beyond what he had outlined in his Hetch Hetchy Report. Instead, he optimistically offered assurances that the city could proceed with building the Hetch Hetchy system and someday supply 400 mgd of mountain water to urban consumers.[18]

For a few decades after passage of the Raker Act, the city held fast to its plans for a 400 mgd system. However, by midcentury the city began to reassess this expectation. In the 1920s, the Modesto and Turlock Irrigation Districts built the 283-foot-high Don Pedro Dam a few miles upstream from La Grange Dam. This concrete curved gravity dam had a storage capacity of about 290,000 acre-feet, a maximum reservoir elevation of 580 feet above sea level, and a power-generating capacity of 30 megawatts.[19] San Francisco did not object to this large reservoir (which lay well downstream from the Hetch Hetchy Aqueduct intake), with O'Shaughnessy reporting that in the years following passage of the Raker Act, the city's "relations with the irrigations of Turlock and Modesto have been more than friendly."[20] Carrying on this "friendly" relationship, in the 1940s the city allied with the districts in the marketing of hydroelectric power generated at Moccasin Creek. This interaction soon paved the way for a significant restructuring of the city's Hetch Hetchy system.[21]

In Freeman's original plan, reservoirs built across Cherry Creek and at Lake Eleanor would connect into the Hetch Hetchy reservoir via an eight-mile-long tunnel. With this, the flow of all three watersheds (Cherry Creek, Eleanor Creek, and the main Tuolumne) would have fed into a single main storage reservoir at Hetch Hetchy. But a different idea emerged after the end of World War Two. With the city and the districts finding common ground in the arena of marketing hydroelectric power, in the 1950s they initiated a new plan to develop the Cherry Creek and Lake Eleanor watersheds. In this reconfigured system, the two watersheds would not be connected to the Hetch Hetchy reservoir but would instead conjointly store their streamflow in a large reservoir impounding Cherry Creek (named Lake Lloyd); this reservoir would feed into the Holm Powerhouse (with a capacity of 165 megawatts) located a short distance downstream from the aqueduct's Early Intake. With completion of the 315-foot-high Cherry Creek Dam in 1956 (and opening of the Holm Powerhouse in 1960), operation of the Hetch Hetchy system entered a new phase in which the water available for diversion into the aqueduct was significantly diminished. As a historian for the city later explained, "Lake Lloyd water storage has a dual purpose—conservation and power production—but was not designed for contribution to [the] San Francisco domestic water supply. Water releases are for power generation at Holm Powerhouse and to meet irrigation priorities under the Raker Act."[22]

During the 1960s, the two irrigation districts began planning for a major enhancement of their joint storage capacity along the lower Tuolumne. The result was the 585-foot-high New Don Pedro Dam that, superseding (and inundating) the original Don Pedro Dam built in the 1920s, provided an enlarged storage capacity of more than 2 million acre-feet at a maximum surface elevation of 855 feet above sea level. With this tremendous dam, the districts confirmed their hydraulic dominance over the watershed.[23]

Through the latter years of the twentieth century, San Francisco maintained a desire to draw 300 mgd from the Hetch Hetchy Aqueduct, but by the 1990s the expected annual draw from the system had been reduced to 240 mgd, a figure that has remained relatively steady into the twenty-first century.[24] In Freeman's March 1914 report analyzing the impact of Raker Act stipulations, he had optimistically assessed that a draw of 400 mgd could be possible by increasing storage in the upper Tuolumne watershed. But it was never to be. And here we can see the likely impact of the obligation to provide the irrigation districts an extra 1,650 cfs for sixty days every spring, which equates to 175 mgd spread out over an entire year; added to the 240 mgd generally supplied by the Hetch Hetchy Aqueduct, this flow closely approximates the 400 mgd that Freeman envisioned in his 1912 Hetch Hetchy Report.

The East Bay's Mokelumne Aqueduct

In his report, Freeman had conceived of a metropolitan water supply district for the Bay Area that would serve San Francisco, the communities surrounding Palo Alto and San Jose to the south, and the growing cities of Oakland, Berkeley, and Richmond to the east. As it turned out, the East Bay communities centered around Oakland and Berkeley chose not to join with San Francisco in the development of the Hetch Hetchy project, instead opting for their own regional water supply system and their own aqueduct drawing water from the slopes of the Sierra Nevada. Formed in 1923, the East Bay Municipal Utility District (EBMUD) set its sights on the lower Mokelumne River as a source of supply.[25] This was the same river that San Francisco and Freeman had deemed unsuitable as an alternative to the upper Tuolumne watershed and the same river that Eugene Sullivan of the Sierra Blue Lakes Water and Power Company had promoted as a possible source for the city at a House committee hearing in the summer of 1913. And this company's proposal had subsequently attracted the attention of Robert Underwood Johnson, prompting him to proclaim it as a possible source for San Francisco (with a supposed capacity of some 432 mgd). In contrast, Freeman's *Hetch Hetchy Report* had projected a possible supply of about 200 mgd from the Mokelumne, far less than the 400 mgd promised by the Tuolumne/Hetch Hetchy system (see chapter 6).

A full recounting of the history surrounding EBMUD's Mokelumne Aqueduct is beyond the scope of this book. But it is important to note that the Mokelumne and Hetch Hetchy aqueducts are not directly comparable, although it has been speculated that construction of EBMUD's system offers evidence that in 1913 San Francisco could have turned to the Mokelumne as a viable alternative to the Tuolumne/Hetch Hetchy system.[26] Before assuming this supposition to be true, it is worth considering how the two systems differ.

The biggest variance between the two regional water supply regimes is that the storage reservoir impounded by the Hetch Hetchy/O'Shaughnessy Dam (with a capacity of 360,000 acre-feet) lies more than 3,500 feet above sea level; in contrast, EBMUD's Pardee Dam on the Mokelumne (with a capacity of about 200,000 acre-feet) lies at an elevation only about 560 feet above sea level. And this difference in elevation equates to a much smaller hydroelectric power-generating capacity at Pardee Dam (about 24 megawatts) compared to San Francisco's power plants in the upper Tuolumne (with a capacity exceeding 200 megawatts). In addition, water delivered by the Mokelumne Aqueduct needs to be treated and filtered before being piped to household consumers. While water carried by the Hetch Hetchy Aqueduct also requires some treatment, the city benefits from how, in the words of city officials, "water quality in the Tuolumne River [stored at Hetch Hetchy] is so high that the San Francisco Public Utility Commission maintains a filtration avoidance permit for its delivery [to consumers in the Bay Area]."[27]

At the beginning of the twenty-first century, EBMUD's Mokelumne system provided on average about 210 mgd to the greater East Bay (about what Freeman projected as the Mokelumne River's capacity in his *Hetch Hetchy Report*).[28] Supplying a population of 1.4 million residents, the Pardee/Mokelumne system stands as a major component of Greater San Francisco's hydraulic infrastructure. But it could not readily serve as a water supply for the entire Bay Area.

Of course, San Francisco's modern-day Hetch Hetchy system—which on average delivers little more than 240 mgd and, in concert with facilities formerly operated by the Spring Valley Water Company (providing about 40 mgd to supplement the Hetch Hetchy supply), serves a population of over 2.5 million people—has also proven inadequate to meet the needs of the entire Bay Area.[29] In this, we can see how the demands of the Turlock and Modesto Irrigation Districts on the flow of the Tuolumne has impacted the ability of the city to ever meet Freeman's vision of a 400 mgd Tuolumne system. In the end, the lion's share of the Tuolumne goes to the irrigation districts, while San Francisco receives enough of the river's flow to provide for its citizens and for many in greater Silicon Valley. However, that does not include the 1.4 million people served by EBMUD.

Although the Hetch Hetchy Dam and Aqueduct was built in general accord with Freeman's plans, his larger vision of a single metropolitan water district dependent upon the upper Tuolumne watershed never came to be. Viewed more broadly, a regional system divided into two civic regimes—and relying upon both the Tuolumne and the Mokelumne River—has come to provide the greater Bay Area with steady supply of more than 400 mgd. As a corollary, the Modesto and Turlock Irrigation Districts stand as big winners in terms of controlling the region's hydraulic resources.

Hetch Hetchy in the Twenty-First Century

In 2012 the voters of San Francisco rejected a ballot referendum that would have initiated serious planning for the removal of O'Shaughnessy Dam and restoration of Hetch Hetchy Valley as a part of Yosemite National Park. This did not deter anti-dam preservationists—many acting through the organization Restore Hetch Hetchy—from pursuing court action to bring about removal of the dam. But by 2018 these legal efforts foundered, with the California Supreme Court refusing to hear appeals of lower court rulings that affirmed both the constitutionality of the Raker Act and San Francisco's operation of O'Shaughnessy Dam as a centerpiece of the city's municipal water supply system (see introduction).

At some future time, political or legal action might be taken to remove the dam and restore the reservoir area. However, that time is not now; as of 2025 the Hetch Hetchy Aqueduct remains a vital source of water for over 2.5 million people. In a time of climate change and hydrological uncertainty, the importance of water supply reservoirs in California has not diminished but in fact has increased, as periods of extended drought are punctuated with the onslaught of intense storms dumping huge volumes of precipitation across the Sierra Nevada. In response, some water engineers and planners see heightened dams and enlarged reservoirs as valuable components of a regional waterscape designed to sustain the state's economic viability during the decades that lie ahead.[30] In a 2023 article published by the San Francisco broadcaster KQED, the former general manager of the San Francisco Public Utilities Commission, Susan Leal, makes this point explicit in regard to possibly raising the storage level of O'Shaughnessy Dam fifty-five feet: "We may need to build the dam higher to impound more water, we have to consider this because we never planned for the extreme storms."[31]

The idea that San Francisco might raise the dam at Hetch Hetchy and encroach ever further into Yosemite National Park might seem almost impossible to contemplate. But if John Freeman could somehow appear on the scene almost a century after his death, he would likely embrace the enlargement of Hetch Hetchy reservoir and, in tandem, advocate for restructuring the water allocation agreements embedded in the Raker Act. He would almost certainly want more Tuolumne water to be stored high in the Sierra rather than in the comparatively low-lying reservoir formed by the New Don Pedro Dam. As he pointed out in his 1912 *Hetch Hetchy Report,* there was nothing sacrosanct in calling for the Hetch Hetchy reservoir to be capped at 3,800 feet above sea level. Recognizing that the hydropower capacity of his system could be enhanced by increased storage capacity, he had provocatively observed, "If the remarkable water power opportunities . . . were to be a controlling feature of the [Hetch Hetchy Dam] design, regard for the principles of conservation would direct one to make this dam at least 25 feet and possibly 50 feet taller than shown in the drawings, and the possibility that this height may be called for under the development of future years should not be lost sight of."[32]

Is it far-fetched or crazy to propose enlarging the storage capacity of the Hetch Hetchy reservoir? Perhaps. For protectors of the park who look to John Muir for inspiration, it would likely constitute an abomination. But for someone adopting the perspective and progressive vision of John Freeman, it is not so crazy. And it might well be projected as a way to obtain "something better for the city" and for the millions of people who rely on the water and the power brought to them by the Hetch Hetchy Aqueduct.

Freeman after Hetch Hetchy

After passage of the Raker Act in December 1913, Freeman continued his labors as an insurance executive, businessman, investor, and engineering consultant.[33] Most notably, he retained his presidency of the Manufacturers Mutual Fire Insurance Company, kept his position as consulting engineer for New York City's Catskill Aqueduct (which began delivering water to the metropole in 1916), and in 1915 joined the board of directors of the corporation that controlled the Great Western Power Company. In the spring of 1914, he also helped plan Hartford's Nepaug water system in central Connecticut and served as a consultant/

advisor for the Denver Union Water Company in negotiating the sale of its water supply assets to a municipal authority. However, the start of World War One in August 1914 jolted the American economy, delayed the financing of many large industrial initiatives, including water supply and electric power projects, and slowed his consulting practice. Once the United States entered the war in April 1917, Freeman devoted much time to leading the municipally owned Providence Gas Company in Rhode Island and to aiding the war effort as a member of both the National Research Council and the National Advisory Committee on Aeronautics.[34]

In November 1918 the war in Europe came to an end. The next summer, the sixty-four-year-old Freeman embarked on a nine-month trip to Japan and China, where he studied flood control problems along the Yellow River. Upon returning stateside, he undertook a study of San Diego's water supply and, in 1922, served as president of the American Society of Civil Engineers (at that time he visited the Hetch Hetchy damsite in the company of M. M. O'Shaughnessy). In addition, he carried out an extensive analysis of water levels in the Great Lakes (published in 1926 as *Regulation of the Great Lakes and Effect of Diversions of Chicago Sanitary District*) and advocated for establishment of a National Hydraulic Laboratory.[35]

In terms of advancing his own work as a consulting hydraulic engineer, he had great hopes of partnering with his son Roger, envisaging a future where he would design dams and hydroelectric projects and his son would serve as builder/contractor. After graduating from MIT in 1914, Roger joined the Turner Construction Company and began learning the business of construction contracting. A quick study, by 1924 he was in charge of building a hydroelectric dam on the Tippecanoe River in northern Indiana and, at least in the mind of his father, had established himself as qualified to build major hydraulic engineering projects.[36] The senior Freeman was designing an enlargement of the Great Western Power Company's Big Meadows Dam, and he planned for thirty-two-year-old Roger to take responsibility for building the enlarged earthen dam. A long-time member of Great Western's board of directors, Freeman reckoned that his corporate influence and professional standing would help secure a lucrative contract for the nascent father-son partnership.

But the plan collapsed in January 1925, when Roger suddenly died from a ruptured appendix. Father John was devastated and, in many ways, never recovered from this unexpected loss. He soon began squabbling with other members of the Great Western leadership over reimbursement for the time and energy he had put into his Big Meadows design. Angry at being rebuffed in his search for

compensation, Freeman resigned from the board, but not until he had antago-
nized many of its members. With this precipitous action, his standing as a leading
engineer and executive in the electric power industry came to an inglorious end.[37]

The year Roger died, Freeman turned seventy years old. Although his travel-
ing days as a project consultant were largely over, in 1927 he enjoyed a summer
automobile trip through Europe in the company of his daughter Elsie, the youngest of his seven children, and in the fall of 1929 (again accompanied by Elsie) he
crossed the Pacific to attend the World Engineering Congress in Tokyo, Japan.[38]

During the later years of his life, he became interested in seismology, especially in how the risks posed by earthquakes could be ameliorated by improved
structural design and, perhaps, be covered by factory insurance policies. Seeking
to bring the fruits of his seismological research to a wide audience, he assembled a massive 904-page book published by McGraw-Hill in 1932. Although
the perspective and structure of *Earthquake Damage and Earthquake Insurance*
proved very different than his earlier *Hetch Hetchy Report,* they both feature
scores of photos, maps, and tables to support their respective texts. As in his
1912 report for San Francisco, Freeman's treatment of seismology reflected an
appreciation of how the display of visual data and imagery could help make
arguments and analysis more accessible to a broad readership.[39]

In the spring of 1931—as his work on earthquakes and seismology was coming to a close—a testimonial banquet was held in his honor at the Biltmore Hotel
in Providence. More than two hundred friends and professional colleagues gathered to pay tribute to his career as an engineer and insurance executive; many
others who could not attend sent letters attesting to his stature and professional
influence.[40]

During the last years of his life, he also began writing an autobiography,
starting with his youth in the Maine backcountry and his first steps as an engineering assistant for the Essex Company in the Merrimack River valley. On
October 6, 1932, with his autobiography only partly complete, he died following a brief illness. He was buried in a family plot at Providence's Swan Park
Cemetery, joining his deceased sons Nat and Roger.[41]

It would be too strong to say that Freeman was a beloved member of America's engineering fraternity when he died at age seventy-seven. But he was
respected. As Vannevar Bush, dean and vice president at MIT (and later one of
the leaders of the Manhattan Project, which created the atomic bomb), acknowledged, "[Freeman's] professional life of fifty-six years was crammed with useful and largely original work—eagerly undertaken, thoroughly and brilliantly
completed—which won for him the respect and likings of colleagues, the

loyalty of co-workers, and grateful public appreciation. . . . He set a stiff pace for others, but spared not himself. If he exacted hard stints from his helpers, he inspired them too, by enthusiasm and example; in the midst of a demanding job he kept no working hours, and scarcely knew how to call a day a day."[42]

Freeman the Progressive

Although Freeman remained professionally active through the 1920s, little in his later consulting portfolio matched the scale or significance of his work advocating for the Hetch Hetchy Dam and Aqueduct. Considering his work more broadly, however, it is difficult to identify any engineering study written by any engineer that compares with the *Hetch Hetchy Report* that Freeman crafted during the "strenuous" summer of 1912. It stands on its own as a document of engineering advocacy and design, not because it offers a flawless analysis of an engineering issue (it does not) but because of how Freeman understood the political dimensions of the Hetch Hetchy project and intuited the political maneuvering necessary to win approval of the Raker Act. Although no one ever used the term during his lifetime, with the report and his subsequent politicking in the nation's capital, he proved himself a master in the art of technopolitics.[43]

In the context of Progressivism and the conservation of natural resources, Freeman's involvement in (and leadership of) the Hetch Hetchy project is especially intriguing because of his ties to both the privately owned Spring Valley Water Company and northern California's two largest private electric power companies. In the midst of a battle frequently characterized in terms of "public" versus "private" interests, Freeman's relationship with Spring Valley, the Great Western Power Company, and Pacific Gas and Electric highlights more than simply his skills as an entrepreneurial expert. When he expanded his consulting practice to the West Coast in the years after 1905, Freeman advanced his financial and professional aspirations. Continuing along this path, he likely could have garnered a substantial consulting fee from Spring Valley to oppose the Hetch Hetchy project—he was more than pleased when the private company hired him to formulate a design for Calaveras Dam—but his actions were not determined solely in terms of monetary gain. He was also energized by the opportunity to conjure a municipal water supply project of breathtaking scale. The possibility of providing San Francisco with a tremendous supply of mountain water (and electric power) that could sustain its growth for decades into the future was professionally intoxicating. He saw what was possible if the constraints of the Garfield Permit could be shed, and once Marsden Manson had failed to deliver on a multitude of promises, he could champion a plan to make "something better for the city." With this, a seemingly professional assignment also became very personal.

To see Freeman as a progressive engineer is not to deny that he was, in the words of Donald Worster, a "contract engineer" serving the interests of a client while advancing a particular vision of technological progress. But to believe that Progressivism was all about—or only about—providing society with answers devoid of financial or political or personal self-interest is to misunderstand the impulses that motivated many people who worked to create what they perceived to be a better world. John Muir and his followers advocated an idea of what they thought would provide the greatest good vis-à-vis Hetch Hetchy, and they prided themselves that they worked without financial inducement. That did not mean, however, that they were Progressive in their worldview and Freeman was not.

After publishing his *Hetch Hetchy Report* and joining battle with the "nature lovers," Freeman derived great pleasure in countering their arguments and portraying their positions in the worst possible light. Money mattered to Freeman, but so did pride. Establishing himself as the key advocate of the Hetch Hetchy Dam and Aqueduct, he abhorred the notion that the project might be blocked by people seeking to preserve the valley in opposition to his plans. Failure was unacceptable; should success require weeks of seventeen-hour days far distant from wife and family, that was a price he willingly paid. Should we need it, the story of Hetch Hetchy offers yet one more example of how history can be driven by the actions and imaginations of individuals acting on the public stage.

Knowing Nature, Knowing Culture

In his book *The Organic Machine,* environmental historian Richard White posits the concept of "knowing nature through labor" as a way to study how humans experience the natural world. Focusing on the Columbia River, White describes how people came to know the nature of the waterway through their labors: Indians netting salmon while perched above rocky shoals; fur traders navigating treacherous sandbars; trappers paddling against a surging current; and—perhaps most provocatively—workers clearing streambeds in preparation for massive hydroelectric power plants.[44]

While White illustrates how thousands of laborers engaged in the building of huge hydroelectric power plants knew nature through their labor, his analysis does not encompass the hydraulic engineers who conceived and designed these water control projects. But make no mistake, the knowledge used by engineers to construct water-powered textile mills, navigation canals, municipal aqueducts, irrigation projects, and hydroelectric power plants is derived from observant study—knowing—of the natural world. Knowing nature, and investigating the character and attributes of the physical landscape, can come in the form of surveying river gradients and the elevations of riparian escarpment, measuring streamflow and seasonal floods, and studying geological strata and formations. Similarly, measurement of water flow over weirs and through penstocks, pipes, and turbines—skills a young Freeman honed during his early years with the Essex Company in Lawrence—rely upon labor expended in search of how nature acts and how it reacts to human artifice and stimuli.[45]

The knowledge that engineers have of the natural world may seem far removed from the experiences that environmentalists often seek in studying wilderness and ecological systems, understandings that stand in opposition to a worldview in which river basins are evaluated in terms of merchantable commodities, kilowatt hours, and the ability to supply millions of gallons of water per day to urban consumers. But engineers undoubtedly know nature through their labor; in fact, it is precisely because of such knowledge that hydraulic engineering projects can dramatically transform river basins. But is knowledge of nature alone enough to bring a large hydro-project to completion? The story of Hetch Hetchy offers a compelling answer: No.

John Freeman stood as the greatest of the park invaders that Muir vilified, and although he came to the Hetch Hetchy battle relatively late in the game (almost a decade after Mayor Phelan first filed for water rights at the valley's outlet), he ultimately proved to be the city's most important advocate. A preeminent engineer with wide experience in water control technology, Freeman possessed the technical skills to design a major water supply and hydropower system. However, his triumph in advocating what Muir derided as a "water tank" required more than a technical understanding of California's topography, geology, and hydrography. There was also a political dimension of his work that proved vital to its success. While Freeman needed to "know nature" in planning his Hetch Hetchy system, to bring success to the city he also needed to "know culture"—to know how to cultivate public support for the proposed project and to bring this support (or at least the perception of such support) to bear upon politicians and bureaucrats responsible for managing the nation's public domain.

Freeman's knowledge of America's political economy and the culture that sustained it did not come by happenstance. It accrued through years of work that encompassed water-powered textile manufacture, fire safety, civic park improvement, hydroelectric power, and—last but not least—municipal water supply. And his leadership in the world of mutual factory insurance, which included lobbying for favorable tax treatment at the state and federal level, further honed his political skills. In the context of Hetch Hetchy, the cultural acumen that Freeman amassed in his earlier career proved as important as the technical skills necessary to calculate, for example, the capacity of high-pressure water supply conduits or the sustainable yield of a rural watershed subject to seasonal floods.

Passage of the Raker Act depended upon a politically adept campaign to win support for an expansive vision of a municipal water supply and hydropower system. In carrying out this campaign, Freeman relied upon a knowledge of nature, but all would have been for naught had he not been able to navigate the political labyrinth raised by forces confronting the city.

Although City Clerk J. S. Dunnigan was unable to attend the testimonial dinner held in Providence almost twenty years after the city's battle over Hetch Hetchy had been won, he soon wrote Freeman in playful reaction to the published accolades: "I read with great pleasure proceedings of the testimonials given you by your engineering and other friends. [But] I did not find any reference to your pernicious activities in 1913 in and about the corridors of the National Capitol—no reference whatsoever to your wonderful accomplishments as a large and pernicious lobby. There is no doubt about your great engineering ability, your business sagacity, educational attainments and general all around good citizenship. I cannot understand why you suppressed one of your greatest abilities—lobbyist."[46]

Lobbyist. It was a skill that Muir and his supporters likely never imagined within the provenance of an engineer, even one as prominent as John R. Freeman. But an understanding of culture, energized through politically adroit counsel and arm-twisting, is what turned the tide for Hetch Hetchy and, to Freeman's mind, brought "something better for the city." By knowing culture, not simply the technical intricacies of hydraulic engineering, Freeman masterminded the outcome of one of the epic confrontations in American environmental history. It was John Freeman, and not John Muir, who held the day and materialized the future of Hetch Hetchy Valley.[47]

Acknowledgments

In the fall of 1997, I held a research fellowship at the Dibner Institute for the History of Science and Technology, then located on the campus of the Massachusetts Institute of Technology in Cambridge, Massachusetts. During my residency at the Dibner, I spent much time at MIT's Institute Archives and Special Collections (now known as Distinctive Collections, MIT Libraries) investigating the John R. Freeman Papers (MC51), a remarkable manuscript collection documenting the work of one of America's most important engineers of the Progressive Era and a prominent MIT graduate, class of 1876.

Previously I had made extensive use of the Freeman papers in my book *Building the Ultimate Dam: John S. Eastwood and the Control of Water in the West* (paperback ed., University of Oklahoma Press, 2005), in which Freeman's influential opposition to the technology of multiple arch dams—and his advocacy of massive gravity dams—constitutes an analytic centerpiece. Building upon this experience, I believed that writing a broader book on Freeman's career as a dam designer and hydraulic engineer would be a logical next step. As I worked my way through the voluminous paper trail that documents Freeman's professional life from the 1870s through his death in October 1932, I confronted the incredible amount of work Freeman had undertaken in the fields of water power and hydroelectricity, municipal water supply, civic park design, fire suppression, and factory insurance. The more I explored his papers, including his revealing professional diaries, the more I worried that it would be difficult to write a book that would do justice to—or at least comprehensively investigate—the myriad facets of his professional career. At the same time, I came to see that one particular project stood out as something special: the Hetch Hetchy Dam and San

Francisco's early-twentieth-century campaign to build a municipal water supply reservoir in Yosemite National Park.

By the end of my Dibner fellowship, I realized that I wanted to craft a history of the Hetch Hetchy controversy that placed Freeman—in all his professional complexity—at center stage and not lurking somewhere at the periphery. But life is filled with twists and turns; not long after my return to Lafayette College, I joined up with David Billington at Princeton to write a history of major dams built by the Bureau of Reclamation and the Army Corps of Engineers; that became *Big Dams of the New Deal Era: A Confluence of Engineering and Politics* (University of Oklahoma Press, 2006). I also embarked on an exploration into how dams and their place in American culture are reflected in postcard imagery, resulting in *Pastoral and Monumental: Dams, Postcards, and the American Landscape* (University of Pittsburgh Press, 2013). And I began collaborating with Norris Hundley on a history of the St. Francis Dam disaster. Although delayed by Norris's untimely passing, this effort eventually spawned *Heavy Ground: William Mulholland and the St. Francis Dam Disaster* (Huntington Library Press, 2015; paperback ed., University of Nevada Press, 2020). As work on *Heavy Ground* wound to a close, my focus returned to Freeman and the damming of Hetch Hetchy. Finally—after navigating the headaches and delays brought by the Covid-19 crisis—I was able to complete *The Man Who Dammed Hetch Hetchy.*

In researching the Freeman/Hetch Hetchy nexus during my time as a Dibner research fellow, I benefited greatly from the assistance of Elizabeth Andrews and the staff of MIT's Institute Archives and Special Collections. I also gratefully acknowledge the help of Randal Brandt at the Bancroft Library, University of California, Berkeley, whom I first met when he was a librarian at the Water Resources Center Archives.

At the University of Oklahoma Press, the help of Chuck Rankin, Steven Baker, Helen Robertson, Riley Hines, Amy Hernandez, and especially acquisitions editor Joe Schiller has proven invaluable in bringing this book to publication. With pleasure I acknowledge the support and friendship of Mark Feige and Jared Orsi, editors of Oklahoma's Public Lands History series. Thanks also to Sally Bennett Boyington for her skillful copyediting and to Erin Greb for her cartographic work in creating the map of the Hetch Hetchy Aqueduct and related sites.

At Lafayette College I enjoyed a sabbatical leave in the spring of 2023 that allowed me to complete the manuscript of *The Man Who Dammed Hetch Hetchy.* During the time spent on this project since the 1990s, I have been supported by

History Department heads Arnie Offner, Andy Fix, Deborah Rosen, Josh Sanborn, Paul Barclay, Rebekah Pite, and Rachel Goshgarian; departmental secretaries Kathy Ankaitus, Tammy Yeakel, and Rebecca Stocker; and Lafayette College provosts June Schlueter, Wendy Hill, Abu Rizvi, John Meier, and Laura McGrane. I have also enjoyed support from funds generously provided by the Cornelia F. Hugel endowed professorship.

Over the years I benefited from conversations regarding Freeman and Hetch Hetchy with my colleagues David P. Billington, Donald J. Pisani, and Norris Hundley. They stand as giants in the fields of western water history and the history of American engineering and technology. It has been my extraordinary good fortune to know and work with them. I also gratefully acknowledge the longstanding friendships of environmental and technological historians Mark Harvey, David Soll, Jim Sherow, Bonnie Lynn-Sherow, Marty Melosi, Char Miller, Bill Rowley, Gray Fitzsimons, Pat Malone, Dick Wiltshire, and Patty Limerick. Thanks to all for their sage counsel. And special thanks to Carol.

As the years passed by while I worked on *The Man Who Dammed Hetch Hetchy*, Berry The Dog, Ottohound, Rosie, Carrie, Tucker, Tango, and Nelson exhibited great patience as they plotted to nudge me away from the ubiquitous computer screen and out into the wondrous world beyond the front door. No doubt, more than a few insights related to John R. Freeman and Hetch Hetchy came to mind while I was ruminating in the midst of a quiet evening dog walk.

Appendix

Raker Act Water Allocation—4,000 CFS for 60 Days

In the spring of 1913, representatives of both San Francisco and the Turlock and Modesto Irrigation Districts negotiated terms to be included in the Raker Act that would guide how the city and the districts would divide the flow of the Tuolumne River. The city recognized that the districts possessed a right to a year-round, constant flow of 2,350 cubic feet of water per second (cfs). But the irrigation districts wanted a guarantee of more water during spring flood time and demanded that the Raker Act include, as soon codified in paragraph C, section 9, explicit acknowledgment that the city "recognize the rights of the irrigation districts to the extent of four thousand second-feet of water out of the natural daily flow of the Tuolumne River for combined direct use and collection into storage reservoirs as may be provided by said irrigation districts during the period of sixty days immediately following April fifteenth of each year." This stipulation required that for sixty days every spring the districts would be guaranteed an additional 1,650 cfs of flow beyond their previously recognized right to 2,350 cfs (thus, a total of 4,000 cfs during those sixty days).

The question then arises: Exactly how much water equates to a flow of 1,650 cfs for a period of sixty days? Or, considered more practically, what did San Francisco concede to the Turlock and Modesto Irrigation Districts to facilitate passage of the Raker Act? And did this additional allocation have an impact on the city's subsequent ability to divert 400 million gallons per day (mgd) into the Hetch Hetchy Aqueduct? The following calculations provide context for understanding the implications of the allocation of 1,650 cfs for sixty days:

1 cubic foot of water = 7.48 gallons (approximately)
There are 60 seconds in a minute, or 86,400 seconds in a day, so1 cfs for a
    day = 7.48 gallons × 86,400 seconds = 646,272 gallons per day
There are 325,851 gallons in an acre-foot, so
646,272 gallons ÷ 325,851 gallons = 1.983 acre-feet
Thus, the daily volume of 1 cfs = 1.983 acre-feet

When the Raker Act stipulated an extra flow of 1,650 cfs for sixty days, how much water was required to meet this demand? Calculating this amount is a simple matter of multiplication:

1,650 cfs × 1.98 acre-feet = 3,272 acre-feet per day
1,650 cfs for 60 days = 3,272 acre-feet × 60 days = 196,320 acre-feet
196,320 acre-feet ÷ 365 days = 537.8 acre-feet per day, for every day of
    the year

The question now becomes: How many gallons are in 537.8 acre-feet? And what does that amount equate to if distributed over a full year?

1 acre = 43,560 square feet, so 1 acre-foot = 43,560 cubic feet
1 acre-foot = 43,560 cubic feet × 7.48 gallons per cubic foot = 325,851
    gallons
A flow of 1,650 cfs for 60 days = 537.8 acre-feet per day stretched out for
    a year, thus 537.8 acre-feet per day × 325,851 gallons = 175,242,667
    gallons per day, such that 1,650 cfs for 60 days = 175.242 mgd dis-
    tributed over the course of a full year

Freeman's Hetch Hetchy plan called for supplying San Francisco with 400 mgd. So how does this compare with the sixty-day, 1,650 cfs entitlement granted to the irrigation districts? It is a matter of simple division to calculate this ratio:

175.242 mgd ÷ 400 mgd = .438 or 43.8%

The Raker Act stipulation calling for the irrigation districts to be supplied a flow of 4,000 cfs for sixty days every spring might not seem like much, just some minor adjustment of no great consequence. Wrong. In truth, it guaranteed an extra flow beyond 2,350 cfs that, in aggregate, approached 45 percent of the entire annual capacity of Freeman's planned 400 mgd system. Year after year in perpetuity, this flow represents, from the city's perspective, an enormous amount of water.

Notes

Introduction

1. For the full text, see the Proposition F page on the *Ballotpedia* website, https://ballotpedia.org/San_Francisco_Hetch_Hetchy_Reservoir_Initiative,_Proposition_F_(November_2012).

2. John Muir, "The Endangered Valley," in *Let Everyone Help to Save the Famous Hetch Hetchy Valley*, November 1909, p. 16, National Archives, College Park, MD, Record Group 95, Entry 22, box 4; see also John Muir, *The Yosemite* (New York: Century, 1912), 262.

3. Muir, "Endangered Valley," 16.

4. The law authorizing construction of a dam at Hetch Hetchy was sponsored by US Representative John Raker of California. The full text of the Raker Act is available at Museum of the City of San Francisco website, http://www.sfmuseum.org/hetch/hetchy.html.

5. John McPhee (in "Farewell to the 19th Century," *The New Yorker*, September 21, 1999) describes the Edwards Dam removal and the broader movement to breach dams. In the 1950s, US Bureau of Reclamation plans to inundate part of Dinosaur National Monument with a hydropower dam were successfully opposed by the Sierra Club (Mark Harvey, *A Symbol of Wilderness: Echo Park and the American Conservation Movement* [Seattle: University of Washington Press, 2000]). But in 1956 the bureau won approval for the Glen Canyon Dam across the Colorado River above Grand Canyon, and by the 1960s the Sierra Club was attacking the need for this huge dam (Eliot Porter and David Brower, *The Place No One Knew: Glen Canyon on the Colorado* [San Francisco: Sierra Club, 1963]). After Glen Canyon Dam's completion in 1966, *The Place No One Knew* became a touchstone for the growing "anti-dam" movement.

6. Anti-dam activism sparked the creation of Restore Hetch Hetchy, an organization established in 1999 and dedicated to the removal of O'Shaughnessy Dam. See the organization's website, https://www.hetchhetchy.org/.

7. Ed Lee and David Chui, "Ballot Measure on S.F. Water Supply Unnecessary," *SFGate,* March 2, 2012.

8. Quotation is from Louis Sahagun, "Lungren, Feinstein Spar over Hetch Hetchy Valley Restoration," *Los Angeles Times,* December 13, 2011. See also Aaron Sankin, "Hetch Hetchy Ballot Measure: Environmental Group Seeks to Tear Down SF's Primary Source of Drinking Water," *Huffington Post,* July 11, 2012, https://www.huffingtonpost .com/2012/07/11/hetch-hetchy-ballot-measure_n_1665596.html?ir=San+Francisco.

9. "Endorsements 2012: San Francisco Propositions," *San Francisco Bay Guardian,* October 3, 2012; "Prop F for Fail," *SFGate*, October 5, 2012.

10. After this defeat, Restore Hetch Hetchy sought to remove O'Shaughnessy Dam through court action. But in 2018 the California Supreme Court refused to review lower court rulings affirming San Francisco's right to use the valley for a municipal reservoir. Paige Blankenbuehler, "Why Hetch Hetchy Is Staying Under Water," *High Country News,* May 30, 2016; Bob Egelko, "State High Court Rejects Berkeley Group's Suit to Drain Hetch Hetchy Reservoir," *San Francisco Chronicle,* October 17, 2018. For discussion of how restoration of Hetch Hetchy Valley might ultimately occur, see Kenneth Brower, *Hetch Hetchy: Undoing a Great American Mistake* (Berkeley, CA: Heyday, 2013).

11. Holway Jones, *John Muir and the Sierra Club: The Battle for Yosemite* (San Francisco: Sierra Club, 1965), 92–95; Roderick Nash, *Wilderness and the American Mind*, 5th ed. (New Haven, CT: Yale University Press, 2014), 161–64, 171; Donald Worster, *A Passion for Nature: The Life of John Muir* (New York: Oxford University Press, 2008), 426–29, 450. Pinchot's advocacy is also featured in Robert W. Righter, *The Battle over Hetch Hetchy: America's Most Controversial Dam and the Birth of Modern Environmentalism* (New York: Oxford University Press, 2005), 60–83; and John Warfield Simpson, *Dam!: Water, Power, Politics and Preservation in Hetch Hetchy and Yosemite National Park* (New York: Pantheon, 2005), 126–28, 141–43, 153–56.

12. Pinchot's involvement with Hetch Hetchy is discussed in chapters 1 and 6.

13. Freeman's family origins, education, and career prior to Hetch Hetchy are described in chapter 2. Key primary sources documenting Freeman's life and career are held in the John Ripley Freeman (1855–1932) Papers (MC 51), Distinctive Collections, Massachusetts Institute of Technology, Cambridge (hereafter, JRF Papers). This archive encompasses 120 boxes of letters, reports, diaries, and other manuscript material covering his life from the 1860s until his death in October 1932.

14. John R. Freeman to Walter Fisher, April 7, 1911, JRF Papers, box 67.

15. John R. Freeman, *On the Proposed Use of a Portion of the Hetch Hetchy, Eleanor and Cherry Valleys Within or Near to the Boundaries of the Stanislaus U.S. National Forest Reserve and the Yosemite National Park as Reservoirs for Impounding Tuolumne River Flood Waters and Appurtenant Works for the Water Supply of San Francisco, California and Neighboring Cities* (San Francisco: by Authority of the Board of Supervisors, Rincon Publishing, 1912). Hereafter, Freeman, *Hetch Hetchy Report.*

16. Righter, *Battle over Hetch Hetchy,* 102; Kevin Starr, *Endangered Dreams: The Great Depression in California* (Berkeley: University of California Press, 1996), 282.

Freeman's role in the planning of the Hetch Hetchy project is also noted in Ray W. Taylor, *Hetch Hetchy: The Story of San Francisco's Struggle to Provide a Water Supply for Her Future Needs* (San Francisco: Ricardo J. Orozco, 1926), 111–17; and M. M. O'Shaughnessy, *Hetch Hetchy: Its Origin and History* (San Francisco: privately printed, 1934), 26, 40–41.

17. Donald Worster, *Rivers of Empire: Water, Aridity and the Growth of the American West* (New York: Pantheon, 1985); Norris Hundley, *The Great Thirst: Californians and Water, 1770–1990* (Berkeley: University of California Press, 1992); Marc Reisner, *Cadillac Desert: The American West and Its Disappearing Water* (New York: Viking Press, 1986). Recent publications describing the Hetch Hetchy controversy usually note Freeman's work as a consulting engineer, although they provide little insight regarding his pre–Hetch Hetchy career or the precise nature of his work for the city; see Jones, *John Muir*; Kendrick A. Clements, "Politics and the Park: San Francisco's Fight for Hetch Hetchy, 1908–1913," *Pacific Historical Review* (May 1979): 185–215; and Righter, *Battle over Hetch Hetchy*. Bruce Sinclair's essay "Engineering the Golden State: Technics, Politics, and Culture in Progressive Era California" (in *Where Minds and Matter Meet: Technology in California and the West*, ed. Volker Janssen [San Marino, CA: Huntington Library Press, 2012], 43–70) discusses Freeman's work on Hetch Hetchy in a broad context, comparing and contrasting his work with that of City Engineers Marsden Manson and M. M. O'Shaughnessy. Histories of the Los Angeles Aqueduct usually note Freeman's participation on the city's board of engineers; see William L. Kahrl, *Water and Power: The Conflict over Los Angeles' Water Supply in the Owens Valley* (Berkeley: University of California Press, 1982), 50–52. Freeman's professional influence looms large in my treatment of the dam engineering career of John S. Eastwood; see Donald C. Jackson, *Building the Ultimate Dam: John S. Eastwood and the Control of Water in the West,* paperback ed. (Norman: University of Oklahoma Press, 2005), 109–33, 227–35.

18. In placing Freeman as a central figure who guided the outcome of the Hetch Hetchy controversy in 1912–13, my intent is not to deny that other historical actors also played a role in either supporting or opposing the Raker Act. Clearly, Freeman's work advocating for, and socially constructing, the Hetch Hetchy Dam took place within a broader network of historical actors (including anti-dam activists aligned with Muir, San Francisco city officials, Spring Valley Water Company leaders, farmers in the Turlock and Modesto Irrigation Districts, various secretaries of the interior, US congressmen and senators, the Advisory Board of Army Engineers, and others). Freeman operated within a dynamic social environment in which no single person, collective, or organization absolutely controlled if, when, or how a dam was to be built at Hetch Hetchy. But Freeman was not simply one among many actors in the story of Hetch Hetchy; he was the key and essential participant who, as of 1912–13, exercised dramatic influence over how, for decades to come, the valley was to be used as a municipal reservoir. For more on the complexities of "actor network theory," see Bruno Latour, *Reassembling the Social: An Introduction to Actor-Network-Theory* (New York: Oxford University Press, 2005). For more on the social construction of technology, see Donald Mackenzie and Judy

Wajcman, eds., *The Social Shaping of Technology,* 2nd ed. (Buckingham, UK: Open University Press, 1999).

19. Historical studies of Progressivism and the Progressive Era are extensive and include Richard Hofstadter, *The Age of Reform* (New York: Vintage Books, 1955); Robert H. Wiebe, *The Search for Order, 1877–1920* (New York: Hill and Wang, 1967); Maureen A. Flanagan, *America Reformed: Progressives and Progressivisms, 1890s–1920s* (New York: Oxford University Press, 1907); Michael McGerr, *A Fierce Discontent: The Rise and Fall of the Progressive Movement in America, 1870–1920* (New York: Free Press, 2014); and Daniel Rodgers, "In Search of Progressivism," *Reviews in American History* 10 (December 1982): 113–32. For a review of the subject with scores of sources and useful references, see "Progressive Era," *Wikipedia,* https://en.wikipedia.org/wiki/Progressive_Era.

20. Samuel F. Hays, *Conservation and the Gospel of Efficiency: The Progressive Conservation Movement, 1890–1920* (Cambridge, MA: Harvard University Press, 1959), quotations from 2–3.

21. John R. Freeman to Albert E. Pillsbury, September 29, 1913, JRF Papers, box 94.

22. Jackson, *Building the Ultimate Dam,* esp. chapter 6, "Confrontation at Big Meadows: John R. Freeman and the 'Psychology' of Dam Design (1911–1913)," 109–33.

Chapter 1

1. Quotations are from N. King Huber, *The Geologic Story of Yosemite National Park* (Washington, DC: Government Printing Office, 1987), 11.

2. The origin of the name Hetch Hetchy is discussed in Robert W. Righter, *The Battle over Hetch Hetchy: America's Most Controversial Dam and the Birth of Modern Environmentalism* (New York: Oxford University Press, 2005), 16. It is most likely derived from the Ahwahneechee or Tuolumne word *hatchatchie,* referencing a species of grass.

3. Char Miller, *Hetch Hetchy: A History in Documents* (Peterborough, ON: Broadview Press, 2020), 5–6. Also see Righter, *Battle over Hetch Hetchy,* 14–16; and M. Kat Anderson, *Tending the Wild: Native American Knowledge and Management of California's Natural Resources* (Berkeley: University of California Press, 2005).

4. John Muir, "The Hetch Hetchy Valley," *Sierra Club Bulletin,* January 1908, 216.

5. Stephen G. Hyslop, *Contest for California: From Spanish Colonization to the American Conquest* (Norman, OK: Arthur H. Clark, 2012).

6. There are many histories of the California Gold Rush; among the most notable is J. S. Holliday, *The World Rushed In: The California Gold Rush Experience* (New York: Simon and Schuster, 1981).

7. The Anglo-American "discovery" of Yosemite and Hetch Hetchy Valleys and attacks on Native tribes are described in Jen Huntley, *The Making of Yosemite: James Mason Hutchings and the Origin of America's Most Popular National Park* (Lawrence: University Press of Kansas, 2011), 55–59; Righter, *Battle over Hetch Hetchy,* 16–17; and John Warfield Simpson, *Dam!: Water, Power, Politics and Preservation in Hetch Hetchy and Yosemite National Park* (New York: Pantheon, 2005), 4–13. In the early

twentieth-century battle over the use of Hetch Hetchy as either park or reservoir, the interests of Euro-Americans dominated the political discourse surrounding the controversy, and the rights of indigenous peoples with long-standing ties to the Sierra Nevada landscape were ignored. It is now widely recognized that the creation of national parks in the nineteenth and early twentieth centuries purposefully excluded Native tribes from park lands. Only in recent years has this important subject begun to receive serious attention; for more on the relationship of indigenous people to Yosemite National Park, see Miller, *Hetch Hetchy.* See also Christina Gish Hill, Matthew J. Hill, and Brooke Neely, eds., *National Parks, Native Sovereignty: Experiments in Collaboration* (Norman: University of Oklahoma Press, 2024).

8. Huntley, *Making of Yosemite.*

9. Huntley, *Making of Yosemite,* 115–16.

10. Huntley, *Making of Yosemite,* 120–26. See S. 203, Public Act no. 159, "An Act Authorizing a Grant to the State of California of 'the Yo-Semite Valley' and of the Land Embracing the Mariposa Big Tree Grove."

11. Frederick Law Olmsted, *Yosemite and the Mariposa Grove: A Preliminary Report, 1865* (Yosemite National Park: Yosemite Association, 1995). Olmsted's report is described in Holway Jones, *John Muir and the Sierra Club: The Battle for Yosemite* (San Francisco: Sierra Club, 1965), 30–33; and Huntley, *Making of Yosemite,* 113–14.

12. Huntley, *Making of Yosemite,* 1.

13. Preceding Muir's visit, in 1868 geologist Josiah D. Whitney noted how "the walls of this [Hetch Hetchy] Valley are not quite so high as those of the [Merced] Yosemite; but, still anywhere else than in California, would be considered as wonderfully grand." J. D. Whitney, *The Yosemite Book: A Description of the Yosemite Valley and the Adjacent Region of the Sierra Nevada* (New York: published by the authority of the legislature, 1868), 98–99. Also see Jones, *John Muir,* 84.

14. John Muir, "The Hetch Hetchy Valley," *Boston Weekly Transcript,* March 25, 1873.

15. Muir, "Hetch Hetchy Valley."

16. Muir, "Hetch Hetchy Valley."

17. Muir soon followed up his piece in the *Boston Weekly Transcript* with another description of his 1871 visit to the valley. See John Muir, "Hetch Hetchy Valley," *Overland Monthly,* July 1873.

18. Bierstadt's paintings are discussed in Righter, *Battle over Hetch Hetchy,* 19. Lieutenant Montgomery Macomb's 1878–79 survey of the West, including Hetch Hetchy, is discussed in Jones, *John Muir,* 85. The 1885 article is Xenos Clark, "The Hetch Hetchy Valley: A New Yosemite," *Outing: An Illustrated Monthly Magazine of Recreation* (May 1885). For a map illustrating the 720 acres of valley land held in private ownership prior to 1890, see Ray W. Taylor, *Hetch Hetchy: The Story of San Francisco's Struggle to Provide a Water Supply for Her Future Needs* (San Francisco: Ricardo J. Orozco, 1926), 110.

19. John Muir, "Living Glaciers of California," *Harper's New Monthly Magazine,* November 1875, 769–76. See Donald Worster, *A Passion for Nature: The Life of John Muir* (New York: Oxford University Press, 2008), 241.

20. Worster, *Passion for Nature,* 276–304, 309.

21. Robert Underwood Johnson, *Remembered Yesterdays* (Boston: Little Brown, 1923).

22. Worster, *Passion for Nature,* 310–14; Johnson, *Remembered Yesterdays,* 278–83. Righter (in *Battle for Hetch Hetchy*, 22) notes that Johnson had corresponded with Muir as early as 1877; it appears that the two men did not meet in person until 1889.

23. Richard Orsi, *Sunset Limited: The Southern Pacific Railroad and the Development of the American West* (Berkeley: University of California Press, 2005), 363–64.

24. Worster, *Passion for Nature,* 312–16; Jones, *John Muir,* 43–47; Johnson, *Remembered Yesterdays,* 287–89.

25. "Fisher Hearing, Day 1," p. 126, in John Ripley Freeman (1855–1932) Papers (MC 51), Distinctive Collections, Massachusetts Institute of Technology, Cambridge (hereafter, JRF Papers), box 72.

26. John Muir, "Treasures of the Yosemite," *Century Magazine,* August 1890, 483–500; John Muir, "Features of the Proposed Yosemite National Park," *Century Magazine,* September 1890, 656–67.

27. Jones, *John Muir,* 44.

28. Fifty-First Cong., 1st sess., chap. 1263, "An act to set apart certain tracts of land in the State of California as forest reservations," October 1, 1890. See Jones, *John Muir,* 43–47.

29. "An act to set apart certain tracts of land . . ." October 1, 1890.

30. The Sierra Club's origins are detailed in Jones, *John Muir,* 3–23; and Worster, *Passion for Nature,* 328–31.

31. The movement to expand irrigation in the arid West is described in Donald J. Pisani, *From the Family Farm to Agribusiness: The Irrigation Crusade in California and the West, 1850–1931* (Berkeley: University of California Press, 1984). Also see Donald J. Pisani, *To Reclaim a Divided West: Water, Law, and Public Policy* (Albuquerque: University of New Mexico Press, 1992).

32. Charles Hillman Brough, *Irrigation in Utah* (Baltimore, MD: Johns Hopkins University Press, 1898); George Thomas, *The Development of Institutions under Irrigation with Special Reference to Early Utah Conditions* (New York: Macmillan, 1920).

33. Alan M. Peterson, *Land, Water and Power: A History of the Turlock Irrigation District, 1887–1987* (Spokane, WA: Arthur H. Clark, 1989), 20–36.

34. Pisani, *Family Farm,* 250–83.

35. "Brief of L.L. Bennett, on Behalf of the Modesto Irrigation District," n.d. (ca. July 1907), in *Reports on the Water Supplies of San Francisco, 1900 to 1908 Inclusive,* published by authority of the board of supervisors (San Francisco: Britton & Rey, 1908), 170B. Righter (in *Battle for Hetch Hetchy,* 251) notes that this 230-page compendium of letters, reports, briefs, and various filings related to Hetch Hetchy is available at the San Francisco Public Library. I also have a copy in my personal library.

36. For more on La Grange Dam and the irrigation districts, see Peterson, *Land, Water and Power,* 68–83. Also see James D. Schuyler, *Reservoirs for Irrigation, Water Power and Domestic Water Supply,* 2nd ed. (New York: John Wiley & Sons, 1909), 256–62.

37. "Brief of L.L. Bennett," 170B.

38. John Wesley Powell, *Report on the Lands of the Arid Region of the United States* (Washington, DC: Government Printing Office, 1879).

39. A. H. Thompson, "Report on the Location and Survey of Reservoir Sites during the Fiscal Year Ended June 30, 1891," in US Geological Survey, *Twelfth Annual Report, Part 2—Irrigation* (Washington, DC: Government Printing Office, 1891), 36–38.

40. "Tuolumne River—Hetch Hetchy Valley," in *Twenty-First Annual Report of the USGS (1899–1900), Part 4—Hydrography* (Washington, DC: Government Printing Office, 1901), 449–65.

41. USGS, *Twenty-First Annual Report, Part 4,* 450, 457.

42. USGS, *Twenty-First Annual Report, Part 4,* 465.

43. USGS, *Twenty-First Annual Report, Part 4,* 459.

44. "Demographics of San Francisco," *Wikipedia,* https://en.wikipedia.org/wiki /Demographics_of_San_Francisco#cite_note-2020CensusFF-1.

45. For the pre-1910 history of San Francisco's water supply, see Marsden Manson, "Outline of the History of the Water Supply of the City of San Francisco," in *Water Supplies of San Francisco,* 3–13; Hermann Schussler, *The Water Supply of San Francisco: Before, during, and after the Earthquake of April 18th, 1906, and the Subsequent Conflagration* (New York: Martin D. Brown Press, 1906); and Taylor, *Hetch Hetchy,* 3–102.

46. Schussler, *Water Supply,* 8–26; Schuyler, *Reservoirs for Irrigation,* 267–73.

47. Schussler, *Water Supply,* 20–22.

48. Donald J. Pisani, "'Why Shouldn't California Have the Grandest Aqueduct in the World?': Alex Von Schmidt's Lake Tahoe Scheme," *California Historical Quarterly* (Winter 1974): 347–60.

49. Gray Brechin, *Imperial San Francisco: Urban Power, Earthly Ruin* (Berkeley: University of California Press, 1999), 99–100; James P. Walsh and Timothy J. O'Keefe, *Legacy of a Native Son: James Duval Phelan and Vila Montalvo* (Los Gatos, CA: Forbes Mill Press, 1993).

50. Edward F. Treadwell, annotator, *The Charter of the City and County of San Francisco* (San Francisco: Bancroft-Whitney, 1899), 338.

51. Franklin K. Lane's career is outlined in Anne Wintermute Lane and Louise Herrick Wall, eds., *The Letters of Franklin K. Lane, Personal and Political* (Boston: Houghton Mifflin, 1922), 31–37.

52. Treadwell, *Charter,* 340, 343.

53. Fifty-Sixth Cong., 2nd sess., chap. 372, "An Act Relating to rights of way through certain parks, reservations, and other public lands," September 15, 1901.

54. Fifty-Sixth Cong., 2nd sess., chap. 372. Exactly how the terms of the 1901 Right-of-Way Act came to include the Yosemite reservation is uncertain, with historian Kendrick Clements noting a "possible collusion between the city and the law's author, Representative Marion DeVries of Stockton." However, DeVries disavowed any such interaction, and Clements accepts this denial, noting the lack of "concrete evidence to support this speculation." Kendrick A. Clements, "Politics and the Park: San Francisco's Fight for Hetch Hetchy, 1908–1913," *Pacific Historical Review* (May 1979): 188. In its final provision, the act stipulated "that any permission given by the secretary of the interior under the provisions of this act may be revoked by him or his successor in his

discretion, and shall not be held to confer any right, or easement, or interest in, to, or over any public land, reservation, or park."

55. Jones (in *John Muir,* 90–91) discusses how preservationists were unaware of provisions in the 1901 Right-of-Way Act relating to Yosemite, Sequoia, and General Grant National Parks.

56. "Filings of James D. Phelan for Reservoir Rights of Way, Notice of Appropriation of Water, July 29, 1901"; James D. Phelan, "Right of way application for a Tuolumne River reservoir site," October 15, 1901; James D. Phelan, "Right of way map for the Lake Eleanor reservoir site," October 15, 1901; all in *Water Supplies of San Francisco,* 106–8. Taylor (in *Hetch Hetchy,* 46–47) notes that J. B. Lippincott was engaged to do survey work for Phelan and the city related to planning for the Hetch Hetchy Dam; this is also referenced in Jones, *John Muir,* 49. Also see C. E. Grunsky to the Honorable Board of Public Works, July 28, 1902, in *Water Supplies of San Francisco,* 38–39.

57. Phelan's 1901 Right-of-Way Act filings are documented on pages 109–11 of *Water Supplies of San Francisco.*

58. C. E. Grunsky, "Report on the Tuolumne River Water Supply Project," July 28, 1902, in *Water Supplies of San Francisco,* 36–81. Grunsky was born near Stockton, California, in 1855 and educated at Stockton High School. In the 1870s he trained as an engineer at the Realschule and Polytechnikum at Stuttgart, Germany. Upon returning from Europe in 1877, he worked as a topographer for the California State Engineering Department. From 1900 through 1904, he served as San Francisco's city engineer, whereupon he became a consulting engineer based out of San Francisco. As discussed in chapter 4, he undertook work investigating alternative sources of water supply for San Francisco that was included in Freeman's *Hetch Hetchy Report,* published in 1912. For biographical information, see his obituaries "Carl Ewald Grunsky," *Science* 79 (June 22, 1935), 556; and "Carl Ewald Grunsky," *New York Times,* June 10, 1934,

59. C. E. Grunsky, "Report on the Availability of Water Supply Sources for San Francisco, California," November 24, 1902, in *Water Supplies of San Francisco,* 14–35.

60. E. A. Hitchcock to Commissioner of the General Land Office, December 22, 1903, in *Water Supplies of San Francisco,* 129–30.

61. Franklin K. Lane, "Petition to the Secretary of the Interior for Review of the Matter of the Application of Jas. D. Phelan for Rights-of-Way in Hetch Hetchy Valley and Lake Eleanor within Yosemite National Park," ca. April 1904, in *Water Supplies of San Francisco,* 112–27; also see Lane and Wall, *Letters of Franklin K. Lane,* 41–42.

62. E. A. Hitchcock, "Letter of the Hon. Secretary of the Interior, to the President, Denying the Application of San Francisco," February 20, 1905, in *Water Supplies of San Francisco,* 128–32.

63. "San Francisco Board of Supervisors Resolution No. 6949," adopted January 29, 1906; facsimile in Taylor, *Hetch Hetchy,* 60.

64. "Marsden Manson," in Will Hager, ed., *Hydraulicians in the USA, 1800–2000* (New York: CRC Press, 2015), 2305; "Marsden Manson," *Transactions of the American Society of Civil Engineers* 95 (1931): 1554–56.

65. Pinchot's support is described in Righter, *Battle over Hetch Hetchy,* 60–61, 66–70.

66. Gifford Pinchot to William Colby, February 17, 1905, in Jones, *John Muir,* 92–93.

67. Simon Winchester, *A Crack in the Edge of the World: America and the Great California Earthquake of 1906* (New York: HarperCollins, 2005). The response of the Spring Valley Water Company is described in Schussler, *Water Supply.*

68. Marsden Manson to Gifford Pinchot, May 10, 1908, National Archives, College Park, MD, Record Group 95 (hereafter, NA, RG95), Entry 22, box 4.

69. Gifford Pinchot to Marsden Manson, May 18, 1908, NA, RG95, Entry 22, box 4.

70. The Schmitz/Ruef scandals are described in Walter Bean, *Boss-Ruef's San Francisco* (Berkeley: University of California Press, 1952); the Bay Cities Water Company scandal is described on pages 128 and 140–44. City water supply issues in the Schmitz/Ruef era are also described in Taylor, *Hetch Hetchy,* 59–66.

71. For background on Garfield and on Pinchot's support for his appointment as interior secretary, see Samuel F. Hays, *Conservation and the Gospel of Efficiency: The Progressive Conservation Movement, 1890–1920* (Cambridge, MA: Harvard University Press, 1959), 72–73.

72. A transcript of Garfield's July 24, 1907, Hetch Hetchy hearing in San Francisco is available in *Water Supplies of San Francisco,* 148–210. The irrigation districts' objections are also presented in "Brief of L. L. Bennett, on Behalf of the Modesto Irrigation District," 170A–170G.

73. For more on Muir and Roosevelt's relationship, see Worster, *Passion for Nature,* 366–70; also see John Clayton, *Natural Rivals: John Muir, Gifford Pinchot and the Creation of America's Public Lands* (New York: Pegasus Books, 2019), 42–46.

74. In September 1907, Roosevelt wrote to Muir and acknowledged that "Garfield and Pinchot are rather favorable to the Hetch Hetchy [Dam] plan." While Roosevelt expressed some sympathy for Muir's position against the dam, he pointedly advised Muir, "[S]o far everyone that has appeared has been for [the Hetch Hetchy Dam project] and I have been in the disagreeable position of seeming to interfere with the development of the State [of California] for the sake of keeping a valley, which apparently hardly anyone wanted to have kept, under national control." Theodore Roosevelt to John Muir, September 16, 1907. In Miller, *Hetch Hetchy*, 104.

75. See "Petition of Marsden Manson, City Engineer of San Francisco . . . to Reopen the Matter of Application of James D. Phelan . . . ," May 7, 1908, in *Water Supplies of San Francisco,* 213–16.

76. Gifford Pinchot to Norman Hapgood, April 30, 1908, NA, RG95, Entry 22, box 4.

77. Pinchot to Hapgood, April 30, 1908.

78. Pinchot to Hapgood, April 30, 1908.

79. The "Garfield Permit" was officially issued as "Decision of the Secretary of the Interior Department, Washington, D.C., Granting the City and County of San Francisco, Subject to Certain Conditions, Reservoir Sites and Rights of Way at Lake Eleanor and Hetch Hetchy Valley in the Yosemite National Park, May 11, 1908." The full text of the permit is printed in *Water Supplies of San Francisco,* 217–22.

80. "Decision of the Secretary of the Interior."

81. "Decision of the Secretary of the Interior."

82. Hays, *Conservation,* 128–33. Righter (in *Battle over Hetch Hetchy,* 73) notes that Horace McFarland of the American Civic Association was able to address the conference

and assert that the "Hetch Hetchy Valley belongs to all America and not to San Francisco alone." Nonetheless, his remarks apparently had little impact.

83. Marsden Manson, "Report on the Tuolumne River Water Supply Project, with Plans and a Cost Estimate," September 14, 1908; Marsden Manson, "Report of the City Engineer on the Cost of the Original Construction of Water Works from Hetch Hetchy and Lake Eleanor and Tuolumne River," July 23, 1908; both in *Water Supplies of San Francisco*, 92–98, 99–104.

84. For more on land purchases and the bond referendum, see Taylor, *Hetch Hetchy*, 81–84.

85. Jones, *John Muir*, 94–95.

86. Sierra Club Board of Directors, Report to Secretary Garfield, ca. September 1907; *Sierra Club Bulletin*, June 1908, 264–65; Jones, *John Muir*, 94–95. While the beginning of a defense of Hetch Hetchy was being planned by members of the Sierra Club, preservationists were also concerned that the loosely organized reservations that were designated as "national parks" were vulnerable to use and exploitation because there was no overarching authority in the federal government charged with their administration. This is an issue explored by conservation historian Samuel Hays, who noted that as early as 1904 Pinchot had "recommended that Congress transfer the national parks to the Forest Service so that he could administer them." This sparked a reaction from preservationists fearful of Pinchot; by 1910, preservationists were calling for the creation of a "National Park Bureau" under the control of the Department of the Interior. See Hays, *Conservation*, 195–97. Creation of the National Park Service would not occur until 1916.

87. H. L. Atkinson to William Colby, September 19, 1907, in Jones, *John Muir*, 94–95.

88. John Muir, "The Hetch Hetchy Valley," *Sierra Club Bulletin*, January 1908, 211–20.

89. "Let All the People Speak and Prevent the Destruction of the Yosemite Park," anti-dam pamphlet ca. December 1908/January 1909, copy in JRF Papers, box 70. Holway Jones attributes the pamphlet to "Colby and [Sierra Club member] E.T. Parsons. . . . although in substance it had been approved by Muir" (*Hetch Hetchy*, 102).

90. Public letter signed by John Muir, J. N. LeConte, E. T. Parsons, and W. F. Bade, addressed to "All Lovers of Nature and Scenery," December 21, 1908, printed in the pamphlet "Let All the People Speak and Prevent the Destruction of the Yosemite Park," JRF Papers box 70.

91. Jones, *John Muir*, 102–3.

92. Righter, *Battle over Hetch Hetchy*, 74–75 (includes quotations from Edward McCutcheon, lawyer for the Spring Valley Water Company).

93. For detailed discussion of the House and Senate hearings on the proposed land-swap legislation related to Hetch Hetchy, and the success of preservationist in countering the legislation, see Jones, *John Muir*, 100–105; Righter, *Battle over Hetch Hetchy*, 73–78.

94. Horace McFarland to Alden Sampson, March 11, 1909, referenced in Jones, *John Muir*, 105.

95. Gifford Pinchot to Marsden Manson, May 15, 1909, NA, RG95, Entry 22, box 4.

96. Marsden Manson to Gifford Pinchot, May 22, 1909, NA, RG95, Entry 22, box 4.

97. Horace McFarland to Gifford Pinchot, May 19, 1909, NA, RG95, Entry 22, box 4. In the letter, McFarland complained that "the whole effort to get this water supply is part of the dirty politics of San Francisco rather than an attempt to supply a real public need."

98. Formation of the Society for the Preservation of National Parks is described in Jones, *John Muir,* 97–98.

99. Clements, "Politics and the Park," 192–94; Jones, *John Muir,* 100–107. Also see *Let Everyone Help to Save the Famous Hetch Hetchy Valley,* November 1909, National Archives, College Park, MD, Record Group 95, Entry 22, box 4.

100. For more on the bureaucratic battle between Ballinger and Pinchot, see James Penick Jr., *Progressive Politics and Conservation. The Ballinger-Pinchot Affair* (Chicago: University of Chicago Press, 1968); Char Miller, *Gifford Pinchot and the Making of Modern Environmentalism* (New York: Island Press, 2001), 208–11; and Alpheus Thomas Mason, *Bureaucracy Convicts Itself: The Ballinger-Pinchot Controversy, 1910, and Its Meaning Today* (New York: Viking, 1941).

101. Taft's visit to Yosemite National Park and Ballinger's trip to Hetch Hetchy with Muir are discussed in Worster, *Passion for Nature,* 435–36. Quotations are from John Muir to Marion and Katherine Hooker, October 20, 1909, and John Muir to William Colby, November 1909, both in Worster, *Passion for Nature,* 491.

102. Ballinger's interest in revoking the Garfield Permit is detailed in Clements, "Politics and the Park."

103. George Otis Smith, Director of the US Geological Survey, to Secretary of the Interior February 25, 1912, in *Proceedings before the Secretary of the Interior in Re Use of Hetch Hetchy Reservoir Site in Yosemite National Park by the City of San Francisco* (Washington, DC: Government Printing Office, 1910), 8–9 (hereafter, *Ballinger Hearing*). E. G. Hopson's "Report on Tuolumne River Supply for San Francisco," submitted to Secretary Ballinger, November 23, 1909, with a copy also sent to George Smith, is available in JRF Papers, box 70.

104. Richard Ballinger to Mayor and Board of Supervisors, San Francisco, February 25, 1910, in *Ballinger Hearing,* 6.

105. For Taft's dismissal of Pinchot, see Righter, *Battle over Hetch Hetchy,* 100.

106. The January 1910 referendums on Hetch Hetchy and the Spring Valley purchase, including vote totals, are described in Taylor, *Hetch Hetchy,* 97–101; for the Building Trades quotation, see page 101. Following the authorization of Hetch Hetchy project bonds, the city acted to purchase water and power rights filings that a private power syndicate (controlled by William Hammond Hall and John Hayes Hammond) had made on Eleanor and Cherry Creeks. These filings and the subsequent purchase by San Francisco in April 1910 are described in Righter, *Battle over Hetch Hetchy,* 79–80. The Lake Eleanor filings cost the city $400,000, and the Cherry Creek filings cost $600,000. These purchases were negotiated prior to when John R. Freeman became a consultant for the city.

Chapter 2

1. See chapters 3 and 4 regarding payments Freeman received from San Francisco. The largest of these ($34,000) came in October 1912. This was a time in which the average American worker earned less than $1,000 per year.

2. For Muir's early years in rural Wisconsin, see Donald Worster, *A Passion for Nature: The Life of John Muir* (New York: Oxford University Press, 2008), 42–66. Muir did not leave the family farm until 1860 at age 22.

3. John's father, Nathaniel Dyer Freeman, was born in January 1824 in Portland, Maine; his mother, Mary Elizabeth Morse, was born in Portland in October 1825. For basic biographical/genealogical data on the Freeman family, see the *Find a Grave* website (www.findagrave.com) records of the West Bridgton Cemetery.

4. John R. Freeman, "Autobiography," unpublished typescript, ca. 1930–32, pp. 13, 21, in John Ripley Freeman (1855–1932) Papers (MC 51), Distinctive Collections, Massachusetts Institute of Technology, Cambridge (hereafter, JRF Papers), box 1. In the last years of his life, starting about 1930, Freeman began writing an autobiography. He died before its completion, but the surviving partial draft of some four hundred typed pages offers valuable insight into his early years and career through the early twentieth century.

5. Freeman, "Autobiography," 30.

6. Freeman, "Autobiography," 9.

7. In his autobiography, Freeman makes no reference to any siblings. However, data on internments at the West Bridgton Cemetery available at www.findagrave.com indicates that two sons of "Nathan D. & Mary Elizabeth Morse Freeman" are buried in the cemetery. One is denoted as "Infant son Freeman" who died in 1853 "aged less-than one year," and the other is Frederic Henry Freeman, who died on December 8, 1863, "aged 1y. 11m. 8d." While the first child was born prior to John, Frederic was born when John was six years old, and it is impossible to believe that John had no knowledge of his younger brother's life and death. However, he makes no reference to any siblings in his autobiography.

8. Freeman, "Autobiography," 17–18.

9. Freeman, "Autobiography," 19.

10. Louis Hunter, *Water Power: A History of Industrial Power in the United States, 1780–1930* (Charlottesville: University Press of Virginia, 1979), 210–11.

11. Theodore Steinberg, *Nature Incorporated: Industrialization and the Waters of New England* (Cambridge: Cambridge University Press, 1991), 47–95. Also see Francois Weil, "Capitalism and Industrialization in New England, 1815–1845," *Journal of American History* 84 (March 1998): 1334–54.

12. Patrick M. Malone, *Waterpower in Lowell: Engineering and Industry in Nineteenth-Century America* (Baltimore, MD: Johns Hopkins University Press, 2009), 8–99; Hunter, *Water Power,* 211–17; Steinberg, *Nature Incorporated,* 50–95.

13. Peter Malloy, "Nineteenth Century Hydropower: Design and Construction of Lawrence Dam, 1845–1848," *Winterthur Portfolio* 15 (Winter 1980): 315–43. Also see Peter A. Ford, "'Father of the Whole Enterprise': Charles S. Storrow and the Making of Lawrence, Massachusetts, 1845–1860," *Massachusetts Historical Review* 2 (2000):

76–117; and George Swain, "Report on the Water-Power of the Streams of New England," in Census Office, *Reports of the Water Power of the United States,* in *Tenth Census,* vol. 16 (Washington, DC: Government Printing Office, 1885), 1–109.

14. Peter A. Ford, "Charles S. Storrow, Civil Engineer: A Case Study of European Training and Technological Transfer in the Antebellum Period," *Technology and Culture* 34 (April 1993): 271–99; Charles S. Storrow, *A Treatise on Water-Works for Conveying and Distributing Supplies of Water; with Tables and Examples* (Boston: Hilliard, Gray, 1835).

15. Maurice Dorman, *Lawrence Yesterday and Today, 1845–1918* (Lawrence, MA: Dick & Trumpold, 1918).

16. Donald Cole, *Immigrant City: Lawrence, Massachusetts, 1845–1921* (Chapel Hill: University of North Carolina Press, 1963). Population statistics for nineteenth-century Lawrence are available at the Lawrence History Center's "Lawrence History Timeline" web page, http://www.lawrencehistory.org/history/timeline.

17. Freeman (in "Autobiography," 80–82) noted Storrow's frequent visits to the Essex Company's offices in Lawrence ("once or twice a week") and commented, "Mr. Storrow was to me in various ways the most wonderful engineer and the finest example of an 'old school gentleman' that I have known."

18. Freeman, "Autobiography," 33.

19. Freeman, "Autobiography," 34.

20. Freeman, "Autobiography," 66–67. Also see "Personal Accounts, Oct. 1903," and "Summary of Cash Paid, Jan. 1st to Dec. 31st 1878," both in JRF Papers, box 95.

21. Samuel C. Prescott, *When M.I.T. Was "Boston Tech"* (Cambridge, MA: Technology Press, 1954), 29, 331–36.

22. Prescott, *"Boston Tech,"* 45–67.

23. Freeman, "Autobiography," 63.

24. Freeman, "Autobiography," 55–62.

25. For a biography of Howison, see "Guide to the George Holmes Howison Papers, ca. 1862–1917," BANC MSS C-B 1037, Bancroft Library, University of California, Berkeley.

26. Freeman, "Abstracts of Notes from Notebooks of J.R.F., 1872," 6–7, JRF Papers, box 1.

27. In 1912–13, Freeman's criticism of multiple-arch dam technology made pointed reference to what he termed the "psychological" effect produced by the downstream face of multiple-arch buttress dams. See Donald C. Jackson, "Confrontation at Big Meadows: John R. Freeman and the 'Psychology' of Dam Design (1911–1913)," in *Building the Ultimate Dam: John S. Eastwood and the Control of Water in the West,* paperback ed. (Norman: University of Oklahoma Press, 2005), 109–33.

28. Freeman to Percy Long, September 2, 1912, JRF Papers, box 67.

29. For acknowledgment of Freeman's enduring relationship to the school, see Prescott, *"Boston Tech,"* 209, 261, 264, 271. Freeman's work in planning MIT's Cambridge campus is noted in Mark M. Jarzombek, *Designing MIT: Bosworth's New Tech* (Cambridge, MA: MIT Press, 2017).

30. Freeman, "Autobiography," 69, 75.

31. "Hiram Francis Mills," *ASCE Transactions* 87 (1924): 1299–1302.

32. For discussion of "mill powers," see Malone, *Waterpower in Lowell,* 39–41.

33. The definition and pricing of "mill powers" and "surplus water" at Lawrence is in Swain, "Report on Water-Power," 26–29. Also see "Quantity of Water per Mill-Power Corresponding to Fall," January 1884, JRF Papers, box 95.

34. Freeman, "Autobiography," 71, 75, 88.

35. The Cochituate Aqueduct is described in Sarah S. Elkind, *Bay Cities and Water Politics: The Battle for Resources in Boston and Oakland* (Lawrence: University Press of Kansas, 1998), 17–29; and Fern L. Ness, *Great Waters: A History of Boston's Water Supply* (Hanover, NH: University Press of New England, 1983), 1–15.

36. Elkind, *Bay Cities,* 49; Ness, *Great Waters,* 11. In 1878 the Boston Water Board (successor to the Cochituate Water Board) drew almost 9.4 million gallons per day from the Merrimack watershed. See *Boston Water Board Annual Report 1879* (Boston: Rockwell and Churchill, City Printers, 1879), 9, 28.

37. Freeman, "Autobiography," 70–71.

38. Freeman, "Autobiography," 73.

39. Freeman's fieldwork could bring him into close contact with the natural world and the dangers posed by rapidly flowing rivers. In later life he well remembered how, while surveying Sewell's Falls along the upper Merrimack River in December 1881, he "narrowly escaped drowning . . . [when] the boat from which [he] was making soundings in the rapids became detached. . . , compelling [him] without warning to run the rapids. Upon [his] boat striking a submerged rock [he] found [himself] overboard in deep swift water, compelled to swim ashore with high rubber boots and heavy winter clothing." He was lucky to survive this perilous drenching. Freeman, "Autobiography," 78, 85.

40. Freeman also engaged in the updating of old equipment, describing his work "under Mr. Mills in designing the settings, for the new turbines, new headgates, and new penstocks, to replace the cumbrous old high-breasted water wheels at Washington Mills, which had been the first water wheels started at Lawrence in 1848. Later I supervised this reconstruction." Freeman, "Autobiography," 77, 87.

41. Freeman, "Autobiography," 79.

42. John R. Freeman to Theodore E. Schwartz, May 23, 1881; "Cash Received, 1881," JRF Papers, both in box 95. See also Freeman, "Autobiography," 79.

43. Freeman, "Autobiography," 75–76, 88–90.

44. Freeman, "Autobiography," 88–89.

45. Freeman, "Autobiography," 90–91.

46. "Commodification" of the Merrimack River is a prominent theme in Steinberg, *Nature Incorporated.*

47. In 1886 Mills was elected chairman of the Massachusetts State Board of Public Health's Committee on Water Supply and Sewerage; under his charge "the famous Lawrence Experimental Station was developed, which . . . is probably the leading laboratory in the world devoted solely to the study of the purification of water and sewerage" ("Hiram Francis Mills," 1300–1301). Also see Steinberg, *Nature Incorporated,* 235–39.

48. For the importance of the Lawrence Experimental Station, see Barbara Rosenkrantz, *Public Health and the State: Changing Views in Massachusetts, 1842–1936* (Cambridge, MA: Harvard University Press, 1972), 101.

49. *The Factory Mutuals: 1835–1935* (Providence, RI: Manufacturers Mutual Fire Insurance Company, 1935), 24–49.

50. *Factory Mutuals,* 51–56, 88, 96.

51. *Factory Mutuals,* 93.

52. Harold F. Williamson, *Edward Atkinson: The Biography of an American Liberal* (New York: Arno Press, 1972).

53. Freeman, "Autobiography," 141–43.

54. Freeman, "Autobiography," 147.

55. Freeman, "Autobiography," 149–50.

56. Freeman, "Autobiography," 152; *Factory Mutuals,* 101.

57. John R. Freeman, "Experiments Relating to the Hydraulics of Fire Streams," *ASCE Transactions* 21 (1890): 303–482; John R. Freeman, "The Nozzle as an Accurate Water Meter," *ASCE Transactions* 24 (1891): 492–527. For more on his sprinkler and other patents, see Freeman, "Autobiography," 158.

58. Freeman's interest in networking and in furthering his professional stature dated back to at least 1881 when he "became a member of the Boston Society of Civil Engineers and attended a majority of its monthly meetings in Boston. Although this roundtrip required a 60 mile rail journey and a midnight return [to Lawrence], it was one of the best investments of my life [because of] the broader acquaintance that it gave to [me] with the leading engineers of Boston at that time." Freeman, "Autobiography," 79–80.

59. Freeman, "Autobiography," 124.

60. *Factory Mutuals,* 120–21; Freeman, "Autobiography," 276–77.

61. His courtship of Bessie Clark and their marriage is described in Freeman, "Autobiography," 262–64.

62. *Factory Mutuals,* 354. In 1903 his annual salary as president of six factory mutual companies was $20,000. JRF Diary, January 31, 1903, JRF Papers, box 3.

63. John R. Freeman, "Autobiography—Insurance Providence," 15–20, 79–90.

64. Nat *Brandt, Chicago Death Trap: The Iroquois Theatre Fire of 1903* (Carbondale: Southern Illinois University Press, 2003); Scott Knowles, *The Disaster Experts: Mastering Risk in Modern America* (Philadelphia: University of Pennsylvania Press, 2011), 76–86; John R. Freeman, "On the Safeguarding of Life in Theaters," *ASME Transactions* 27 (1906): 71–170. In 1906 the ASME reprinted the report verbatim, but the page numbers were changed with the text in the reprint starting on page 7 (instead of page 71).

65. Freeman, "On the Safeguarding of Life," 81–82.

66. Freeman, "On the Safeguarding of Life," 91–100; Freeman, "Autobiography," 297–98.

67. According to Knowles (*Disaster Experts,* 85), Freeman's report "was printed and reprinted, widely cited (and it is still cited) in newspapers and technical journals, and established Freeman's credentials as a fire safety expert with interests far broader than simply inspecting New England factories for insurance companies."

68. Knowles, *Disaster Experts,* 76.

69. Freeman, "Autobiography," 284–88.

70. Freeman, "Autobiography," 244–57. The other two original members of the board were chairman Henry Sprague (a Boston lawyer) and Wilmot R. Evans (a banker from

Everett). *First Annual Report of the Metropolitan Water Board, January 1, 1896* (Boston: Wright & Potter, 1895), 5.

71. Ness, *Great Waters*, 15–35; Elkind, *Bay Cities*, 99–117. Consideration was given to filtering water drawn from the Merrimack River. However, taking water from the "polluted Merrimack" was deemed less desirable than drawing water from a large storage reservoir (Ness, *Great Waters*, 19–20).

72. Freeman, "Autobiography," 247.

73. John R. Freeman, "Memo to the Metropolitan Water Board, Presented Orally Oct. 24th 1895; Revised Nov. 10th–16th, 1896," 2, JRF Papers, box 93.

74. JRF Diary, October 23, 1895.

75. Freeman, "Memo to the Metropolitan Water Board." Emphasis in original.

76. "Frederic Pike Stearns," *ASCE Transactions* 83 (1920): 2132–38. Stearns's design for the Nashua/Wachusett system is detailed in *Report of the Massachusetts State Board of Health upon a Metropolitan Water Supply* (1895).

77. Freeman, "Autobiography," 247–48; JRF Diary, November 8, 1895.

78. Freeman, "Autobiography," 249. Freeman claimed he wished "to avoid an ugly controversy with the Chief Engineer [Stearns]" and thus "[gave] up the contest for [his] plan."

79. Freeman, "Autobiography," 283.

80. Freeman, "Autobiography," 278–80.

81. Freeman's long relationship with New York City is documented in JRF Papers, boxes 101–6.

82. J. Hampden Dougherty to John R. Freeman, April 2 and May 6, 1902; John R. Freeman to J. Hampden Dougherty, May 19, 1902; both in JRF Papers, box 102.

83. For a description of the city's nineteenth-century water supply history, see Charles H. Weidner, *Water for a City: A History of New York City's Problem from the Beginning to the Delaware River System* (New Brunswick, NJ: Rutgers University Press, 1974), 3–139.

84. Weidner, *Water for a City*, 58, 84.

85. David Soll, *Empire of Water: An Environmental and Political History of New York City Water Supply* (Ithaca, NY: Cornell University Press, 2013), 16.

86. Weidner, *Water for a City*, 147.

87. F. C. Moore to John R. Freeman, September 24 and October 4, 1895, JRF Papers, box 93. Freeman's introduction to Coler is referenced in F. C. Moore to John R. Freeman, August 21, 1899, JRF Papers, box 101.

88. John R. Freeman, *Report upon New York's Water Supply with Particular Reference to the Need to Procure Additional Sources . . . Made to Bird S. Coler Comptroller* (New York: Martin B. Brown, 1900).

89. Freeman, *Report upon New York's Water Supply* (quotation from page 9). For a discussion of Freeman's desire to dam the Housatonic, see Weidner, *Water for a City*, 151–52.

90. By 1902, city officials had determined "that it would be unwise to attempt to obtain additional water from interstate sources." Weidner, *Water for a City*, 166.

91. Weidner, *Water for a City*, 149. By the time Freeman began his investigations in the late summer of 1899, the Ramapo Water Company's proposal was already in trouble.

However, not until March 1901 did the state legislature formally revoke the company's charter.

92. William Burr, Rudolph Hering, and John R. Freeman, *Report of the Commission on Additional Water Supply for the City of New York* (New York: Martin B. Brown Press, 1904).

93. For the "origin and early activity" of the board of water supply, see Weidner, *Water for a City*, 176–90. Freeman's August 1905 appointment as a consulting engineer is noted on page 180. During the years when the Catskill Aqueduct was being planned and built, Freeman received an annual stipend of $7,500. Freeman, "Autobiography," 301.

94. For an excellent history of the Catskill Aqueduct system through the course of the twentieth century, see Soll, *Empire of Water.* Also see Weidner, *Water for a City*, 191–283.

95. William Kahrl, *Water and Power: The Conflict over Los Angeles' Water Supply in the Owens Valley* (Berkeley: University of California Press, 1982).

96. Kahrl, *Water and Power,* 149.

97. Kahrl, *Water and Power,* 149–53.

98. "Report of the Board of Consulting Engineers on the Project of the Los Angeles Aqueduct . . . Dec. 22, 1906," in *First Annual Report of the Chief Engineer of the Los Angeles Aqueduct to the Board of Public Works, March 15th, 1907*, 117–32.

99. For Freeman's work on the Los Angeles Aqueduct, see JRF Papers, box 64.

100. For Freeman's work on the Charles River Dam project, see JRF Papers, boxes 91–92. Also see Karl Haglund, *Inventing the Charles River* (Cambridge, MA: MIT Press, 2003), 174–76.

101. Michael Rawson, *Eden on the Charles: The Making of Boston* (Cambridge, MA: Harvard University Press, 2010), 226–27.

102. Rawson, *Eden on the Charles,* 227.

103. "Resolves of 1901, chapter 105," in *Report of the Committee on Charles River Dam* (Boston: Wright & Potter Printing, 1903), 5.

104. John R. Freeman's "Report of the Chief Engineer" constitutes most of the *Report of the Committee on Charles River Dam*, 38–572.

105. Freeman, "Report of the Chief Engineer," 64, 108.

106. William B. De Las Casas to John R. Freeman, July 23, 1903. The commission's history is described in *A History and Description of the Boston Metropolitan Parks* (Boston: Wright & Potter Printing, 1900). Freeman's interaction with Frederick Law Olmsted Jr. is documented in Olmsted Brothers to John R. Freeman, January 19, 1905, JRF Papers, box 92.

107. John R. Freeman to W. B. De Las Casas, September 16, 1903, JRF Papers, box 92. By that time, sanitation engineers recognized mosquitos as disease vectors; eradication of marshy wetlands was seen as a way to reduce mosquito populations. L. O. Howard, *Mosquitoes: How They Live; How They Carry Disease; How They Are Classified; How They May Be Destroyed* (New York: McClure, Phillips, 1902).

108. John R. Freeman, *Report on Improvement of the Upper Mystic River and Alewife Brook by Means of Tide Gates and Large Drainage Channels* (Boston: Wright & Potter Printing, 1904). Also see Hiram Mills to John R. Freeman, April 14 and June 2,

1904; and W. Lyman Underwood to John R. Freeman, July 25, 1904; both in JRF Papers, box 92.

109. Freeman, *Report on . . . Upper Mystic River*, 3, 5; Freeman, "Autobiography," 418–20.

110. John R. Freeman, *On the Proposed Use of a Portion of the Hetch Hetchy, Eleanor and Cherry Valleys Within or Near to the Boundaries of the Stanislaus U.S. National Forest Reserve and the Yosemite National Park as Reservoirs for Impounding Tuolumne River Flood Waters and Appurtenant Works for the Water Supply of San Francisco, California and Neighboring Cities* (San Francisco: by Authority of the Board of Supervisors, Rincon Publishing, 1912), 56, 148, 152.

111. Construction of the dam was overseen by Hiram A. Miller. See Edward C. Sherman, "The Closure of the Charles River Dam," *Engineering News* 60 (November 5, 1908): 498–99.

112. The historiography of electric power is voluminous and includes Thomas P. Hughes, *Networks of Power: Electrification in Western Society, 1880–1930* (Baltimore, MD: Johns Hopkins University Press, 1983); David Nye, *Electrifying America: Social Meanings of a New Technology* (Cambridge, MA: MIT Press, 1992); and Richard Hirsh, *Technology and Transformation in the American Electric Utility Industry* (New York: Cambridge University Press, 1989). For a discussion of hydroelectric power with a focus on California, see Jackson, *Building the Ultimate Dam.*

113. During his career, Freeman participated in a wide range of hydroelectric power initiatives, including the Keokuk Dam across the Mississippi River, the Great Western Power Company's Feather River system, and Holter and Hauser Lake dams in Montana, as well as schemes for the Pittsburgh Hydroelectric Company, Puget Sound Power Company, Western Canada Power Company, Calgary Power Company, and Mexican Northern Power Company.

114. Freeman's early relationship with the Pittsburgh Reduction Company is documented in JRF Papers, box 106. Also see Roy A. Hunt, *The Aluminum Pioneers* (New York: Newcomen Society, 1951).

115. For more on Freeman's consulting work for the aluminum industry, see JRF Papers, boxes 106–8, 111.

116. Charles Wetmore to John R. Freeman, May 24 and 28, 1909; John R. Freeman to Charles Wetmore, June 1, 1909; all in JRF Papers, box 82.

117. For example, Freeman purchased seventeen shares of Pittsburgh Reduction Company stock. Alfred E. Hunt to John R. Freeman, October 16, 1891, JRF Papers, box 106.

118. "Mississippi Power [Investment Prospectus], July 11, 1910"; John R. Freeman to Charles A. Stone, July 12, 1910; Charles A. Stone to John R. Freeman, July 14, 1910; all in JRF Papers, box 83.

119. The origins of the Great Western Power Company are described in Charles M. Coleman, *P. G. and E. of California: The Centennial History of Pacific Gas and Electric Company, 1852–1952* (New York: McGraw-Hill, 1952), 211–24, 267–76. For Freeman's relationship with the Great Western Power Company, see Jackson, *Building the Ultimate Dam,* 109–33.

120. John R. Freeman to A. W. Bullard, February 23, 1912, JRF Papers, box 63. Although the company was publicly known as Great Western Power Company, legally it was a subsidiary of the Western Power Company. Freeman's stock was in the latter company.

121. JRF Diary, April 9–18, 1910.

122. Freeman, "Autobiography," 431–32; Theodore Roosevelt to John R. Freeman, December 28, 1908, JRF Papers, box 122. In Roosevelt's words, Taft was to "look over the canal matters and report to me exactly what the present status of the canal work is."

123. Jackson, *Building the Ultimate Dam*, 109–13.

124. For a description of how, after the death of President Edwin Hawley on February 1, 1912, Freeman convinced Great Western Power's corporate leadership to stop construction on John Eastwood's Big Meadows multiple-arch dam, see Jackson, *Building the Ultimate Dam*, 115–33.

Chapter 3

1. Marsden Manson to John R. Freeman, March 19, 1910, John Ripley Freeman (1855–1932) Papers (MC 51), Distinctive Collections, Massachusetts Institute of Technology, Cambridge (hereafter, JRF Papers), box 70.

2. Ray Taylor reports that Freeman was engaged by the city "at the suggestion" of President Taft but provides no source for that information; Ray W. Taylor, *Hetch Hetchy: The Story of San Francisco's Struggle to Provide a Water Supply for Her Future Needs* (San Francisco: Ricardo J. Orozco Publishers, 1926), 109. Kendrick Clements references Taylor in reporting that Taft had recommended Freeman but offers no corroborating source; Kendrick Clements, "Politics and the Park: San Francisco's Fight for Hetch Hetchy, 1908–1913," *Pacific Historical Review* [1979]: 185–215). Taft's supposed recommendation is not corroborated by documents held in the Freeman Papers, and exactly how a "suggestion" from Taft might have reached Manson or other San Francisco leaders in March 1910 is unclear. But regardless of what might have prompted him, Manson did reach out on March 19, initiating Freeman's engagement with San Francisco.

3. John R. Freeman to Marsden Manson, March 21, 1910, JRF Papers, box 67; JRF Diary, April 16–18, 1910, JRF Papers, box 3.

4. His fees were referenced in his March 21 telegram to Manson. The Washington meeting is referenced in Marsden Manson to John R. Freeman, April 28, 1910, JRF Papers, box 67.

5. John R. Freeman to Charles H. Lee, May 13, 1910, JRF Papers, box 70.

6. "Brief by John R. Freeman, Hydraulic Engineer, San Francisco, Future Water Supply from Hetch Hetchy, May 8, 1910," JRF Papers, box 70. For his 1912 report, see John R. Freeman, *On the Proposed Use of a Portion of the Hetch Hetchy, Eleanor and Cherry Valleys Within or Near to the Boundaries of the Stanislaus U.S. National Forest Reserve and the Yosemite National Park as Reservoirs for Impounding Tuolumne River Flood Waters and Appurtenant Works for the Water Supply of San Francisco, California and Neighboring Cities* (San Francisco: by Authority of the Board of Supervisors, Rincon Publishing, 1912). Hereafter, Freeman, *Hetch Hetchy Report*.

7. "Brief by John R. Freeman."

8. "Brief by John R. Freeman."

9. "Brief by John R. Freeman."

10. "Brief by John R. Freeman."

11. JRF Diary, May 15–19, 1910.

12. JRF Diary, May 23, 1910.

13. This Big Bend Dam meeting is described in Donald C. Jackson, *Building the Ultimate Dam: John S. Eastwood and the Control of Water in the West,* paperback ed. (Norman: University of Oklahoma Press, 2005), 112–13.

14. Clements, "Politics and the Park," 185–215. Manson's knowledge of Harroun's previous employment with the Spring Valley Water Company, and his subsequent participation in the preparation of the Hopson-Hill Report on Hetch Hetchy, is described in Marsden Manson to Percy Long, March 26, 1910, Carton B, Marsden Manson Papers, Bancroft Library. After May 1910, Harroun played no further role in the Hetch Hetchy controversy.

15. Ballinger's request for President Taft's authorization of the Army Board is noted in *Proceedings before the Secretary of the Interior in Re Use of Hetch Hetchy Reservoir Site in Yosemite National Park by the City of San Francisco* (Washington, DC: Government Printing Office, 1910), 10–11 (hereafter, *Ballinger Hearing*).

16. *Ballinger Hearing,* 15.

17. Freeman's absence from the May 25 meeting was briefly referenced by City Attorney Long when he informed Ballinger, "[Mr. Freeman] was compelled to return to New York on account of a previous engagement. He said he could return here if you wished." *Ballinger Hearing,* 34.

18. *Ballinger Hearing,* 61.

19. "Order, in the Matter of the Permit of May 11, 1908 to San Francisco, Relating to Hetch Hetchy Valley," by R. A. Ballinger, May 27, 1910, in *Ballinger Hearing,* 69.

20. As described in chapter 4, this deadline would subsequently be pushed back to the summer of 1912.

21. James Rolph and Percy V. Long to The Advisory Board, United States Army Engineers, July 14, 1912, in Freeman, *Hetch Hetchy Report,* 5.

22. Holway Jones, *John Muir and the Sierra Club: The Battle for Yosemite* (San Francisco: Sierra Club, 1965), 126.

23. See Marsden Manson to Col. John Biddle, June 18 and August 5, 1910; and John R. Freeman to Marsden Manson, June 25, 1910; all in JRF Papers, box 67.

24. Submittal of bill noted in JRF Diary, June 25, 1910. The amount is documented in Marsden Manson to John R. Freeman, July 26, 1910, JRF Papers, box 67.

25. JRF Diary, August 9–21, 1910.

26. JRF Diary, August 21–24, 1910.

27. John R. Freeman to Marsden Manson, September 3, 1910, JRF Papers, box 67.

28. Freeman to Manson, September 3, 1910. The Thirlmere controversy is described in Harriet Ritvo, *The Dawn of Green: Manchester, Thirlmere, and Modern Environmentalism* (Chicago: University of Chicago Press, 2009).

29. Freeman to Manson, September 3, 1910.

30. JRF Diary, August 27–September 2, 1910.

31. Manson to Freeman, July 28, 1910, JRF Papers, box 67.

32. John R. Freeman to Marsden Manson, September 3, 1910, JRF Papers, box 58 (this letter, which concerns the bill he submitted to the city in June, is different than the September 3 letter from Freeman to Manson located in JRF Papers, box 67).

33. John R. Freeman to Marsden Manson, telegram, September 19, 1910, JRF Papers, box 69.

34. JRF Diary, October 2, 1910.

35. JRF Diary, October 3, 1910.

36. JRF Diary, October 4–8, 1910. Unfortunately for historians, Freeman's entry for October 4 is very brief and simply reads, "Exploring Hetch Hetchy Valley."

37. Marsden Manson to John R. Freeman, October 13, 1910, JRF Papers, box 58.

38. H. H. Wadsworth [assistant engineer to Army Board] to Marsden Manson, October 26, 1910; Marsden Manson to John R. Freeman, October 28, 1910; both in JRF Papers, box 69.

39. Marsden Manson to John R. Freeman, December 17, 1910, JRF Papers, box 69.

40. Marsden Manson to John R. Freeman, December 21, 1910, JRF Papers, box 67. For Freeman's 1900 New York City report, see John R. Freeman, *Report upon New York's Water Supply with Particular Reference to the Need to Procure Additional Sources . . . Made to Bird S. Coler Comptroller* (New York: Martin B. Brown, 1900).

41. John R. Freeman to Marsden Manson, December 27 and 28, 1910, JRF Papers, box 67.

42. Marsden Manson to John R. Freeman, January 4, 1911, JRF Papers, box 67.

43. JRF Diary, January 6–21, 1911.

44. John R. Freeman to Marsden Manson, January 27, 1911, JRF Papers, box 67.

45. Marsden Manson to John R. Freeman, February 2, 1911, JRF Papers, box 67.

46. JRF Diary, October 8, 1910.

47. James Phelan to J. P. Tumulty, December 26, 1913, in Clements, "Politics and the Park," 185.

48. Samuel Storrow to S. P. Eastman, October 8, 1910, JRF Papers, box 61. Samuel Storrow was a Los Angeles–based civil engineer who had graduated from MIT in 1890.

49. S. P. Eastman to John R. Freeman, August 19, 1910, JRF Papers, box 67.

50. John R. Freeman to S. P. Eastman, ca. August 23, 1910, JRF Papers, box 67.

51. John R. Freeman to S. P. Eastman, September 3, 1910; S. P. Eastman to John R. Freeman, September 8, 1910; both in JRF Papers, box 67.

52. Long's willingness to allow Freeman to consult with Spring Valley is noted in JRF Diary, October 8, 1910. The Calaveras Dam trip is noted in JRF Diary, October 9, 1910.

53. S. P. Eastman to John R. Freeman, October 11, 1910, JRF Papers, box 60.

54. S. P. Eastman to John R. Freeman, second letter, October 11, 1910, JRF Papers, box 60.

55. Freeman to Manson, September 3, 1910, JRF Papers, box 58.

56. John R. Freeman to S. P. Eastman, November 12, 1910, JRF Papers, box 60.

57. John R. Freeman to A. P. Davis, November 22, 1910, and January 3, 1911; A. P. Davis to John R. Freeman, December 28, 1910, and January 9 and 21, 1911; John R.

Freeman to J. B. Lippincott, November 22 and December 15, 1910, and January 3, and February 2, 1911; J. B. Lippincott to John R. Freeman, December 9 and 22, 1910, and January 14 and 24, 1911; all in JRF Papers, box 60.

58. John R. Freeman to F. C. Herrmann, January 2, 1911; S. P. Eastman to John R. Freeman, January 11, 1911; both in JRF Papers, box 60. Herrmann was the "construction engineer" for Spring Valley.

59. S. P. Eastman to John R. Freeman, February 18, 1911, JRF Papers, box 60.

60. John R. Freeman to S. P. Eastman, April 11, 15, and 24, 1911, JRF Papers, box 60.

61. S. P. Eastman to John R. Freeman, May 8, 1911, JRF Papers, box 60.

62. The Yuba/Spaulding Dam project is documented in JRF Papers, box 72.

63. Biographical information in Walter L. Fisher Papers Finding Aid, Manuscript Division, Library of Congress; https://hdl.loc.gov/loc.mss/eadmss.ms012073.

64. James Penick, *Progressive Politics and Conservation: The Ballinger-Pinchot Affair* (Chicago: University of Chicago Press, 1968).

65. Marsden Manson to John R. Freeman, March 14, 1911, JRF Papers, box 67.

66. John R. Freeman to Marsden Manson, March 22, 1911, JRF Papers, box 67.

67. John R. Freeman to Walter Fisher, April 7, 1911, JRF Papers, box 67. Also see John R. Freeman to Michael Casey, April 7, 1911, JRF Papers, box 67.

68. With Ballinger's resignation, city officials apparently considered lobbying Congress for a new law authorizing the Lake Eleanor/Hetch Hetchy system. Such action is intimated in a March 25 telegram from Long to Freeman reporting, "[M]y plan is to secure legislation at special session if Taft agreeable." JRF Papers, box 67. The Freeman papers include no subsequent reference to such legislation. Evidence that California Congressman Raker submitted a bill appears in *Congressional Record: Proceedings and Debates of the 62nd Congress, First Session,* vol. 47, 625. On April 25, 1911, Raker introduced H.R. 7275 to allow "San Francisco to construct storage reservoirs . . . in Lake Eleanor and Hetch Hetchy Valleys." There is no evidence of further action on the bill by the 62nd Congress.

69. Percy Long to John R. Freeman, April 26, 1911, JRF Papers, box 67.

70. Marsden Manson to John R. Freeman, May 2, 1911; Allen Hazen to John R. Freeman, May 4, 1911; John R. Freeman to Allen Hazen, May 6, 1911; John R. Freeman to Percy Long, May 6, 1911; all in JRF Papers, box 67.

71. JRF Diary, April 20–June 23, 1911.

72. JRF Diary, June 23, 1911.

73. JRF Diary, June 28–30, 1911.

74. JRF Diary, July 1, 1911.

75. JRF Diary, July 2 and 4, 1911.

76. JRF Diary, July 6–8, 1911.

77. John R. Freeman to Percy Long, July 9, 1911, JRF Papers, box 67.

78. JRF Diary, July 10–16, 1911.

79. JRF Diary, July 16–30, 1911. He spent six days in British Columbia (July 18–24) and three days in Idaho (July 26–28).

80. John R. Freeman to William Mulholland, August 5, 1911; William Mulholland to John R. Freeman, August 9, 1911; both in JRF Papers, box 67. JRF Diary, July 30–August 23, 1911.

81. John R. Freeman to Marsden Manson, September 11, 1911, JRF Papers, box 67.

82. Secretary Fisher's trip to Hetch Hetchy in the company of both Manson and the anti-dam preservationist Horace McFarland is described in Jones, *John Muir*, 127–30. In McFarland's view, Manson did not make a good impression, reporting that "Mr. Manson's stock [with Fisher] was at a low ebb. . . . Mr. Manson has in no sense made good" (Horace McFarland to William Colby, September 28, 1911, quoted in Jones, *John Muir*, 129). City Attorney Long also expressed his belief that, from the city's perspective, Manson's interaction with Fisher was "rather unsatisfactory." See Percy Long to John R. Freeman, October 3, 1911, JRF Papers, box 70. Fisher's trip to Hetch Hetchy is also described in Robert W. Righter, *The Battle over Hetch Hetchy: America's Most Controversial Dam and the Birth of Modern Environmentalism* (New York: Oxford University Press, 2005), 110–12. However, Righter mistakenly and incongruously dates this visit to the fall of 1912.

83. John R. Freeman to Marsden Manson, September 18, 1911, JRF Papers, box 67.

84. *Let Everyone Help to Save the Famous Hetch Hetchy Valley*, November 1909, National Archives, College Park, MD, Record Group 95, Entry 22, box 4.

85. John R. Freeman to Percy Long, September 25, 1911, JRF Papers, box 67.

86. Percy Long to John R. Freeman, October 3, 1911.

87. Percy Long to John R. Freeman, October 3, 1911.

88. Walter Fisher to Marsden Manson, October 25, 1911; Marsden Manson to John R. Freeman, October 10, 1911; both in JRF Papers, box 67.

89. John R. Freeman to Marsden Manson, October 2, 1911, JRF Papers, box 67.

90. Freeman's work in Mexico is documented in JRF Papers, boxes 117–18. Also see Freeman to Manson, October 2, 1911.

91. John R. Freeman to Marsden Manson, November 16, 1911, JRF Papers, box 67.

92. JRF Diary, June 24–30, 1911.

93. John R. Freeman to Marsden Manson, December 11, 1911, JRF Papers, box 67.

94. John R. Freeman to Walter Fisher, February 15, 1912, JRF Papers, box 67.

95. For example, see John R. Freeman to S. P. Eastman, November 17, 1911; S. P. Eastman to John R. Freeman, November 29, 1911; John R. Freeman to S. P. Eastman, December 6, 1911; and John R. Freeman to F. C. Herrmann, December 8, 1911; all in JRF Papers, box 61. On January 18, 1912, Eastman "acknowledge[d] receipt of your letter of Jan. 6th enclosing your bill in full for the period from August 1st to December 31st, 1911 in the amount of $1,030.20. I have approved the bill" (JRF Papers, box 61). This appears to be the last payment Freeman received from Spring Valley.

96. In the early spring Freeman received a check from the city for $9,879.24 to cover recent "professional services." Marsden Manson to John R. Freeman, March 25, 1912, JRF Papers, box 67.

97. William Mulholland and J. B. Lippincott, "Report on the Development of Groundwaters of the Livermore Valley," February 2. 1912, in Spring Valley Water Company, *The Future Water Supply of San Francisco from the Conservation and Use of Its Present Resources* (San Francisco: Rincon Publishing, 1912), 178–86.

98. No mention of Mulholland's and Lippincott's work for Spring Valley appears in Clements, "Politics and the Park"; Jones, *John Muir*; or Righter, *Battle over Hetch Hetchy*.

99. John R. Freeman to Thomas Haven [assistant city attorney], February 20, 1912, JRF Papers, box 67.

100. John R. Freeman to Marsden Manson, February 23, 1912, JRF Papers, box 67.

101. John R. Freeman to Percy Long, March 1, 1912, JRF Papers, box 67.

102. JRF Diary, January 15 and February 29, 1912. On February 29, he spent "from 9:30 to 12:50 with Dr. Maclaurin President of Technology studying plans for the new institute."

103. JRF Diary, February 16 and March 22–23, 1912.

104. Freeman's reentry into Great Western Power Company affairs is discussed in Jackson, *Building the Ultimate Dam,* 116–18.

105. Marsden Manson to John R. Freeman, January 24, 1912, JRF Papers, box 70.

106. Marsden Manson to John R. Freeman, March 25, 1912, JRF Papers, box 67.

107. Percy Long to Walter Fisher, January 20, 1912, JRF Papers, box 70.

108. Manson to Freeman, January 24, 1912.

109. John R. Freeman to Percy Long, February 15 and March 25, 1912, JRF Papers, box 67.

110. As discussed in chapter 6, by late April Assistant Engineer Max Bartell had submitted a report to the city engineer, "Acting on your verbal instructions to investigate the water resources of the Mokelumne River." Max Bartell to Marsden Manson, April 24, 1912, Max Bartell Papers, Water Resources Collections and Archives, University of California, Riverside.

111. Percy Long to John R. Freeman, March 5, 1912, JRF Papers, box 67.

112. Col. John Biddle to Marsden Manson, April 26, 1912, JRF Papers, box 67. In January, Manson had requested that these four sources "be omitted" (Marsden Manson to Col. John Biddle, January 6, 1912, JRF Papers, box 70).

113. Marsden Manson to John R. Freeman, April 29, 1912, JRF Papers, box 70.

114. JRF Diary, May 10–12, 1912. Even prior to meeting with Long, Freeman recognized the daunting task that the city (and Manson) faced in meeting the government's "show cause" demand; he also recognized the pressure that he himself was operating under, as he shared with Manson in a "personal" letter dated April 22: "I am in an almost desperate state of mind over finishing the two principal pieces of work expected of me within the prescribed dates and in anything like a creditable manner." But he was nonetheless preparing to fulfill these tasks, and he advised Manson that he was planning to go to San Francisco by early June: "My best guess is that I shall start for the Coast about a month or six weeks hence." John R. Freeman to Marsden Manson, April 22, 1912, JRF Papers, box 67.

115. John R. Freeman to Percy Long, May 11, 1912; and "Progress on Studies under Supervision of J.R. Freeman, Consulting Engr.," May 11, 1912; both in JRF Papers, box 67. In his diary entry for Saturday, May 11, Freeman noted that he had spent "all day" with Long "reviewing progress until 8pm[.] Dictated report of progress — In eve had dinner at University Club." The next day he "review[ed] certain Hetchy data. . . . [and] saw Long to RR station at 5:30." Two days after Long's departure, Freeman noted in his diary (May 14) that he spent "all day on Hetch Hetchy studies mainly on aqueduct location."

116. John R. Freeman to Marsden Manson, telegram, May 15, 1912, JRF Papers, box 67.

117. Mayor Rolph to John R. Freeman, May 18, 1912, JRF Papers, box 67. Freeman spent Sunday, May 19, "planning [the] Hetchy campaign." JRF Diary, May 19, 1912.

118. John R. Freeman and Percy Long to James Rolph, May 22, 1912, JRF Papers, box 67. The final terms set out by Fisher are detailed in John R. Freeman to Marsden Manson, May 25, 1912, JRF Papers, box 67.

Chapter 4

1. John R. Freeman to Marsden Manson, May 25, 1912, John Ripley Freeman (1855–1932) Papers (MC 51), Distinctive Collections, Massachusetts Institute of Technology, Cambridge (hereafter, JRF Papers), box 67.

2. John R. Freeman to William Mulholland, May 29, 1912, JRF Papers, box 67.

3. Mulholland was willing to help Freeman with "our cost reports on tunnels" but made no commitment on being a consultant; William Mulholland to John R. Freeman, June 4, 1912, JRF Papers, box 67.

4. Horace Ropes's early work with Freemen is noted in John R. Freeman, *Report upon New York City's Water Supply with Particular Reference to the Need to Procure Additional Sources . . . made to Bird S. Coler Comptroller* (New York: Martin B. Brown, 1900), 199. Also joining the entourage was the Denver engineer George Prince; John R. Freeman to George Prince, May 25, 1912, JRF Papers, box 67. Once in California, Freeman also engaged the services of J. H. Dockweiler, who had served as city engineer of Los Angeles in the 1890s, and Cyril Williams, a San Francisco–based hydraulic engineer.

5. JRF Diary, June 4–9, 1912, JRF Papers, box 3; John R. Freeman, *On the Proposed Use of a Portion of the Hetch Hetchy, Eleanor and Cherry Valleys Within or Near to the Boundaries of the Stanislaus U.S. National Forest Reserve and the Yosemite National Park as Reservoirs for Impounding Tuolumne River Flood Waters and Appurtenant Works for the Water Supply of San Francisco, California and Neighboring Cities* (San Francisco: by Authority of the Board of Supervisors, Rincon Publishing, 1912). Hereafter, Freeman, *Hetch Hetchy Report.*

6. "Noted Engineer Quiets Fears over Hetch Hetchy: Great Inland Lake Will Help, not Mar, Beautiful Mountain Scenery," *San Francisco Evening Post,* June 10, 1912; "S.F. Must Have Hetch Hetchy, Says Expert: John R. Freeman, World Famous Engineer, Arrives to Complete Data," *San Francisco Examiner,* June 10, 1912.

7. "Expert Here to Make Report to Government on Hetch Hetchy Plan," *San Francisco Chronicle,* June 10, 1912.

8. "Engineer to Work on Water Problem," *San Francisco Call,* June 10, 1912.

9. JRF Diary, June 10, 1912.

10. JRF Diary, June 11–17, 1912.

11. Charles Whiting Baker, "A Notable Water Supply," *Engineering News* 68 (December 26, 1912): 1207–15.

12. John R. Freeman to Roger Freeman, June 21, 1912, JRF Papers, box 5.

13. JRF Diary, July 27, 1912.

14. Freeman, *Hetch Hetchy Report* , 77–79.

15. For census data on the Bay Area from 1860 through 2010, see the "Historical Data" section of *Bay Area Census,* http://www.bayareacensus.ca.gov/historical
/copop18602000.htm.

16. Freeman, *Hetch Hetchy Report,* 63–64.

17. The enlarged Lake Chabot is described in Freeman, *Hetch Hetchy Report,* 134–35.

18. "Seven East Bay Mayors to Plan Water System—To Buy People's Plant," *San
Francisco Examiner,* July 8, 1912.

19. It would take another 10 years, but in 1923 voters would approve creation of the
East Bay Municipal Utility District (EBMUD).

20. Freeman's *Hetch Hetchy Report* includes three appendixes that focus on a possible metropolitan water district, including "Report on Progress Made by Municipalities
Surrounding San Francisco Bay Toward Cooperation in Future Water Supply," 161–66.

21. See Freeman, *Hetch Hetchy Report,* appendix 17, "Abstract of Report on the
Modesto and Turlock Irrigation Districts . . . ," 357–63; also see pp. 98–107. The La
Grange Water and Power Company possessed a claim to 60 cfs of Tuolumne flow in
the vicinity of La Grange Dam, and this claim is briefly referenced in Freeman's *Hetch
Hetchy Report* (pp. 100–101), but it is given little attention compared to the much
larger claims of the two irrigation districts. This power company was later known as the
Yosemite Power Company; in 1921 its assets were purchased by the Turlock Irrigation
District. See Alan M. Peterson, *Land, Water, and Power: A History of the Turlock Irrigation District, 1887–1987* (Spokane, WA: Arthur H. Clark, 1989), 221. For the deleterious
effect of overwatering on land within the two districts, see Peterson, *Land, Water, and
Power,* 159–60.

22. Freeman, *Hetch Hetchy Report,* 108–28, 221–302; for comparison of Freeman's
pressurized gravity-flow system with the Grunsky and Manson plans, see pp. 138–44.

23. Branner's work with Horace Ropes on surveying and evaluating the aqueduct
right-of-way is noted in Freeman, *Hetch Hetchy Report* (pp. 110–11), which includes
Branner's July 1, 1912, report to Freeman on the geology of the right-of-way and the
statement "as far as the geology is concerned the route selected . . . is entirely feasible."
Appendix 9, "Preliminary Estimate of Cost of the Hetch Hetchy Project" (pp. 239–301),
is one of the most detailed and complete components of Freeman's report; it is credited
to "Horace Ropes C.E. in Consultation with John R. Freeman C.E."

24. William Mulholland to John R. Freeman, August 8, 1911, JRF Papers, box 67.

25. Freeman, *Hetch Hetchy Report,* 252; Freeman also references how recent projects
for the Great Western Power Company, Pacific Gas and Electric, and the Oakdale Irrigation District paid common laborers between $2.00 and $2.75 per day. Also see p. 223 of
the report for more on Los Angeles tunnel work costs.

26. Freeman, *Hetch Hetchy Report,* 118–19. For a detailed cost breakdown of building the dam to an initial spillway height of 3,760 feet above sea level, see pp. 286–88.

27. Freeman, *Hetch Hetchy Report,* 119. During the summer of 1912, when testifying at the House Public Lands Committee hearing, former Mayor Phelan claimed
Freeman "has shown that by planting trees or vines over the dam, the idea of a dam,

the appearance of a dam, is entirely lost; so coming upon it, it will look like a emerald gem in the mountains." See *Hetch Hetchy Dam Site: Hearing before the Committee on the Public Lands, House of Representatives, 63rd Congress, First Session, on H.R. 6281* (Washington, DC: Government Printing Office, 1913), 166; also quoted in Char Miller, *Hetch Hetchy: A History in Documents* (Peterborough, ON: Broadview, 2020), 146. While Freeman describes in his report (p. 119) how his design will be "given more liberal architectural treatment than is common with masonry dams," he makes no mention of covering the downstream face with trees or vines; I am not aware that Freeman ever discussed or postulated such a design feature for his Hetch Hetchy Dam.

28. A cross-sectional drawing of Freeman's Calaveras Dam design appears in Spring Valley Water Company, *The Future Water Supply of San Francisco from the Conservation and Use of Its Present Resources* (San Francisco: Rincon Publishing, 1912), 35.

29. Freeman, *Hetch Hetchy Report,* 29–32.

30. Freeman, *Hetch Hetchy Report,* 29–32.

31. See chapter 5 for the Army Board's valuation of Freeman's power system.

32. John R. Freeman to William Mulholland, June 14, 1912, JRF Papers, box 67.

33. John R. Freeman to Charles W. Baker, August 31, 1912, JRF Papers, box 67.

34. Freeman, *Hetch Hetchy Report,* 10. Freeman was not responsible for originating the "mirrored-lake" image, which appeared earlier in James Phelan, "The Hetch Hetchy and San Francisco," *Out West* 1 (February 1911): 169–77.

35. Freeman, *Hetch Hetchy Report,* 35, 37.

36. Henry H. Sprague to John R. Freeman, June 27, 1912, JRF Papers, box 67.

37. Dexter Brackett to John R. Freeman, August 16, 1912, JRF Papers, box 67.

38. John R. Freeman to Marsden Manson, September 18, 1911, JRF Papers, box 67.

39. Freeman, *Hetch Hetchy Report,* 148.

40. Freeman, *Hetch Hetchy Report,* 148.

41. For example, he advised Hazen, "I believe that just as in Boston we went to the Nashua, although filtered Merrimack water may have figured cheaper, and in New York we went beyond the Hudson to the Catskills, so [too] the citizens of San Francisco have an overwhelming desire to go to the mountains rather than to the San Joaquin [or Sacramento River]." John R. Freeman to Allen Hazen, May 6, 1911, JRF Papers, box 67.

42. Allen Hazen, "Report on Filtered Water Supply for San Francisco," in Freeman, *Hetch Hetchy Report,* 315–25.

43. Freeman, *Hetch Hetchy Report,* 325.

44. Freeman, *Hetch Hetchy Report,* 160o.

45. Freeman, *Hetch Hetchy Report,* 151.

46. Freeman, *Hetch Hetchy Report,* 156–160s.

47. Freeman, *Hetch Hetchy Report,* 159.

48. Freeman, *Hetch Hetchy Report,* 157.

49. Freeman, *Hetch Hetchy Report,* 157.

50. Freeman, *Hetch Hetchy Report,* 156.

51. An Army Board trip to the McCloud River is described in "Engineer Inspects McCloud Project," *Sacramento Bee,* July 29, 1912.

52. Freeman, *Hetch Hetchy Report,* 343.

53. Freeman, *Hetch Hetchy Report,* 343.

54. Freeman, *Hetch Hetchy Report,* 343–44.

55. US Geological Survey streamflow data for the various rivers available at Wikipedia: https://en.wikipedia.org/wiki/Cosumnes_River; https://en.wikipedia.org/wiki/Calaveras_River; https://en.wikipedia.org/wiki/Stanislaus_River; https://en.wikipedia.org/wiki/Mokelumne_River; https://en.wikipedia.org/wiki/Tuolumne_River.

56. Freeman, *Hetch Hetchy Report,* 160d.

57. Freeman, *Hetch Hetchy Report,* 160–160a.

58. Freeman, *Hetch Hetchy Report,* 160a.

59. Freeman, *Hetch Hetchy Report,* 160e.

60. John R. Freeman to Marsden Manson, October 2, 1911, JRF Papers, box 67. Specifically, Freeman admonished Manson, "[Y]ou must earnestly get your hooks and spurs into Grunsky and inject more energy into his collection of data."

61. JRF Diary, August 1, 1912.

62. JRF Diary, August 20, 1912.

63. John R. Freeman to Charles W. Baker, August 31, 1912, JRF Papers, box 67.

64. See chapter 5.

65. Freeman, *Hetch Hetchy Report,* 179.

66. William Mulholland and J. B. Lippincott, "Report on the Development of Ground Waters of the Livermore Valley," in Spring Valley Water Company, *Future Water Supply,* 185.

67. JRF Diary, June 22 and 24, 1912.

68. William B. Bourn to John R. Freeman, July 10, 1912, in Spring Valley Water Company, *Future Water Supply,* 499; John R. Freeman to William B. Bourn, July 10, 1912, JRF Papers, box 67.

69. John R. Freeman, "Regarding the Purchase of [Spring Valley] Works by City" [draft], June 22, 1912, JRF Papers, box 72; JRF Diary, June 23 and July 1, 2, 3, and 15, 1912.

70. JRF Diary, August 7, 1912.

71. "A Proposal for the Purchase of the Spring Valley Water Works by the City of San Francisco," August 9, 1912; Freeman, *Hetch Hetchy Report,* 395–400. As an attachment to the city's offer, Freeman wrote an independent "opinion" in which he declared, "It seems to me that the analysis presented in the accompanying proposal is logical and that its conclusions are sound and fair." See "Opinion of Consulting Engineer upon Proposal to Purchase," John R. Freeman to Mayor Rolph, August 9, 1912, in Freeman, *Hetch Hetchy Report,* 401.

72. "Mayor Bids for Spring Valley—Offers $38,500,000 for Entire Plant," *San Francisco Call,* August 13, 1913; also see "City Makes Offer to Buy Spring Valley," *San Francisco Chronicle,* August 13, 1912. Freeman's endorsement of the city's offer is reported in "Engineer Approves the Plan," *San Francisco Call,* August 13, 1912.

73. "Offer to Purchase Spring Water Company's Plant and Reply of Spring Valley Water Company," September 14, 1912, JRF Papers, box 67.

74. John R. Freeman to J. H. Dockweiler, October 24, 1912, JRF Papers, box 67.

75. John R. Freeman to William Mulholland, September 2, 1912, JRF Papers, box 67.

76. John R. Freeman to William Mulholland, July 27, 1912, JRF Papers, box 67.

77. "Argument for Hetch Hetchy Claims Filed, Engineer's Report Sets Forth the Need for an Absolutely New Permit," *San Francisco Call,* July 17, 1912.

78. "Engineer's Report on Water Supply Is Submitted, John R. Freeman Discards All Previous Studies in His Work," *San Francisco Chronicle,* July 17, 1912; "Hetch Hetchy Must Be City's Main Water Supply," *San Francisco Evening Post,* July 16, 1912.

79. "City's Main Water Supply."

80. "City's Main Water Supply."

81. "Engineer's Report."

82. JRF Diary, July 14, 1912.

83. JRF Diary, July 17, 19, 21, and 24 and August 2, 1912. Freeman also paid close attention both to gathering the illustrations and to how they would appear in the published report. See John R. Freeman to Underwood and Underwood, Photographers, June 18, 1912; and John R. Freeman to Britton & Rey, Engravers, July 5, 1912 (both letters in JRF Papers, box 67). In the latter letter, he advised that the Tuolumne Meadows photograph needed "artist work . . . [with] a little more life and snap put into it, possibly the distant hills need a little retouching to give them more relief."

84. William Dumhoff, *The Bohemian Grove and Other Retreats: A Study in Ruling-Class Cohesiveness* (New York: Harper & Row, 1974).

85. JRF Diary, August 8–10, 1912.

86. "Mayor Rolph Joins Hetch Hetchy Party," *San Francisco Examiner,* August 13, 1912. Planning for this trip is detailed in John R. Freeman to Colonel John Biddle, August 5, 1912, JRF Papers, box 67.

87. JRF Diary, August 12–15, 1912.

88. "Rolph Back and Enthuses over Water Plan," *San Francisco Examiner,* August 17, 1912.

89. "Expert Indorses the Hetch Hetchy," *San Francisco Chronicle,* August 25, 1912; also see "Growth of S.F. Depends on Its Water Supply, Freeman Tells Commonwealth Club Hetch Hetchy Plans Are Based on Future," *San Francisco Examiner,* August 25, 1912.

90. "Rolph Smashes Hetch Hetchy Nature Fakes," *San Francisco Call,* August 25, 1912.

91. JRF Diary, August 24, 1912.

92. JRF Diary, August 23, 1912.

93. John R. Freeman to Albert Northrop, August 31, 1912, JRF Papers, box 67; JRF Diary, August 31–September 3, 1912.

94. JRF Diary, July 3, 1912.

95. "Manson Quits Office When under Fire by Mayor," *San Francisco Evening Post,* August 7, 1912.

96. Freeman, *Hetch Hetchy Report,* 160s.

97. John R. Freeman to Charles W. Baker, editor, *Engineering News,* August 31, 1912, JRF Papers, box 67.

98. JRF Diary, August 22, 1912.

99. For more on O'Shaughnessy's career, see "Michael Maurice O'Shaughnessy," *ASCE Transactions of the American Society of Civil Engineers* 100 (1935): 1710–12. Also see M. M. O'Shaughnessy, *Hetch Hetchy: Its Origin and History* (San Francisco: privately printed, 1934); and M. M. O'Shaughnessy, "Construction of the Morena Rock Fill Dam, San Diego County, California," *ASCE Transactions* 75 (December 1912): 27–67.

100. On August 31, Freeman "had a long talk with the new City Engineer, Mr. O'Shaughnessy"; JRF to Mayor Rolph, August 31, 1912, JRF Papers, box 67. On September 3 he had another "conference with new City Engineer O'Shaughnessy"; JRF Diary, September 3, 1912. For the August 31 meeting, see O'Shaughnessy, *Hetch Hetchy,* 21.

101. After being fired as city engineer, Manson became resentful that Freeman's plan came to define the Hetch Hetchy project going forward, and he complained that Freeman had no authority to devise a system that ignored what had been the basis of the Garfield Permit. This is evident in handwritten notes he inscribed in his copy of Freeman's *Hetch Hetchy Report*; among other complaints, he asks on page 20, "What authority is there to spend Hetch Hetchy bond money for designing works for a 'Metropolitan District'?" and "Why this amount [400 million gallons per day] so far in advance of time?" But despite his criticisms and carping, Manson had little impact on how the Hetch Hetchy story played out from the fall of 1912 through passage of the Raker Act in December 1913. See copy of *Hetch Hetchy Report* in carton 2, Marsden Manson Papers, Bancroft Library, University of California, Berkeley.

102. John R. Freeman to Percy Long, September 23, 1912, JRF Papers, box 67.

103. John R. Freeman to Mayor Rolph, September 23, 1912; John R. Freeman to Percy Long, September 23, 1912; both in JRF Papers, box 67.

104. The total of Freeman's bill is in "Engineer Freeman's Little Bill," *Pacific Underwriter,* October 10, 1912; also see J. S. Dunnigan to John R. Freeman, October 2, 1912, JRF Papers, box 67.

105. For Freeman's billing of "seven-hour days" and his billing of 68 "days" in July, see "Experts Bill Criticized," *San Francisco Call,* October 26, 1912.

106. Freeman to Long, September 23, 1912.

107. Freeman to Long, September 23, 1912.

108. "Auditor Holds Up Freeman Account," *San Francisco Call,* October 25, 1912; "Experts Bill Criticized," *San Francisco Call,* October 26, 1912.

109. Support for paying Freeman in full appears in "Freeman's Bill Should Be Paid without Hesitation," *San Francisco Bulletin,* November 1, 1912.

110. Percy Long to John R. Freeman, October 29, 1912, JRF Papers, box 67.

111. Freeman to Rolph, September 23, 1912. Emphasis in original.

112. John R. Freeman to Andrew Y. Wood, September 2, 1912, JRF Papers, box 67.

113. John R. Freeman to Percy Long, September 2, 1912, JRF Papers, box 67.

Chapter 5

1. Donald C. Jackson, *Building the Ultimate Dam: John S. Eastwood and the Control of Water in the West,* paperback ed. (Norman: University of Oklahoma Press, 2005), 18–19.

2. John R. Freeman, *On the Proposed Use of a Portion of the Hetch Hetchy, Eleanor and Cherry Valleys Within or Near to the Boundaries of the Stanislaus U.S. National Forest Reserve and the Yosemite National Park as Reservoirs for Impounding Tuolumne River Flood Waters and Appurtenant Works for the Water Supply of San Francisco, California and Neighboring Cities* (San Francisco: by Authority of the Board of Supervisors, Rincon Publishing, 1912). Hereafter, Freeman, *Hetch Hetchy Report.*

3. JRF Diary, September 13–17, 1912, John Ripley Freeman (1855–1932) Papers (MC 51), Distinctive Collections, Massachusetts Institute of Technology, Cambridge (hereafter, JRF Papers), box 3.

4. JRF Diary, September 18–19, 1912. Freeman's work on the new campus is detailed in Mark M. Jarzombek, *Designing MIT: Bosworth's New Tech* (Cambridge, MA: MIT Press, 2017), 40–52, 69–70, 79–84, 116–31.

5. Jarzombek, *Designing MIT,* 38; JRF Diary, January 15, 1912.

6. JRF Diary, September 19 and October 9, 1912.

7. For Freeman's dismissive view of the traditional talents of architects, see Jarzombek, *Designing MIT,* 41–44.

8. JRF Diary, October 31 and November 21, 1912.

9. John R. Freeman to Percy Long, September 2, 1912, JRF Papers, box 67; also see "Distribution List of the Hetch Hetchy Reports," JRF Papers, box 70.

10. See "Distribution List of Hetch Hetchy Reports"; John R. Freeman to Robert Collier, November 6, 1912; John R. Freeman to Lyman Abbott, October 21, 1912; John R. Freeman to Hiram Mills, October 23, 1912; John R. Freeman to Henry Pritchett, October 23, 1912; John R. Freeman to Louis Brandeis, October 28, 1912; John R. Freeman to A. Lawrence Lowell, November 5, 1912; all in JRF Papers, box 70.

11. John R. Freeman to Henry Sharpe, November 5, 1912, JRF Papers, box 70.

12. Freeman to Pritchett, October 23, 1912.

13. Freeman to Brandeis, October 28, 1912.

14. Freeman to Mills, October 23, 1912.

15. John R. Freeman to Henry Barker, October 30, 1912, JRF Papers, box 70.

16. Freeman to Lowell, November 5, 1912.

17. Freeman to Lowell, November 5, 1912.

18. John R. Freeman to Desmond Fitzgerald, October 24, 1912, JRF Papers, box 67.

19. John R. Freeman to M. M. O'Shaughnessy, October 10, 1912, JRF Papers, box 67.

20. Spring Valley Water Company, *The Future Water Supply of San Francisco from the Conservation and Use of Its Present Resources* (San Francisco: Rincon Publishing, 1912).

21. Spring Valley Water Company, *Future Water Supply,* 1.

22. Freeman, *Hetch Hetchy Report,* 392.

23. Freeman, *Hetch Hetchy Report,* 83–86. Freeman's critique of the long-term groundwater capacities of the Alameda/Livermore watershed is presented on pp. 86–90.

24. Freeman, *Hetch Hetchy Report,* 191. This overarching point is reinforced on page 97, where Freeman avers, "The only source of any particular account in providing water for the future growth of San Francisco, owned by the Spring Valley Water Company, is whatever may now be available in the Alameda Creek source. Estimates

provided by the Spring Valley Water Company grossly over-estimate what could possibly be stored and drawn from that source."

25. John R. Freeman to Percy Long, October 17, 1912. Freeman later cast further aspersions on Mulholland's work for Spring Valley, privately confiding, "It is past my own comprehension how Mulholland could run amuck so badly or have his eyes so blinded by the retainer of the Spring Valley Company." John R. Freeman to M. M. O'Shaughnessy, December 26, 1912. Both letters in JRF Papers, box 67.

26. Percy Long to John R. Freeman, October 24, 1912, JRF Papers, box 67.

27. John R. Freeman to Allen Hazen, November 20, 1912, JRF Papers, box 67.

28. E. R. Jones (attorney for Modesto Irrigation District) and P. H. Griffin (attorney for Turlock Irrigation District) to Advisory Board, United States Army Engineers, November 1, 1912, JRF Papers, box 68.

29. Burton Smith and H. S. Crowe, "Report of the Turlock and Modesto Irrigation District . . ." October 24, 1912, JRF Papers, box 68.

30. Smith and Crowe, "Report," 7.

31. Smith and Crowe, "Report," 6.

32. Donald Worster, *A Passion for Nature: The Life of John Muir* (New York: Oxford University Press, 2008), 441–47.

33. "Hetch Hetchy Must Be City's Main Supply," *San Francisco Evening Post,* July 16, 1912; Robert U. Johnson to William Colby, September 12, 1912, quoted in Holway Jones, *John Muir and the Sierra Club: The Battle for Yosemite* (San Francisco: Sierra Club, 1965), 134.

34. Jones, *John Muir,* 134–35. Also see "Brief of Sierra Club in Opposition to Grant of Hetch Hetchy Valley to San Francisco for a Water Supply," ca. November 1, 1912, copy in JRF Papers, box 70.

35. "Brief of Sierra Club," 1.

36. "Brief of Sierra Club," 1–2.

37. For a discussion of how preservationists promoted plans for hotels and camping within Hetch Hetchy, see Robert W. Righter, *The Battle over Hetch Hetchy: America's Most Controversial Dam and the Birth of Modern Environmentalism* (New York: Oxford University Press, 2005), 208–9.

38. "Brief of Sierra Club," 9, 25 (original emphasis).

39. Harriet Monroe to Secretary of the Interior, October 12, 1912, JRF Papers, box 70.

40. Whitman's anti-dam leadership is documented in Jones, *John Muir,* 97, 102–3, 119–20, 157–58.

41. Edmund Whitman, "In the Matter of the Order to Show Cause to the City of San Francisco Why the Hetch Hetchy Valley Should Not Be Eliminated from the Permit of 1908," ca. November 1, 1912, JRF Papers, box 70.

42. Whitman, "Order to Show Cause," 1–2, 15.

43. Whitman, "Order to Show Cause," 14.

44. Whitman, "Order to Show Cause," 6–7.

45. Whitman, "Order to Show Cause," 15.

46. Born in 1860 in Lawrence, Kansas, Edmund A. Whitman was the youngest son of Edmund G. Whitman (1812–83), a Harvard College graduate, class of 1838. After the

Civil War, the elder Whitman served as the US Army's assistant quartermaster in charge of cemeteries. The younger Edmund practiced law from 1886 through the 1920s; his work with the law firm Elder, Whitman, Weyburn & Crocker is noted in *The American Bar: A Biographical Directory of Contemporary Lawyers of the United States and Canada* (New York: James C. Fifield, 1921), 427. For biographical material on the younger Whitman, see "Hearing before the Committee on the Public Lands, House of Representatives, 63rd Congress, First Session, on H.R. 6281" (Washington, DC: Government Printing Office, 1913), 170–71; and Albert N. Marquis, *Who's Who in New England* (Chicago: A. N. Marquis, 1916), 1174. For more on the elder Whitman, see the online finding aid for the Edmund Whitman Papers at the University of Michigan Clements Library: https://findingaids.lib.umich.edu/catalog/umich-wcl-M-1738whi.

47. The Gloucester lawsuit is documented in JRF Papers, box 94.

48. JRF Diary, October 31, 1912.

49. John R. Freeman to Percy Long, November 2, 1912, JRF Papers, box 67.

50. Freeman to Long, November 2, 1912.

51. JRF Diary, November 16–23, 1912.

52. See the Wikipedia article on the 1912 presidential election, https://en.wikipedia .org/wiki/1912_United_States_presidential_election_in_California.

53. JRF Diary, November 22, 1912.

54. "Ready to Present Water Case," *San Francisco Chronicle,* November 23, 1912.

55. JRF Diary, November 23, 1912.

56. A list of attendees appears in "Hearings Before the Secretary of the Interior In Re Use of Hetch Hetchy Reservoir Site, in the Yosemite National Park by the City of San Francisco," Day 1, November 25, pp. 39–45. A typed stenographic record of the five-day meeting is available in JRF Papers, boxes 69–70. Hereafter, this transcript is referenced as "Fisher Hearing, Day _."

57. "Fisher Hearing, Day 1," 34–35.

58. "Fisher Hearing, Day 1," 83–86.

59. John R. Freeman to Mayor Rolph, February 21, 1913, JRF Papers, box 67.

60. "Fisher Hearing, Day 1," 87–91.

61. "Fisher Hearing, Day 1," 94–95. As a follow-up to the wage discrepancy issue, William Bade called attention to a cost variation in how steel penstocks were priced in Freeman's Hetch Hetchy design compared to Hazen's Sacramento River proposal (see pp. 99–105).

62. William Bade to William Colby, November 25, 1912, quoted in Jones, *John Muir,* 146.

63. "Fisher Hearing, Day 1," 124.

64. "Fisher Hearing, Day 1," 131–33. Johnson's testimony relative to the park's origin is quoted in chapter 1.

65. "Fisher Hearing, Day 1," 134–35.

66. William Bade to William Colby, November 29, 1912, quoted in Jones, *John Muir,* 146.

67. Fisher asked Horace McFarland if he shared Johnson's perspective: "Is your view as extreme as that of Mr. Johnson as to the alternative source of supply? Do you

exclude cost from consideration?" McFarland curtly replied, "No sir." "Fisher Hearing, Day 1," 146.

68. "Fisher Hearing, Day 1," 185–86.

69. "Fisher Hearing, Day 1," 187–88.

70. "Fisher Hearing, Day 1," 198. Also see "Fisher Hearing, Day 2," 141.

71. "Fisher Hearing, Day 3," 174–203.

72. Although Mulholland and Lippincott were absent, Freeman noted in his diary (Friday, November 28, 1912) that "Spring Valley got into the game in P.M. and situation became very intense and aggressive."

73. For example, the city's consulting engineer, Myron Fuller, described why he believed Mulholland and Lippincott had erred in their evaluation of the Livermore gravels. See "Fisher Hearing, Day 5," 60–69.

74. "Fisher Hearing, Day 5," 1–60.

75. "Fisher Hearing, Day 5," 93–190.

76. Freeman, *Hetch Hetchy Report,* 160h–160i, 342–44.

77. Freeman, *Hetch Hetchy Report,* 327–30; The McCloud River proposal is described in US Advisory Board of Army Engineers, *Hetch Hetchy Valley Report . . . to the Secretary of the Interior on Investigations Relative to Sources of Water Supply for San Francisco and Bay Communities* (Washington, DC: Government Printing Office, 1913), 22–25 (hereafter, Army Board, *Hetch Hetchy Valley Report*).

78. "Fisher Hearing, Day 2," 81, 88.

79. "Fisher Hearing, Day 2," 94, 224, 227.

80. "Fisher Hearing, Day 5," 224, 227.

81. "Fisher Hearing, Day 5," 228.

82. "Fisher Hearing, Day 5," 248.

83. John R. Freeman to M. M. O'Shaughnessy, December 9, 1912, JRF Papers, box 67. It is also worth recognizing that Freeman had spent a long week in Washington representing the city, and this came at significant personal cost given that he missed being at home with his family for Thanksgiving. On this point, his diary entry for November 28 reads, "Thanksgiving Day—Worked all day and until midnight. Had a glass of milk for dinner and a piece of Rhode Island turkey for supper in company with O'Shaughnessy. A long and lonesome day."

84. Freeman to O'Shaughnessy, December 9, 1912.

85. John R. Freeman, "Summary of Hearing before Secretary Fisher on the Hetch Hetchy Water Supply," *Engineering News* 68 (December 26, 1912): 1215.

86. JRF Diary, December 1, 1912.

87. The results of O'Shaughnessy's study to determine the "cost of the McCloud River Project fully developed to deliver 400,000,000 gallons daily" is summarized in M. M. O'Shaughnessy to United States Board of Army Engineers, January 13, 1913, JRF Papers, box 67. By O'Shaughnessy's calculation, a 400 mgd McCloud River aqueduct would have cost at least $45 million more than a comparable Hetch Hetchy aqueduct.

88. JRF Diary, December 1, 1912.

89. JRF Diary, December 11, 1912.

90. JRF Diary, December 14, 1912.

91. JRF Diary, December 17, 1912.

92. JRF Diary, December 22, 1912.

93. John R. Freeman to M. M. O'Shaughnessy, December 26, 1912, JRF Papers, box 67.

94. Colonel John Biddle to Secretary of the Interior, December 26, 1912, JRF Papers, box 67.

95. John R. Freeman to M. M. O'Shaughnessy, December 31, 1912, JRF Papers, box 67.

96. John R. Freeman to Percy Long, January 4, 1913, JRF Papers, box 67.

97. John R. Freeman to J. H. Dockweiler, January 6, 1913; later in the month, Freeman expressed further exasperation with the Army Board, averring that "they were leaving all analysis to [their assistant engineer H. H.] Wadsworth and that Wadsworth knew nothing about municipal water supply or aqueduct building, his whole training having been on river navigation improvement." John R. Freeman to Mayor Rolph and Percy Long, January 25, 1913; both letters in JRF Papers, box 67.

98. John R. Freeman to Walter Fisher, January 7, 1913, JRF Papers, box 67.

99. JRF Diary, January 23–24, 1913.

100. John R. Freeman to Walter Fisher, "Letter of Transmittal," January 30, 1913, JRF Papers, box 70. See O'Shaughnessy to Army Board, January 13, 1913, for McCloud River aqueduct cost estimate.

101. JRF Diary, December 27, 1912.

102. JRF Diary, January 6, 1913.

103. JRF Diary, January 17, 1913.

104. John R. Freeman to Richard Maclaurin, February 5, 1913, copy of telegram glued into JRF Diary, February 1913. Sensing what was to come, Freeman expressed his "fears that efficiency may be sacrificed for exterior appearances."

105. Richard Maclaurin to John R. Freeman, ca. February 15, 1913, copy of letter glued into JRF Diary, February 1913. In his diary on Monday, February 17, 1913, Freeman plaintively noted, "[O]ffer to prepare Institute Technology plans was not accepted." For more on the decision to hire Bosworth as architect for MIT's "New Technology" campus, see Jarzombek, *Designing MIT*, 69–70.

106. JRF Diary, February 19, 1913; John R. Freeman to Percy Long, February 21, 1913, JRF Papers, box 67.

107. JRF Diary, February 19, 1913. Ever the multitasker, later that day Freeman also "[s]aw R.I. Congressman O'Shaughnessy & got him to introduce bill for relief [of] Mutual Insurance Co. from income tax."

108. JRF Diary, February 20, 1913; Freeman to Rolph, February 21, 1913.

109. JRF Diary, February 23, 1913. Also see Percy Long to John R. Freeman, February 22, 1913; and Mayor Rolph to John R. Freeman, February 22, 1913; both in JRF Papers, box 67.

110. Army Board, *Hetch Hetchy Valley Report,* 50.

111. Army Board, *Hetch Hetchy Valley Report,* 51.

112. Army Board, *Hetch Hetchy Valley Report,* 15–18.

113. Army Board, *Hetch Hetchy Valley Report,* 49.

114. Army Board, *Hetch Hetchy Valley Report,* 50. Although the Army Board did not claim to have precisely determined all costs associated with the proposed Hetch Hetchy system and various alternative source, it believed its estimates to be "sufficient to permit such comparisons as are necessary to be made for judging the relative cost and merits of the different projects" (p. 51).

115. John R. Freeman to M. M. O'Shaughnessy, December 31, 1912; John R. Freeman to Percy Long, January 4, 1913; John R. Freeman to J. H. Dockweiler, January 6, 1913; all in JRF Papers, box 67.

116. John R. Freeman to Mayor Rolph, February 26, 1913, JRF Papers, box 67.

117. JRF Diary, February 28, 1913.

118. John R. Freeman to Mayor Rolph, March 1, 1913, JRF Papers, box 67.

119. Secretary Walter Fisher to the Mayor and Supervisors of San Francisco, March 1, 1913, "Water Supply, City of San Francisco, Application for Lake Eleanor and Hetch Hetchy Valley Reservoir Sites. Act of February 15, 1901," JRF Papers, box 67.

120. On March 12, 1912, after he had replaced Garfield as interior secretary, Franklin Lane apparently formally revoked the Garfield Permit with the understanding that the bill subsequently submitted to Congress in the spring of 1913 constituted the appropriate means for San Francisco to receive "the relief desired in this matter." See Franklin K. Lane to Hon. Scott Ferris, [House] Committee on the Public Lands, March 12, 1913, reprinted in "Hetch Hetchy Dam Site: Hearing before the Committee on the Public Lands, House of Representatives, 63rd Congress, First Session, on H.R. 6281" (Washington, DC: Government Printing Office, 1913), 20–21.

Chapter 6

1. JRF Diary, March 3, 1913, John Ripley Freeman (1855–1932) Papers (MC 51), Distinctive Collections, Massachusetts Institute of Technology, Cambridge (hereafter, JRF Papers), box 3.

2. John R. Freeman to Myron Fuller, March 4, 1913, JRF Papers, box 67.

3. Lane's political and professional career prior to being appointed secretary of the interior is documented in Anne Wintermute Lane and Louise Herrick Wall, eds., *Letters of Franklin K. Lane, Personal and Political* (Boston: Houghton Mifflin, 1922), 17–128.

4. Franklin Lane to John H. Wigmore, May 9, 1903, in Lane and Wall, *Letters,* 42.

5. Lane to Wigmore, May 9, 1903.

6. Lane and Wall, *Letters,* 129–30.

7. Charles Seymour, ed., *The Intimate Papers of Colonel House,* vol. 1 (New York: Houghton Mifflin, 1926), 83–113. Lane's selection as interior secretary (and the absence of Hetch Hetchy as a factor) is noted in Robert W. Righter, *The Battle over Hetch Hetchy: America's Most Controversial Dam and the Birth of Modern Environmentalism* (New York: Oxford University Press, 2005), 118–19.

8. Richard Watrous to William Bade, March 10, 1913, quoted in Holway Jones, *John Muir and the Sierra Club: The Battle for Yosemite* (San Francisco: Sierra Club, 1965), 153.

9. "63rd United States Congress," *Wikipedia,* https://en.wikipedia.org/wiki/63rd _United_States_Congress.

10. Wilson described the purpose of the summer session in a speech delivered to Congress on April 8, 1913, as "[t]ariff reform" and "reform of our banking and currency laws"; he made no mention of Hetch Hetchy. Woodrow Wilson, "April 8, 1913: Message Regarding Tariff Duties," *Presidential Speeches,* University of Virginia, Miller Center, https://millercenter.org/the-presidency/presidential-speeches/april-8-1913-message -regarding-tariff-duties.

11. Various versions of the Raker bill are noted in "Hetch Hetchy Grant to San Francisco," 63rd Cong., 1st sess., H.R. Rep. No. 41, August 5, 1913.

12. JRF Diary, March 3–April 1, 1913.

13. Donald C. Jackson, chapter 4 in *Building the Ultimate Dam: John S. Eastwood and the Control of Water in the West*, paperback ed. (Norman: University of Oklahoma Press, 2005).

14. JRF Diary, April 2–May 2, 1913.

15. On April 17, he "discussed matters and lunched with Long at Bohemian Club." A week later he met with O'Shaughnessy. JRF Diary, April 17, 25, and 26, 1913.

16. Percy Long to John R. Freeman, May 15, 1913, JRF Papers, box 67.

17. John R. Freeman to Percy Long, May 21, 1913, JRF Papers, box 67.

18. JRF Diary, June 10–19, 1913.

19. JRF Diary, June 20–July 8, 1913.

20. Freeman's trip into Russia is described in JRF Diary, July 14–August 11, 1913.

21. Jones, *John Muir,* 148–49.

22. William Colby to Horace McFarland, March 14, 1913, quoted in Jones, *John Muir,* 154.

23. Colby to McFarland, March 14, 1913.

24. For more on the irrigation crusade, see Donald J. Pisani, *From the Family Farm to Agribusiness: The Irrigation Crusade in California and the West, 1850–1931* (Berkeley: University of California Press, 1984).

25. David Igler, *Industrial Cowboys: Miller & Lux and the Transformation of the Far West, 1850–1920* (Berkeley: University of California Press, 2005).

26. Percy Long to John R. Freeman, August 21, 1913, JRF Papers, box 67.

27. Percy Long told Freeman, "The Modesto-Turlock Districts were represented by three attorneys including the Superior Judge of Stanislaus County and several engineers." Long to Freeman, August 21, 1913.

28. Raker Act, sect. 9, paragraph C ; sect. 5, paragraph B; Army Board report, 105.

29. M. M. O'Shaughnessy to John R. Freeman, August 30, 1913, JRF Papers, box 67.

30. "Hetch Hetchy Dam Site: Hearing before the Committee on the Public Lands, House of Representatives, 63rd Congress, First Session, on H.R. 6281" (Washington, DC: Government Printing Office, 1913), 3 (hereafter, "Hetch Hetchy: House Hearing").

31. According to Raker, "The original bill H.R. 112 and the amended bill H.R. 4319 have been supplanted by the amended bill introduced on June 23, 1913, H.R 6281"; "Hetch Hetchy: House Hearing," 4.

32. Lane clarified, "I am not entirely without partisanship in this matter . . . because some 10 or 11 years ago, when I was the city attorney of San Francisco . . . I was requested

by the Board of supervisors to come here and make an argument before [Secretary Hitch-cock] on the city's behalf . . . [and] since that time I have always advised in sympathy with the purpose of this bill." Franklin Lane testimony, "Hetch Hetchy: House Hearing," 5.

33. Lane testimony, 15–19.

34. Franklin K. Lane to Hon. Scott Ferris, June 24, 1913, in "Hetch Hetchy: House Hearing," 21–22.

35. David Houston testimony, 22–25; George Smith testimony, 34–41; Henry Graves testimony, 42–44; and Frederick Newell testimony, 44–50; all in "Hetch Hetchy: House Hearing."

36. Chairman Ferris introductory remarks, "Hetch Hetchy: House Hearing," 4.

37. Gifford Pinchot testimony, "Hetch Hetchy: House Hearing," 31.

38. Pinchot testimony, 32.

39. Pinchot testimony, 25.

40. Colonel John Biddle testimony, "Hetch Hetchy: House Hearing," 51–94; quotation is from 51.

41. Biddle testimony, 64.

42. Colonel Spencer Cosby and Lt. Colonel Harry Taylor testimony, "Hetch Hetchy: House Hearing," 92–93.

43. Percy Long testimony, "Hetch Hetchy: House Hearing," 109, 114.

44. M. M. O'Shaughnessy testimony, "Hetch Hetchy: House Hearing," 133–34, 139.

45. O'Shaughnessy testimony, 139.

46. James Needham testimony, "Hetch Hetchy: House Hearing," 268.

47. L. I. Dennent testimony, "Hetch Hetchy: House Hearing," 252–68.

48. To assuage fears that the city might shirk its roadbuilding responsibilities, O'Shaughnessy reassured Raker, "There is a qualifying clause put in the bill that [roads related to the Hetch Hetchy project] shall be built so far as directed by the Secretary of the Interior. We submit ourselves entirely to his direction and discretion as to the character of these improvements." O'Shaughnessy testimony, 159–60.

49. Whitman made clear to the committee that he had twice visited Hetch Hetchy, first in 1904 and later in 1909. See Edmund Whitman testimony, "Hetch Hetchy: House Hearing," 176.

50. Whitman testimony, 226–27.

51. Whitman testimony, 238.

52. Whitman testimony, 239.

53. "Offer of the Blue Lakes Company" for a Mokelumne supply is referenced in John R. Freeman, *On the Proposed Use of a Portion of the Hetch Hetchy, Eleanor and Cherry Valleys Within or Near to the Boundaries of the Stanislaus U.S. National Forest Reserve and the Yosemite National Park as Reservoirs for Impounding Tuolumne River Flood Waters and Appurtenant Works for the Water Supply of San Francisco, California and Neighboring Cities* (San Francisco: by Authority of the Board of Supervisors, Rincon Publishing, 1912), 160b–160d, 369–72 (hereafter, Freeman, *Hetch Hetchy Report*). Freeman projects the Mokelumne's ultimate capacity to be 200 mgd.

54. Eugene J. Sullivan to Scott Ferris, June 27, 1913, in "Hetch Hetchy: House Hearing," 179.

55. Taggart Aston to William Kent, June 14, 1913, in "Hetch Hetchy: House Hearing," 180.

56. Jones, *John Muir*, 157–58.

57. Jones, 158; Righter, *Hetch Hetchy*, 121–22; John Warfield Simpson, *Dam!: Water, Power, Politics and Preservation in Hetch Hetchy and Yosemite National Park* (New York: Pantheon, 2005), 169–70.

58. Max Bartell to Marsden Manson, City Engineer, Letter with Appendices and Maps, April 24, 1912, Max Bartell Papers, box 2, Water Resources Center Archive, University of California, Riverside (hereafter, Bartell Report).

59. Bartell reported that "the cost . . . necessary to deliver 60 million gallons daily to San Francisco [from the Mokelumne], is $40,978,680." Bartell to Manson, April 24, 1912.

60. Bartell to Manson, April 24, 1912, Appendix A, 14.

61. Freeman, *Hetch Hetchy Report*, 160c–160d, 368. For a 200 mgd Mokelumne system, the report estimated a power generating capacity of 24,000 kilowatts, or a little more than 32,000 horsepower. Freeman also stated that in postulating a 200 mgd supply to be drawn from the Mokelumne, Grunsky found that the system "would interfere seriously with irrigation needs principally because of lack of sufficient storage at low elevations on the North Fork [of the Mokelumne]."

62. Freeman, *Hetch Hetchy Report*, 160e. In preparing his report on the Mokelumne River, Grunsky may not have purposely suppressed Bartell's report, but he did not make much, if any, use of it.

63. US Advisory Board of Army Engineers, *Hetch Hetchy Valley Report . . . to the Secretary of the Interior on Investigations Relative to Sources of Water Supply for San Francisco and Bay Communities* (Washington, DC: Government Printing Office, 1913), 21, 99.

64. Eugene Sullivan testimony, "Hetch Hetchy: House Hearing," 278–363; Aston's absence is noted on 284 and 289.

65. Sullivan testimony, 278–363; questioning in regard to Maud Treadwell, 349–61.

66. Freeman, *Hetch Hetchy Report*, 160d–160e, 365–72.

67. John R. Freeman testimony, Senate Public Land Committee Hearing, September 24, 1912, 76.

68. JRF Diary, August 1, 1912.

69. Jones, *John Muir*, 158. For the "suppressed" Bartell report, see "A National Park Threatened," *New York Times*, July 12, 1913.

70. For Johnson's awareness of Sullivan's claims, see "The Hetch Hetchy Plan: Robert Underwood Johnson Gives Advice on How to Defeat It," *New York Times*, July 27, 1913.

71. Robert Underwood Johnson, "The Hetch Hetchy Scheme, Why It Should Not Be Rushed Through the Extra Session, An Open Letter to the American People," August 1, 1913, copy in JRF Papers, box 67.

72. George Whipple testimony, "Hearings before the Secretary of the Interior In Re Use of Hetch Hetchy Reservoir Site, in the Yosemite National Park by the City of San Francisco" (typed stenographic record for Fisher Hearing), Day 2, 185–98, JRF Papers, box 70. See chapter 5.

73. Freeman, *Hetch Hetchy Report*, 160d, 371–72.

74. "Hetch Hetchy Grant to San Francisco," *Congressional Record,* August 5, 1913, 63rd Cong., 1st sess., Report No. 41.

75. "Hetch Hetchy Grant to San Francisco," 33–35.

76. For the deliberations, see *Congressional Record,* August 29–September 2, 1913.

77. *Congressional Record,* August 29, 1913, 3896.

78. *Congressional Record,* August 29, 1913, 3896.

79. *Congressional Record,* August 30, 1913, 3976.

80. *Congressional Record,* August 30, 1913, 3973–74.

81. *Congressional Record,* August 30, 1913, 3974.

82. *Congressional Record,* August 29, 1913, 3915.

83. City Attorney Long explained this point to Freeman: "If the Senate should decide ... to eliminate the irrigation conditions of the Bill, we would, of course[,] be satisfied herewith, but that suggestion must not come from us as we are morally bound to this agreement and it is so thoroughly understood in Congress." Percy Long to John R. Freeman, August 25, 1913, JRF Papers, box 67.

84. *Congressional Record,* August 30, 1913, 3986–88.

85. *Congressional Record,* August 30, 1913, 3984–4112.

86. *Congressional Record,* September 2, 1913, 4112; September 3, 1913, 4151–52.

Chapter 7

1. JRF Diary, August 21–30, 1913, John Ripley Freeman (1855–1932) Papers (MC 51), Distinctive Collections, Massachusetts Institute of Technology, Cambridge (hereafter, JRF Papers), box 3.

2. Percy Long to John R. Freeman, August 21 and 25, 1913; both in JRF Papers, box 67.

3. Long to Freeman, August 21, 1913.

4. John R. Freeman to Percy Long, September 2, 1913, JRF Papers, box 68.

5. John R. Freeman to M. M. O'Shaughnessy, September 8, 1913, JRF Papers, box 68.

6. John R. Freeman to J. S. Dunnigan, August 29, 1913, JRF Papers, box 67.

7. J. S. Dunnigan to John R. Freeman, August 31, 1913, JRF Papers, box 67.

8. JRF Diary, September 3, 1913.

9. JRF Diary, September 4–5, 1913.

10. JRF Diary, September 10, 1913. Although the committee did not meet, Freeman spent time with Senators Tilman of South Carolina and Lippitt of Rhode Island.

11. JRF Diary, September 22–24, 1913. The senators referenced as meeting with Freeman only include those specifically noted in his diary; there may well have been other senators (or staff members) of both parties with whom he conferred during his time on Capitol Hill.

12. John R. Freeman to Mayor James Rolph, September 11, 1913, JRF Papers, box 68.

13. Freeman to O'Shaughnessy, September 8, 1913. For evidence of the numerous newspapers that published articles and editorials in opposition to the city's plans, see "The Hetch Hetchy 'Grab,'" *Literary Digest,* September 27, 1913, JRF Papers, box 68.

14. "The Hetch Hetchy Water Grab," *Providence Journal,* September 5, 1913, JRF Papers, box 68.

15. John R. Freeman, Letter to the Editor, *Providence Journal,* September 9, 1913, JRF Papers, box 68.

16. Freeman, Letter to the Editor.

17. "Hetch Hetchy Reservoir Site: Hearing before the Committee on Public Lands, United States Senate, Sixty Third Congress, First Session, on H.R. 7207," September 24, 1913 (Washington, DC: Government Printing Office, 1913). Hereafter, "Hetch Hetchy: Senate Hearing."

18. "Change of View Impresses Lane," *Los Angeles Times,* September 4, 1913.

19. "Secretary Lane Stricken," *Washington Post,* September 10, 1913; "Lane Able to Sit Up," *Los Angeles Times,* September 12, 1913.

20. "Secretary Lane Back, Recovered," *New York Times,* October 11, 1913.

21. For more on Smoot's career, see Milton R. Merrill, *Reed Smoot: Apostle in Politics* (Logan: Utah State University Press, 1990).

22. For a biography of Works, see "Works, John Downey," *Biographical Directory of the United States Congress,* https://bioguide.congress.gov/search/bio/W000743.

23. "Oppose Hetch-Hetchy Bill," *Los Angeles Times,* September 3, 1913.

24. Richard Watrous testimony, "Hetch Hetchy: Senate Hearing," 30–31; includes transcript of a letter from Horace McFarland to Richard Watrous, ca. September 20, 1913.

25. Herbert Parsons testimony, "Hetch Hetchy: Senate Hearing," 13.

26. Edmund Whitman testimony, "Hetch Hetchy: Senate Hearing," 21–22.

27. Whitman testimony, 23, 26–27.

28. Whitman testimony, 27.

29. Robert Underwood Johnson testimony, "Hetch Hetchy: Senate Hearing," 36–37.

30. Johnson testimony, 42.

31. Johnson testimony, 43–44.

32. W. C. Lehane testimony, "Levi Winklebeck to Chairman Myers, Sept. 23, 1913," "Hetch Hetchy: Senate Hearing," 47.

33. Lehane testimony, "Thomas Carswell to Chairman Myers, Sept. 23, 1913," and "W. P. Witten to Chairman Myers, Sept. 23, 1913," 47.

34. Lehane testimony, 56.

35. Lehane testimony, 57.

36. Denver Church testimony, "Hetch Hetchy: Senate Hearing," 64.

37. "C. S. Abbott et al. to Congressman Church, August 13, 1913," in Church testimony, 64.

38. "H. C. Hoskins to Congressman Church, August 12, 1913," in Church testimony, 65–66.

39. Church testimony, 66.

40. John Raker testimony, "Hetch Hetchy: Senate Hearing," 66, 69.

41. John Hart, *Muir Woods National Monument* (San Francisco: Golden Gate National Parks Conservancy, 2011).

42. William Kent testimony, "Hetch Hetchy: Senate Hearing," 70.

43. Kent testimony, 71.

44. Alexander Vogelsang testimony, "Hetch Hetchy: Senate Hearing," 71–72.

45. John Dunnigan remarks, "Hetch Hetchy: Senate Hearing," 74.

46. John R. Freeman testimony, "Hetch Hetchy: Senate Hearing," 74.

47. Freeman testimony, 74.

48. Freeman testimony, 74–75.

49. Freeman testimony, 76–77.

50. Freeman testimony, 77.

51. "Hetch Hetchy: Senate Hearing," 78.

52. JRF Diary, September 25, 1913.

53. John R. Freeman to M. M. O'Shaughnessy, September 27, 1913, JRF Papers, box 68.

54. Freeman to O'Shaughnessy, September 27, 1913.

55. J. S. Dunnigan to John R. Freeman, ca. September 30, 1913, JRF Papers, box 68.

56. *Congressional Record,* 63rd Cong., 1st sess., October 4, 1913, 5443–44.

57. *Congressional Record,* October 4, 1913, 5447.

58. *Congressional Record,* October 4, 1913, 5448, 5463.

59. *Congressional Record,* October 4, 1913, 5462.

60. *Congressional Record,* October 4, 1913, 5449. Senator Gronna wondered "if it is fair . . . to try and railroad this bill through without giving those of us who are perhaps ignorant of the conditions and surroundings and opportunity to investigate the matter?" 5448.

61. *Congressional Record,* October 4, 1913, 5474–75.

62. *Congressional Record,* October 4, 1913, 5475.

63. *Congressional Record,* October 7, 1913, 5483–74.

64. Freeman was a multitasker par excellence, and he was remarkably busy during the month of October. A far-ranging trip he took starting on October 8 is described in a letter to his son Roger: "I managed to look in on seven or eight different jobs in the course of three weeks and when I reached home [I discovered] I had spent thirteen out of fourteen consecutive nights on sleeping cars." John R. Freeman to Roger Freeman, November 6, 1913, JRF Papers, box 5.

65. J. S. Dunnigan to John R. Freeman, October 7, 1913, JRF Papers, box 68.

66. Muir's correspondence with Johnson through the summer and fall of 1913 is described in Dean King, *Guardians of the Valley: John Muir and the Friendship That Saved Yosemite* (New York: Scribner, 2023), 324–44.

67. John Muir to Robert U. Johnson, September 12, 1913; referenced in Holway Jones, *John Muir and the Sierra Club: The Battle for Yosemite* (San Francisco: Sierra Club, 1965), 159–60. Jones describes how "leaflets and circulars . . . were printed in large lots of 5,000 to 20,000 and were designed to be easily used by editorial writers."

68. See Jones, *John Muir,* 101–4; also see Robert W. Righter, *The Battle over Hetch Hetchy: America's Most Controversial Dam and the Birth of Modern Environmentalism* (New York: Oxford University Press, 2005), 72–78.

69. For "Circular Number Seven," see Jones, *John Muir,* 182.

70. "The Truth about Hetch Hetchy," Eastern Branch, Society for the Preservation of National Parks, ca. September 1913, JRF Papers, box 70.

71. The address for the SPNP Eastern Branch is listed in "The Truth about Hetch Hetchy" pamphlet as "Pemberton Building, Boston," the same address as Whitman's law firm.

72. Robert Underwood Johnson, "Hetch Hetchy," *Collier's,* October 25, 1913. Emphasis in the original.

73. Johnson, "Hetch Hetchy."

74. "The Truth about Hetch Hetchy"; Johnson, "Hetch Hetchy."

75. J. S. Dunnigan to John R. Freeman, November 9, 1913, JRF Papers, box 68.

76. Horace McFarland, "Hetch Hetchy Lobby," *New York Times,* November 8, 1913.

77. John R. Freeman to J. S. Dunnigan, November 12, 1913, JRF Papers, box 68.

78. John R. Freeman to Frederick Law Olmsted, November 13, 1913, JRF Papers, box 68.

79. Freeman to Olmsted, November 13, 1913.

80. Frederick Law Olmsted to John R. Freeman, November 17, 1913, JRF Papers, box 68.

81. Olmsted to Freeman, November 17, 1913.

82. John R. Freeman to Frederick Law Olmsted, November 18, 1913, JRF Papers, box 68.

83. Freeman to Olmsted, November 18, 1913.

84. Frederick Law Olmsted, "Olmsted on Hetch Hetchy," *Boston Evening Transcript,* November 19, 1913, JRF Papers, box 68. Reprinted in F. L. Olmsted, "Hetch Hetchy: The San Francisco Water Supply Controversy," *Landscape Architecture* 4 (January 1914): 37–46.

85. Olmsted, "Olmsted on Hetch Hetchy."

86. Olmsted, "Olmsted on Hetch Hetchy."

87. JRF Diary, November 19–30, 1913.

88. JRF Diary, December 1–6, 1913.

89. John R. Freeman to Desmond Fitzgerald, December 12, 1913, JRF Papers, box 68.

90. John Muir to R. U. Johnson, October 30, 1913; facsimile of telegram in King, *Guardians of the Valley,* 343. Although Muir did not appear in Washington during the Senate debate, Gifford Pinchot appears to have spent at least some time on Capitol Hill during early December. As reported in the *New York Times,* Senator Works claimed that "if you will go out into the corridor of this Capital now, you will find Gifford Pinchot lobbying in favor of the bill; at least he was here this morning." See "Hetch Hetchy 'Lobbies,'" *New York Times,* December 4, 1913.

91. John Muir, *The Yosemite* (New York: Century, 1912), 262.

92. "Hetch Hetchy Meeting: Protest against Bill to Be Voiced Tonight at Natural History Museum," *New York Times,* November 21, 1913.

93. R. U. Johnson, "Hetch Hetchy Campaign," *New York Times,* December 1, 1913.

94. For a summary of the Senate debate, see Richard Lowitt, "Hetch Hetchy Phase Two: The Senate Debate," *California History* 74 (Summer 1995): 190–203. Lowitt notes that "environmental issues raised by John Muir and the throngs of preservationists endorsing his views, upon which historians have focused, actually received relatively short shrift in the Senate debate" (193). While Lowitt highlights the importance of the Army Board report, he makes no mention of Freeman or Freeman's *Hetch Hetchy Report.*

95. Facsimile copies of such material appear in Char Miller, *Hetch Hetchy: A History in Documents* (Peterborough, ON: Broadview Press, 2020), 160–66.

96. *Congressional Record,* December 3, 1913, 96.

97. *Congressional Record,* December 6, 1913, 334.

98. *Congressional Record,* December 2, 1913, 46–73; December 3, 1913, 105–15.

99. *Congressional Record,* December 4, 1913, 198.

100. *Congressional Record,* December 4, 1913, 182.

101. *Congressional Record,* December 4, 1913, 183.

102. *Congressional Record,* 183; December 6, 1913, 333.

103. *Congressional Record,* December 6, 1913, 343, 345.

104. *Congressional Record,* December 6, 1913, 346–47. Norris further exclaimed, "Pass this bill and you relieve the burden of every man on the coast who earns his living by the sweat of his face. Pass it and you give relief to every woman who toils, to every child who suffers. . . . Pass this bill and you will perform the very highest possible act of conservation."

105. *Congressional Record,* December 6, 1913, 367–68.

106. *Congressional Record,* December 6, 1913, 359.

107. *Congressional Record,* December 6, 1913, 359.

108. *Congressional Record,* December 6, 1913, 360.

109. John R. Freeman to Allen Hazen, December 11, 1913, JRF Papers, box 68. Earlier on December 6, each senator reportedly received a copy of the December 2 *San Francisco Examiner,* which included a fanciful illustration showing a beaux art boulevard, replete with automobiles, extending atop a dam at Hetch Hetchy. There is no evidence that Freeman played any role in the creation of this drawing, but it does reflect his championing of a road that would make the valley accessible to automobiles. A facsimile copy is reproduced in Ray W. Taylor, *Hetch Hetchy: The Story of San Francisco's Struggle to Provide a Water Supply for Her Future Needs* (San Francisco: Ricardo J. Orozco, 1926), 126.

110. *Congressional Record,* December 6, 1913, 382.

111. *Congressional Record,* December 6, 1913, 381–82.

112. *Congressional Record,* December 6, 1913, 382–85.

113. *Congressional Record,* December 6, 1913, 385–86. Even if all 11 senators who were not paired had somehow appeared and voted against the Raker Act, the bill would still have been passed 51 yeas to 44 nays.

114. JRF Diary, December 6, 1913.

115. JRF Diary, December 7–20, 1913.

116. "Engineer Defends Hetch Hetchy Plan," *New York Times,* December 12, 1913.

117. J. S. Dunnigan to John R. Freeman, December 12, 1913, JRF Papers, box 68. Dunnigan was not concerned by the delay in Wilson's signing of the bill; he advised Freeman, "Secretary Lane has written a strong approval of the bill and the President has informed the newspapermen that the act is satisfactory to him and that he will sign it."

118. "Against Hetch Hetchy Bill," *New York Times,* December 10, 1913.

119. "Wilson Signs Hetch Hetchy," *Los Angeles Times,* December 20, 1913.

120. "For Hetch Hetchy Repeal," *New York Times,* December 21, 1913.

121. In fact, Brandegee had not really changed his view of the Raker bill. He had voted to keep the legislation off the Senate floor in October because he feared it would encounter problems with the Senate maintaining a quorum, but he always supported the city's plans: "I voted against proceeding to a consideration of this bill at this time because I thought the situation was such that we possibly would wind up with the absence of a quorum in the end. . . . [S]till if it comes to a question on the passage of the bill I shall vote for its passage. I think it is a wise and proper measure." See *Congressional Record,* October 4, 1913, 5466.

122. Taking into account paired votes, the six nay-voting Democrats were Kern and Shively of Indiana, Hollis of New Hampshire, Martine of New Jersey, Lane of Oregon, and Sheppard of Texas; the eleven Republicans who voted yea were Brandegee and McLean of Connecticut, Perkins of California, Sherman of Illinois, Nelson of Minnesota, Norris of Nebraska, Oliver and Penrose of Pennsylvania, Lippitt of Rhode Island, and La Follette and Stephenson of Wisconsin. For documentation of paired votes, see *Congressional Record,* December 6, 1913, 386.

Chapter 8

1. For more on the creation of the National Park Service, see Robert W. Righter, *The Battle over Hetch Hetchy: America's Most Controversial Dam and the Birth of Modern Environmentalism* (New York: Oxford University Press, 2005), 192–95; Donald C. Swain, "The Passage of the National Park Service Act of 1916," *Wisconsin Magazine of History* (Autumn 1966): 4–17; and Robin W. Winks, "The National Park Service Act of 1916: 'A Contradictory Mandate'?" 74 Denver U.L. Rev. 575 (1997). Gifford Pinchot opposed the creation of a national park system based within the Department of the Interior but favored such a system if placed under the authority of the Forest Service/Department of Agriculture; see Samuel F. Hays, *Conservation and the Gospel of Efficiency: The Progressive Conservation Movement, 1890–1920* (Cambridge, MA: Harvard University Press, 1959), 196–97.

2. Beating back attempts to utilize national parks for water development projects did not always come easily. As Mark Harvey describes, after World War Two the Bureau of Reclamation attempted to build a dam across the Green River in eastern Utah's Dinosaur National Monument, but legislation authorizing the planned Echo Park Dam was defeated in Congress in the 1950s. See Mark Harvey, *A Symbol of Wilderness: Echo Park*

and the American Conservation Movement (Seattle: University of Washington Press, 2015). Quote from Righter, *Battle over Hetch Hetchy*, 193.

3. M. M. O'Shaughnessy, *Hetch Hetchy: Its Origin and History* (San Francisco: privately printed), 21, 41.

4. O'Shaughnessy, *Hetch Hetchy*, 26, 54. O'Shaughnessy also praised Freeman's *Hetch Hetchy Report* (*The Hetch Hetchy Water Supply for San Francisco, 1912, Report by John R. Freeman* [San Francisco: Rincon Publishing, 1912]) as being "conceded by all engineers to be one of the ablest brief documents ever prepared for this purpose [of municipal water supply]" (O'Shaughnessy, *Hetch Hetchy*, 40).

5. The roads and trails built by the city in compliance with the Raker Act are described in Righter, *Battle over Hetch Hetchy*, 200–203.

6. The dedication of O'Shaughnessy Dam in July 1923 is described in Righter, *Battle over Hetch Hetchy*, 151–54.

7. Difficulties in financing construction of the Hetch Hetchy system are described in O'Shaughnessy, "Financial Aspects of the Hetch Hetchy Project," *Hetch Hetchy*, 112–18.

8. O'Shaughnessy, *Hetch Hetchy*, 118. The durability and integrity of O'Shaughnessy's work on the Hetch Hetchy Dam and Aqueduct is not something that should be passed over lightly. Specifically, O'Shaughnessy Dam can be contrasted with California's other great municipal water supply dam of the 1920s, which was built by William Mulholland for the City of Los Angeles; completed in 1926 as a component of the Los Angeles Aqueduct system, the concrete gravity St. Francis Dam failed catastrophically in March 1928, killing some 400 people. See Norris Hundley and Donald C. Jackson, *Heavy Ground: William Mulholland and the St. Francis Dam Disaster* (Huntington Library Press, 2015; paperback ed., Reno: University of Nevada Press, 2020).

9. The construction history of the aqueduct is detailed and (profusely) illustrated in Ted Wurm, *Hetch Hetchy and Its Dam Railroad* (Glendale, CA: Trans-Anglo Books, 1973). See also O'Shaughnessy, *Hetch Hetchy*, 54–118; and Nelson A. Eckart, "The Water Supply System of San Francisco," *American Water Works Association Journal* (May 1940): 751–94. The 11-mile-long Canyon Power Tunnel running from O'Shaughnessy Dam to the Dion R. Holm Powerhouse just above Early Intake was not built until the 1960s, although it had been envisaged as part of Freeman's 1912 hydropower plan.

10. Wurm, *Hetch Hetchy*, 193–228.

11. Righter, *Battle over Hetch Hetchy*, 162. Water arrived on October 24, and formal celebration of the aqueduct's completion was held four days later.

12. The vote tallies for the 1910, 1915, 1921, 1927, and 1928 bond referendums are provided in O'Shaughnessy, *Hetch Hetchy*, 64. The 1928 vote supporting the purchase of Spring Valley was overwhelming: 82,490 yeas to 21,175 nays. Also see John Warfield Simpson, *Dam!: Water, Power, Politics and Preservation in Hetch Hetchy and Yosemite National Park* (New York: Pantheon, 2005), 228–29.

13. Beginning in 1913, Spring Valley began building Calaveras Dam under the supervision of consulting engineer William Mulholland. In contrast to Freeman's proposed concrete gravity design for the site, Mulholland opted for a hydraulic-fill earth embankment structure that quickly drew criticism from the Providence engineer. In October

1913, Freeman expressed concern to O'Shaughnessy about the Calaveras design, which he attributed to "Mulholland's defect of early training and his present tendencies of devoting his main efforts to cheapness and speed rather than having things scientifically strong and safe. This is a natural result of his early training and his attitude appears more that of the successful hustling contractor than that of the educated engineer and he shows in many ways the effect of having worked up from the ranks of common laborer without benefit of a scientific education. Appreciation of his many good qualities should not make us forget the elements of training wherein he is weak." O'Shaughnessy shared Freeman's assessment of Mulholland's design and bemoaned "Mulholland and his swollen ideas of accomplishment." Construction of Mulholland's Calaveras Dam stretched out for over four years; in March 1918, the dam suffered from a partial, yet serious, collapse. The rebuilt dam was completed in 1925 and bought by the city in 1930. A completely new Calaveras Dam, built by the city in the twenty-first century, was completed in 2019. See John R. Freeman to M. M. O'Shaughnessy, October 6, 1913, John Ripley Freeman (1855–1932) Papers (MC 51), Distinctive Collections, Massachusetts Institute of Technology, Cambridge (hereafter, JRF Papers), box 68; O'Shaughnessy, *Hetch Hetchy,* 68–69 (reprinting letter to Freeman, dated October 14, 1913); Allen Hazen and Leonard Metcalf, "Middle Section of Upstream Side of Calaveras Dam Slips into Reservoir," *Engineering News-Record* 80 (April 4, 1918): 679–81; "Failure of Part of Calaveras Dam," *Western Engineering* 9 (May 1918): 173–74; and Scott Blair, "Moving Mountains: Replacement Dam near Calaveras Fault Strengthens San Francisco's Water Supply," *Engineering News-Record,* November 20, 2017, 3–9.

14. John R. Freeman to M. M. O'Shaughnessy, September 8, 1913, JRF Papers, box 68.

15. John R. Freeman to M. M. O'Shaughnessy, February 3, 1914, JRF Papers, box 68.

16. John R. Freeman to J. S. Dunnigan, July 13, 1914; and Percy Long to John R. Freeman, September 10, 1914; both in JRF Papers, box 68.

17. John R. Freeman, "Hetch Hetchy Water Supply System—Effect of Increased Water Priorities Granted to Irrigation Districts," March 21, 1914, JRF Papers, box 69. Also see John R. Freeman to M. M. O'Shaughnessy, March 28, 1914, JRF Papers, box 68.

18. Freeman to O'Shaughnessy, March 28, 1914.

19. The history of the Modesto Irrigation District is documented in Dwight H. Barnes, *The Greening of Paradise Valley: Where the Land Owns the Water and the Power; The First 100 Years of the Modesto Irrigation District* (Modesto, CA: Modesto Irrigation District, 1987), https://www.mid.org. The history of the Turlock Irrigation District is documented in Alan M. Peterson, *Land, Water and Power: A History of the Turlock Irrigation District* (Glendale, CA: Arthur H. Clark, 1987).

20. O'Shaughnessy, *Hetch Hetchy,* 50.

21. The extended controversy involving the city's hydroelectric power system and the Raker Act's stipulation that power from the Hetch Hetchy system was not to be sold to private power companies (such as Pacific Gas and Electric) is covered at length in Righter, *Battle over Hetch Hetchy,* 167–90. Also see O'Shaughnessy, *Hetch Hetchy,* 88–110. During the Senate debate over the Raker bill in 1913, Senator George Norris may have envisioned the Hetch Hetchy project as constituting a great advancement in the

realm of public power, but it never achieved that status; a century later, the city generates power for some municipally owned facilities but still relies on privately owned power distribution lines to serve its residents. Over the years, the city has sought to purchase the PG&E system in the city, but as of 2025, all efforts have failed to win the approval of a two-thirds majority in the required bond referendum.

22. Warren D. Hanson, *A History of the Municipal Water Department and Hetch Hetchy System,* 3rd ed. (San Francisco: City of San Francisco, 1994), 47.

23. For more on the New Don Pedro Dam, see Barnes, *Greening of Paradise Valley.*

24. Hanson, *History,* 52.

25. The history of the East Bay Municipal Utility District is detailed in John Wesley Noble and Gayle B. Montgomery, *Its Name Was M.U.D.: A Story of Water* (Oakland, CA: Eastbay Municipal Utility District, 1999). Also see Righter, *Battle over Hetch Hetchy,* 156–61; and "The $39,000,000 Water Project of the East Bay Cities," *Engineering News-Record* 93 (December 11, 1924): 961.

26. Simpson, *Dam!,* 169.

27. San Francisco Public Utilities Commission, "State of the Regional Water System, October 2020," 22. See Bay Area Water Supply and Conservation Agency (BAWSCA) web page "Water Supply and System," https://bawsca.org/water/supply.

28. Noble and Montgomery, *Its Name Was M.U.D.*, 299, 314; also see Freeman, *Hetch Hetchy Report,* 368–72.

29. For the post–World War Two history of the city's Hetch Hetchy water supply, see Hanson, *History.* Also see San Francisco Public Utilities Commission, "State of the Regional Water System."

30. For example, in recent years the Bureau of Reclamation has seriously investigated the possibility of raising the height of Shasta Dam 18.5 feet and increasing its storage capacity by 634,000 acre-feet. See the bureau's "Shasta Dam and Reservoir Enlargement Project," https://www.usbr.gov/mp/ncao/shasta-enlargement.html.

31. Ezra David Romero, "Celebration and Concern: Hetch Hetchy Reservoir Turns 100, But Climate Change Complicates Its Future," May 2, 2023, KQED Science website, https://www.kqed.org/science/1982551/celebration-and-concern-hetch-hetchy-reservoir -turns-100-but-climate-change-complicates-its-future.

32. Freeman, *Hetch Hetchy Report,* 119.

33. Although not part of his ongoing consulting work, in 1914 Freeman's engagement with lawyer Edmund Whitman finally concluded. For more than a year prior to passage of the Raker Act, Edmund Whitman had played a dual role in Freeman's life. While the Boston attorney remained a stalwart "nature lover" in opposing the Hetch Hetchy Dam, he also loyally served the Providence engineer in defending Freeman in civil court against claims that he had been improperly compensated for work setting a valuation for the Gloucester Water Supply Company (see chapter 5). The long-delayed trial was held in October 1913, with the law firm of Elder, Whitman & Barnum representing Freeman. The court's ruling came on February 20, 1914, whereupon Whitman reported to Freeman, "Today Judge Morton ordered a judgment in your favor." This ruling offered an opportunity for the two men to find common ground after their fierce competition over Hetch Hetchy. But Freeman only begrudgingly acknowledged Whitman's success on his

behalf. After receiving a bill for the law firm's services in April, Freeman advised his old friend Sam Elder (Whitman's senior partner), "Frankly to me the charges are exorbitant and it troubles me all the more to have had so much of this work turned over to Mr. Whitman, for whom, frankly[,] I have no use since hearing his arguments and observing his actions in the Hetch Hetchy case. I feel he was very unjust to me." In total, Freeman was being asked to pay $1,750. By the fall of 1912 he had already paid $750 to the law firm, so the fee in question was only $1,000. The amount was not a burden for Freeman, but the fact that he was in any way beholden to Whitman stiffened his resistance to clear the account. While Elder acknowledged Freeman's dismay concerning how he and Whitman had fought over Hetch Hetchy, he counseled that this should not affect Freeman's obligations to the law firm: "I am sorry about the Hetch Hetchy matter . . . and did not know that you had been criticized [by Whitman]. Have you not allowed your feeling with reference to that tinge your views of the Gloucester matter?" Elder's second request for payment came in May, but the Providence engineer let the matter fester, forcing Elder to resubmit the unpaid bill at the beginning of August. Again, Freeman remained unmoved. Not until the start of the New Year did he finally address the disputed fee. However, he could not bring himself to pay without seeking a concession: "I have paid you already $250 as a retainer and $500 on account. How far can you go on meeting me halfway on the remainder?" Elder refused to meet Freeman "halfway," averring that Freeman had never "appreciated the danger of this case." Nonetheless, he was willing to knock the bill down $100. Mollified at last, Freeman relented and forwarded "a check for $900 in settlement." With that, his professional engagement with one of San Francisco's most tenacious adversaries over Hetch Hetchy came to an end. There appears to be no evidence that Freeman and Whitman ever interacted again. However, from the fall of 1912, when they first realized the dual dynamic of their relationship, through enactment of the Raker Act and onto the civil court decision rendered in February 1914, the two men were parties to the strangest and least probable element of the entire Hetch Hetchy controversy. For the correspondence from which the above quotations are derived, see JRF Papers, box 94.

34. For a synopsis of his professional activities after 1914, see "John Ripley Freeman," *ASCE Transactions* 98 (1933): 1471–76.

35. His work for San Diego in the 1920s is documented in Donald C. Jackson, *Building the Ultimate Dam: John S. Eastwood and the Control of Water in the West,* paperback ed. (Norman: University of Oklahoma Press, 2005); also see JRF Papers, boxes 52–54. His work on the Great Lakes is documented in JRF Papers, boxes 76–80; and in John R. Freeman, *Regulation of the Great Lakes and the Effect of Diversions of Chicago Sanitary District, a Report Submitted to the President and Board of Trustees of the Chicago Sanitary District,* 1926. His advocacy of a national hydraulic laboratory is prominently noted in Vannevar Bush, "Biographical Memoir of John Ripley Freeman, 1855–1932," *National Academy of Sciences Biographical Memoirs* (1935), 177–79. Freeman's October 6, 1922, trip to the Hetch Hetchy damsite was in the company of both O'Shaughnessy and the former city engineer C. E. Grunsky (see photo BANC PIC 1992 058 PIC Oct 6 1922 in the O'Shaughnessy Papers, Bancroft Library, Berkeley). It would have been fitting if Marsden Manson—like Grunsky, a former city engineer—had accompanied the

trio on their excursion, but by that time he was occupied in what he told Freeman was a project to develop "a new system of teaching geography by means of motion pictures." Apparently, this new system did not attract much attention, and when Manson died in 1931, he left his wife in a financially precarious circumstance. In response to an entreaty from Grunsky, Freeman contributed $100 to a fund benefitting Manson's widow. See Marsden Manson to John R. Freeman, May 12, 1924; C. E. Grunsky to John R. Freeman, March 2, 1931; and John R. Freeman to C. E. Grunsky, March 7, 1931; all in JRF Papers, box 68.

36. This project consisted of the Oakdale Dam and Powerplant near Monticello, Indiana; after Roger's death, the reservoir formed by the dam was named Lake Freeman in his honor.

37. Freeman's planning for the enlarged Big Meadows Dam, his son's untimely death, and his dispute with Great Western corporate leadership over compensation are all documented in JRF Papers, boxes 5, 65–67.

38. His trip to Japan is documented in JRF Diary, October 1–December 22, 1929, JRF Papers, box 3.

39. John R. Freeman, *Earthquake Damage and Earthquake Insurance* (New York: McGraw-Hill, 1932). He described how the book began as a pamphlet for "the directors of the Manufacturers Mutual Fire Insurance Company . . . [while] further study resulted in the presentation of a long paper covering these topics, on May 4, 1927, before the Eastern Division of the Seismological Society of America. In September 1929, after further extended investigations, the author briefly urged these matters of research . . . before a conference of eminent seismologists and geophysicists then in session in the Seismological Laboratory at Pasadena Cal[ifornia]; and in May 1930 he earnestly presented this need for research and more accurate data upon earthquake motion, in Washington D.C. before the Eastern Division of the Seismological Society" (p. ix).

40. The testimonial dinner, held on April 21, 1931, is documented in JRF Papers, box 2.

41. For the family gravesite in Providence, see "John Ripley Freeman" on the Findagrave website, https://www.findagrave.com/memorial/129530007/john-ripley-freeman. The primary obituary memorializing his professional life is "John Ripley Freeman," *Transactions of the American Society of Civil Engineers* 98 (1933): 1471–76.

42. Bush, "John Ripley Freeman," 171.

43. In recent years, "technopolitics" has become a phrase commonly used in describing post–World War Two technological systems and the ways in which technology and engineering projects are intertwined with governmental regimes. The term is also relevant in characterizing how Freeman's work in planning and advocating the Hetch Hetchy Dam represented a melding of technology and politics. See Gabrielle Hecht, ed., *Entangled Geographies: Empire and Technopolitics in the Global Cold War* (Cambridge, MA: MIT Press, 2011); and Gabrielle Hecht, *Being Nuclear: Africans and the Global Uranium Trade* (Cambridge, MA: MIT Press, 2013).

44. Richard White, *The Organic Machine: The Remaking of the Columbia River* (New York: Hill and Wang, 1996); chapter 1 is titled "Knowing Nature through Labor: Energy, Salmon, and Society on the Columbia."

45. Carried one step further, even the deployment of mathematical formulae to model the behavior of structures under load represents ways that engineers seek to comprehend nature through work.

46. Dunnigan also fondly reminisced about the fall of 1913, recalling "with great pleasure your sessions with Senator Ben Tillman, Frank Brandegee, Lippit[t], and many others. The sessions we had in Washington brought the lasting admiration and friendship of our dear old friend William Kent, a friendship worthwhile having." J. S. Dunnigan, San Francisco City Clerk, to John R. Freeman, July 15, 1931, JRF Papers, box 2.

47. The observation that Freeman won and Muir lost simply acknowledges that Freeman's vision of a Hetch Hetchy Dam and Aqueduct came to define what was actually built.

Index

Page numbers in italics refer to figures or illustrations. Dams and city/town/place names are in California unless specified otherwise. San Francisco is abbreviated SF.

Union Labor Party (SF), 36, 92

US Congress. *See* US House of Representatives; US Senate

US Forest Service, 5–6, 37, 52, 200, 294n86, 329n1

US General Land Office, 20, 49, 105

US Geological Survey (USGS), 26–29, 32–33, 34, 51

US House of Representatives, 6, 22, 45–46, 52, 166, 192, 195, 197, 199, 205, 210, 214–217, 222, 219–220, 224–25, 227–29, 231, 238, 248, 249, 255, 258, 265, 267, 294n93, 310n27. *See also* House Committee on the Public Lands hearing (1913)

US Senate, 3, 45–46, 169, 192, 195, 213, 217, 219–22, 225, 227–30, 233–37, 239–41, 244–55, *245*, 324n83, 327n90, 328n94. *See also* Senate Committee on Public Lands hearing

United Trades and Labor Council of Pittsburg, Kansas, 247

Vogelsang, Alexander, 230–31, 244–45, 254

Von Schmidt, Alexis, 32

Vyrnwy Dam (Wales), 97

Wachusett Aqueduct, 70–73, 76, 78, 125

Wachusett Dam (Mass.), 70, *71*, 78

Wadsworth, H. H., 107, 319n97

wage rates, discrepancies in, 126–27, 166, 172–73, 183

Walsh, Thomas J., 246

Wapama Falls, *2*, *10*, 20, 258, 264

Washington, D.C., 12–13, 42, 45, 68, 88, 91–92, 94, 99–100, 117, 154–55, 168–70, 182, 184–85, 187, 190–92, 197, 203, 206, 213, 219–20, 222, 227–28, 233, 235, 237, 240, 244, 246, 249, 254, 256, 327n90, 335n46

Washington Mills, 67, 298n40

Washington State Federation of Women's Clubs, 235

Water and Power: The Conflict over Los Angeles' Water Supply in the Owens Valley (Kahrl), 78

Watkins, Carleton, 17–18

Watrous, Richard, 191, 223, 226

Webster, Edwin, 85

West Bridgton (Maine), 55–56, 60, 61

Whipple, George, 170, 175–77, 212

White, Richard, 274

Whitman, Edmund: Ballinger hearing, 93; early years and family, 168, 316n46; Fisher hearing, 170, 172, 177; Freeman, *Hetch Hetchy Report*, critique of, 166–67, 172, 18; Freeman's lawyer, 167–69, 332n33; Freeman's opponent, 168, 181, 234, 238–39, 241, 246, 255; House and Senate committee hearings, 46, 206–7, 223–24, 226, 231

Wilderness and the American Mind (Nash), 5

Williams, Cyril, 309n4

Williams, John S., 253

Wilson, Woodrow, 2, 11, 169, 187, 191–92, *193*, 200, 237, 254–55, 257, 321n10

Winklebeck, Levi, 227

Winnipeg River (MB), 109

Wood, Andrew, 155

Works, John D., 219, 222, 235, 240, 245–46, 248, 251, 254

Worster, Donald, 274

Wright Irrigation District Act, 25–26

Yosemite grant, from federal government, 18

Yosemite National Park: geological history of, 15; Hetch Hetchy Dam, *4*, 14; Hetch Hetchy Valley, included in, 1, 9, 29, *48*; Hitchcock E. A., 35–36, 41; Johnson, Robert Underwood, 21–22; Muir, John, 21–22, 40, 43, 50; Native peoples, exclusion from, 24; origins of, 21–24; private land in, 23, *23*, *24*; Right-of-Way Act, inclusion in, 33–34;

www.ingramcontent.com/pod-product-compliance
Lightning Source LLC
Chambersburg PA
CBHW020447100426
42812CB00036B/3480/J